# Current Topics in Pathology
# 82

Managing Editors

C.L. Berry E. Grundmann

James C.E. Underwood (Ed.)

# Pathology of the Nucleus

Contributors
J. Crocker · P. Dal Cin · R. A. Fallon · H. M. H. Kamel
J. Kirk · P. Quirke · A. A. Sandberg · T. J. Stephenson
G. Terenghi · P. G. Toner · J. C. E. Underwood

Springer Verlag Berlin Heidelberg New York
London Paris Tokyo Hong Kong

J. C. E. UNDERWOOD, Professor, M. D., F. R. C. Path.,
Department of Pathology,
University of Sheffield Medical School,
Beech Hill Road, Sheffield S10 2RX, United Kingdom

C. L. BERRY, Professor, M. D., Ph. D., F. R. C. Path.,
Department of Morbid Anatomy, The London Hospital,
London E1 1BB, United Kingdom

E. GRUNDMANN, Professor Dr.,
Gerhard-Domagk-Institut für Pathologie der
Universität Münster, Domagkstraße 17,
4400 Münster, Federal Republic of Germany

With 104 Figures and 14 Tables

ISBN-13: 978-3-642-74670-3 e-ISBN-13: 978-3-642-74668-0
DOI: 10.1007/978-3-642-74668-0


Softcover reprint of the hardcover 1st edition 1990



2122/3130-543210 – Printed on acid-free paper

# List of Contributors


| | |
|---|---|
| CROCKER, J., Dr. | Histopathology Department, East Birmingham Hospital, Bordesley Green East, Birmingham, B9 5ST, UK |
| DAL CIN, P., Dr. | The Cancer Center of Southwest Biomedical Research Institute, 6401 East Thomas Road, Scottsdale, AZ 85251, USA |
| FALLON, R. A., Dr. | Amersham International, Pollards Wood Laboratories, White Lion Road, Amersham HP7 9LL, UK |
| KAMEL, H. M. H., Dr. | Department of Pathology, The Queen's University of Belfast, Royal Victoria Hospital, Grosvenor Road, Belfast BT12 6BA, Northern Ireland |
| KIRK, J., Dr. | Department of Pathology, The Queen's University of Belfast, Royal Victoria Hospital, Grosvenor Road, Belfast BT12 6BA, Northern Ireland |
| QUIRKE, P., Dr. | Department of Pathology, University of Leeds, Leeds LS2 9JT, UK |
| SANDBERG, A. A., Dr. | The Cancer Center of Southwest Biomedical Research Institute, 6401 East Thomas Road, Scottsdale, AZ 85251, USA |

STEPHENSON, T. J., Dr.
Department of Pathology,
University of Sheffield Medical School,
Beech Hill Road,
Sheffield S10 2RX, UK

TERENGHI, G., Dr.
Histochemistry Department,
The Royal Postgraduate Medical School,
Hammersmith Hospital,
Ducane Road,
London W12 0HS, UK

TONER, P. G., Prof.
Department of Pathology,
The Queen's University of Belfast,
Royal Victoria Hospital, Grosvenor Road,
Belfast BT12 6BA, Northern Ireland

UNDERWOOD, J. C. E., Prof.
Department of Pathology,
University of Sheffield Medical School,
Beech Hill Road, Sheffield
S10 2RX, UK

# Preface

The scope of this book embraces the structure and function of the nucleus in diseased tissues with emphasis on neoplasia because that is the current focus of interest. Molecular aberrations within the nucleus are thought to be fundamental to the genesis of the neoplastic process.

At first, all that was known about tumours were their signs and symptoms, and there was speculation that they due to an excess of "black bile". Then the autopsy disclosed the association of primary malignant tumours with secondary metastases. The advent of microscopy revealed abnormalities of tissue and cellular architecture in tumours and enabled the first glimpse of the nuclear aberrations within these lesions. Now, advances in molecular genetics have caused considerable attention to be directed at the nucleus. It hasn't always been so for two reasons: first, until this century, the primordial state of physics and biochemistry denied investigators the ability to study the nucleus of eukaryotic cells in the way we can today; second, even when the biochemical ingredients of the DNA in the nucleus were ascertained it was not thought possible that complex genetic data could be encoded in combinations of just four bases.

In editing this book I have solicited a series of chapters covering various aspects of nuclear pathology, with some methodological details where appropriate so that readers unfamiliar with these techniques could begin to employ them in their own work.

The first chapter summarises briefly aspects of nuclear morphology in human tumours, which are probably familiar to most histopathologists but are nevertheless diagnostically useful and serve to set the remaining chapters in context. It is to be hoped that further detailed analysis of tumour cell nuclei will reveal the diagnostic usefulness of some of the abnormalities recognised by histopathologists but for which we have, as yet, no biological explanation and can attach no clinical significance. Some of these abnormalities, and others, are explored at an ultrastructural level in the following chapter.

The last few years have seen a spate of publications on nucleolar organiser region enumeration in interphase nuclei in tumours. These intranuclear pools of ribosomal RNA and associated proteins, some of which are argyrophilic, can be identified in sections and cell preparations by a relatively simple silver staining technique within the capabilities of most routine diagnostic laboratories. It brings transcriptional activity and nucleolar organisation within reach of routine histopathology and the diagnostic and prognostic significance of the observations is of considerable interest.

The abnormalities of nuclear size and shape summarised in the first chapter can be quantified, digitised, processed, and subjected to statistical analysis. These procedures and the results derived from them are detailed in the fourth chapter. Because nuclear abnormalities are observed quite consistently in cytological preparations from tumours, it is likely that quantitative and image analysis techniques can be developed to assist in population screening for malignancies accessible to exfoliative cytology. An extension of this approach is flow cytometry, which has the advantage that many cells can be assessed very rapidly using two or more parameters including DNA ploidy and cell markers.

Moving from observations on the whole nucleus to the individual chromosomes, in the sixth chapter it is gratifying to discover that solid tumours are yielding to the techniques of the cytogeneticist. Consistent chromosomal abnormalities are being revealed in certain types of tumour and, from our knowledge of the location of genes on individual chromosomes, the functional significance of these abnormalities is being elucidated.

Finally, to the study of the nucleus and its transcribed products at a molecular level. The relatively new technique of in situ hybridisation enables the abundance and expression of genes to be investigated at a cellular level. This is fundamentally important to the neoplastic process on which most of the work is being done.

In conclusion, I am indebted to the other authors, who shared my enthusiasm for the pathology of the nucleus. Their erudite and well-prepared contributions greatly eased the editing process. On behalf of all authors I thank also Mrs. Ursula Davis and her colleagues at Springer-Verlag for their advice and help inthe preparation of this volume.

Sheffield, UK J. C. E. UNDERWOOD

# Contents



Indexed in ISR

# Nuclear Morphology and Grading in Tumours

J. C. E. UNDERWOOD



## 1 Introduction

VON HANSEMANN (1890) is credited with first postulating that abnormalities of nuclear morphology in tumour cells might be correlated with their biological properties and clinical behaviour. He placed importance on the mitotic rate and the presence of abnormal mitoses (VON HANSEMANN 1892). His postulate has remarkable prescience because it would not be for almost 100 years that evidence would be gathered to provide biochemical support for the notion that the aberrant behaviour of cancer cells was attributable to genetic lesions within their nuclei.

## 2 Structure and Function of the Nucleus

Most of the DNA in eukaryotic cells is in their nuclei; relatively minute quantities of maternally derived DNA reside in mitochondria. In nuclei the DNA is organised into chromosomes in a highly folded configuration. The folded state of nuclear DNA is crucial for not only does it enable the DNA to be accommodated within the limited volume of a nucleus (the unfolded DNA in each chromosome is approximately 5 cm long), but it influences the activity of the genes encoded therein (only a small proportion of the DNA encodes essential proteins). Associated with the DNA and contributing to its folded state are a variety of DNA-binding proteins: these are the histone and non-histone proteins. The non-histone proteins are numerous and subserve many functions. Histones are structural proteins which are, on a basis of mass, as abundant in nuclei as the DNA with which they are associated. The DNA and associated histones constitute the chromatin in interphase nuclei.

Within each nucleus there is at least one subsidiary structure – the nucleolus. The nucleolus is formed around loops of DNA, called nucleolar organiser regions (see Chap. 3), located on the acrocentric chromosomes and coding for rRNA synthesis. The nucleolus contains, in addition to rDNA and rRNA, associated proteins such as RNA polymerase 1 (Alberts et al. 1983).

The shape of the nucleus, frequently abnormal in neoplastic cells, is dictated by the characteristics of the nuclear envelope. The nuclear envelope comprises an inner and an outer nuclear membrane separated by the perinuclear space and perforated by pores. The nuclear aspect of the inner membrane is bordered by the fibrous lamina, but this is often ultrastructurally inconspicuous. The fibrous lamina contains proteins which form the nuclear pores, bind to the inner membrane, and bind to the underlying chromatin. The fibrous lamina is therefore crucial to the structural organisation of the nucleus.

In histological and cytological preparations the nucleus, its chromatin pattern and the nucleoli can be discerned by light microscopy. Their size, shape and staining pattern are of proven value in the diagnosis and grading of neoplasms, although the precise functional significance of some of the observations remains speculative.

## 3 Nuclear Abnormalities in Tumours

Nuclear abnormalities in tumour cells were noted in the early days of microscopy, but the biological explanation of these abnormalities only later achieved its current importance. Although it was realised that the nucleus of cells contained genetic information on the chromosomes which became visible during mitosis, it was not until 1944 that it was demonstrated that DNA carried the code.

The genetic basis for the relatively autonomous and autocrine growth of neoplastic cells is now under active investigation and the role of oncogenes is being elucidated. However, the application of the techniques of molecular biology to the genome of neoplastic cells does not diminish the diagnostic usefulness of the simple time-honoured morphological observations which are summarised in this chapter. It is to be hoped that eventually a molecular explanation will be forthcoming for the wide range of nuclear appearances seen in human tumours. Indeed, it is remarkable that, since the fundamental genetic aberrations in cancer lie within the nuclei of cancer cells, the nuclear abnormalities in these lesions are not more consistent. But it should be borne in mind that morphology is a relatively indirect and insensitive paradigm of function; for example, a chromosomal translocation can result in the expression of a normally unexpressed oncogene, thus enabling autocrine growth, but the total DNA content of the nucleus and its morphology may be unaffected.

## 3.1 Diagnostic Significance

The size, shape and texture of tumour cell nuclei are regarded as important features in the diagnosis and classification of neoplasms (Table 1). Certain features are so common and general that they are helpful in distinguishing between benign and malignant lesions. These include pleomorphism, hyperchromaticism and mitotic figures; these reflect disturbances in nuclear ploidy and DNA synthesis. Of these features, pleomorphism is the least reliable indicator of malignancy. It is often seen, for example, in

**Table 1.** Examples of nuclear morphology restricted to certain tumour types

| Nuclear feature | Diagnostic significance |
|---|---|
| Vacuolated and grooved | Papillary carcinoma of the thyroid |
| Coffee bean nuclei | Brenner tumour |
| Convoluted or cerebriform | T lymphocytes in mycosis fungoides and Sézary syndrome |
| Grooved | Langerhans cells and histiocytosis X |
| Hyaline inclusions | Lymphoplasmacytoid lymphoma (Dutcher-Fahey inclusions) |
| Cleaved | Centrocytic lymphoma |
| Corrugated and blunt ended | Smooth muscle tumour |
| Reniform | Histiocytic neoplasms |

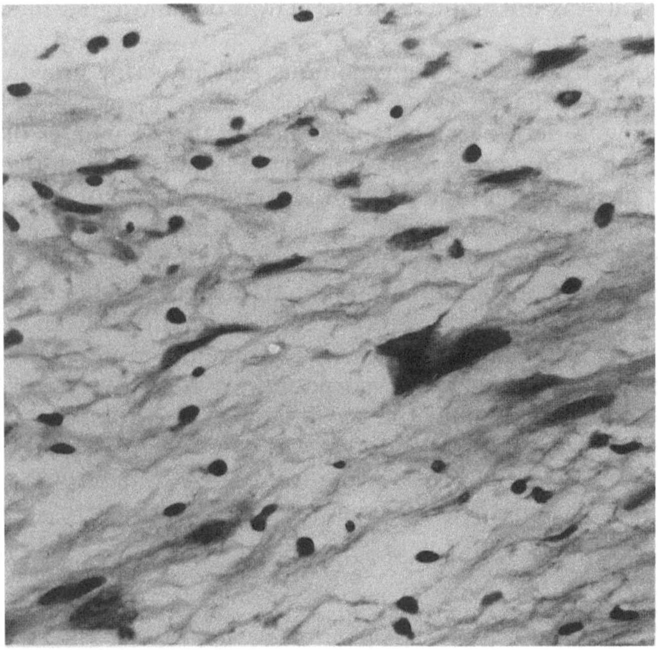

**Fig. 1.** Nuclear pleomorphism in a nerve sheath tumour. In the absence of mitotic activity this abnormality can be attributed to degeneration rather than malignancy. H & E

benign endocrine tumours and in degenerate areas of benign connective tissue neoplasms (Fig. 1). Nuclear pleomorphism is an alarming feature of atypical fibroxanthoma, a benign lesion. Conversely, banal nuclear morphology may be encountered in highly malignant neoplasms. As a rule, it is unwise to attach a malignant significance to nuclear pleomorphism in a neoplasm unless it is accompanied by inappropriate or excessive mitotic activity. Many of the bizarre nuclei seen in neoplasms are probably so genetically deviant as to be effete.

Nuclei may be deformed by cytoplasmic structures in neoplastic cells; the deformability of nuclei may, in itself, be an abnormal feature. Examples include the identation of nuclei by fat vacuoles in lipoblasts in liposarcomas and by mucin vacuoles in the signet-ring cells of adenocarcinomas.

Nuclear morphology is of paramount importance in diagnostic cytology, because information on tissue architecture and organoid growth patterns is usually lacking. Nuclear features favouring malignancy include coarseness of nuclear chromatin, abnormally large or irregular or abundant nucleoli, moulding of nuclear outlines, and hyperchromaticism. Nuclear detail may be less easy to discern in histological sections of solid tissue biopsies due to suboptimal fixation, which often results in artefactual nuclear vesiculation, and to nuclear superimposition. Certain fixatives, especially those containing acetic acid and ethanol (e.g. Carnoy's) which precipitate nucleoproteins, are advocated for preservation of nuclear detail. Thinner sections can be used to improve the rendition of nuclear detail; this may necessitate the use of plastic embedding media.

Some nuclear features are specific to individual tumour types.

### *3.1.1 Clear and Grooved Nuclei: Papillary Carcinoma of the Thyroid*

Papillary carcinoma of the thyroid is characterised by the presence of optically clear nuclei within the tumour cells (Fig. 2). These pale nuclei are often referred to as "Orphan Annie" nuclei because of their resemblance to the empty forlorn eyes of a North American cartoon character of that name. The pallor is attributable to dispersion of the chromatin, giving a ground-glass appearance, and to intracytoplasmic inclusions (GRAY and DONIACH 1969; OYAMA 1989). Another recently recognised diagnostic feature is the presence of grooved nuclei (CHAN and SAW 1986).

These nuclear features are so pathognomonic of this tumour type that their presence virtually overrides any other diagnostic criterion. For example, some thyroid carcinomas have a predominantly follicular growth pattern, but the tumour cells possess clear nuclei; these tumours should be regarded as papillary carcinomas with a follicular pattern because they behave as do the more typical papillary carcinomas (i.e. they are lymphangio-invasive).

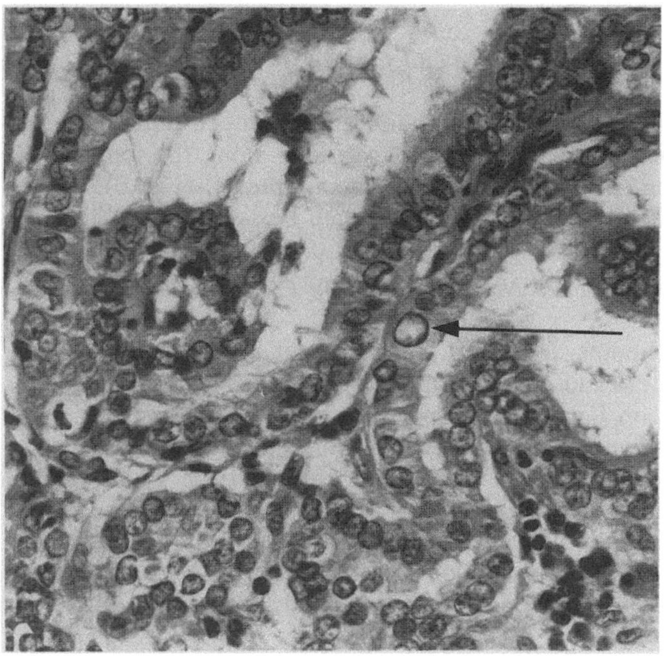

**Fig. 2.** Papillary carcinoma of the thyroid characterised by clear vesicular nuclei (*arrow*). H & E

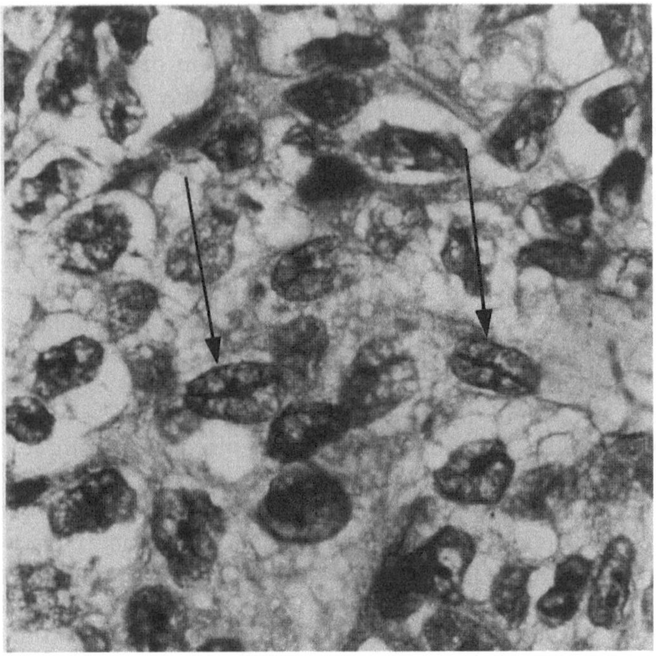

**Fig. 3.** Longitudinally grooved "coffee bean" nuclei (*arrows*) in a Brenner tumour of the ovary. H & E

### *3.1.2 "Coffee Bean" Nuclei: Brenner and Granulosa Cell Tumours*

The nuclei of ovarian Brenner tumours and granulosa cell tumours invariably have ovoid nuclei with a single linear groove along the longitudinal axis, thus resembling coffee beans ("coffee bean" nuclei) (Fig. 3). Similar nuclei are seen in the cells forming Walthard cell nests (AREY 1943).

### *3.1.3 Convoluted and Cleaved Nuclei: Malignant Lymphomas*

T- and B-cell lymphomas are often characterised by distinctive nuclear morphology. Indeed, because the cytoplasm of T and B cells is often minimal and featureless on light microscopy, attention is usually concentrated on the nucleus for the purposes of diagnosis and classification.

Convoluted (or cerebriform) nuclei are characteristic of T-cell lymphomas, including mycosis fungoides and Sézary syndrome. The nuclei of some normal T-cell subpopulations is recognisably convoluted, but the convolutions are exaggerated in the neoplastic state; cells with such nuclei are thus often referred to as "hyperconvoluted" or Lutzner cells (LUTZNER et al. 1971).

Cleaved nuclei are a feature of the B cell-derived centrocytes of germinal centres and their neoplastic counterparts. Recognition that cleaved

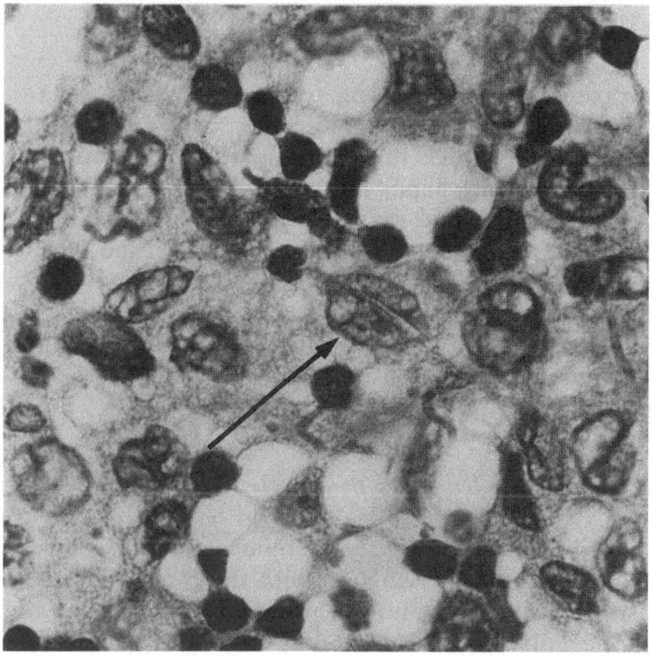

**Fig. 4.** Vesicular nuclei of Langerhans cells, some of which are reniform or grooved (*arrow*), in a case of eosinophilic granuloma in a lymph node. H & E

nuclei are present is important in the subclassification of B-cell lymphomas. When small cleaved cells (centrocytes) are seen in the blood smears from patients with follicular lymphomas they are referred to, by some North American haematologists, as "buttock cells" because of their conspicuous nuclear clefts.

### *3.1.4 Reed-Sternberg Cell Nuclei: Hodgkin's Disease*

The typical Reed-Sternberg cell, the presence of which is virtually a prerequisite for the diagnosis of Hodgkin's disease, has a symmetrical or "mirror image" bilobed nucleus, each lobe of which contains a large round eosinophilic nucleus. It would be thought that such a distinctive nucleus might betray the histogenesis of the Reed-Sternberg cell, but no normal lymphoid cells are candidates on purely morphological grounds. Variants with distinctive nuclei are associated with certain subtypes of Hodgkin's disease, such as the polylobate nucleus of the Reed-Sternberg cell of the lymphocyte and histiocyte rich variant.

### *3.1.5 Reniform Nuclei: Histiocytic Neoplasms*

Normal macrophages possess vesicular elongated nuclei which are frequently angulated or bent so that they resemble the shape of a kidney viewed in anteroposterior profile; thus, this shape is often referred to as reniform. A wide variety of nuclear shapes are seen in neoplasms alleged to be of histiocytic origin; in some the reniform nucleus of the putative parent cell is retained and its recognition aids identification of the type of neoplasm.

The nucleus of Langerhans cells and their neoplastic derivatives, seen for example in some of the histiocytoses, and of interdigitating reticulum cells is characterised by a longitudinal crease or groove which may be evident in a favourable plane of section (Fig. 4).

### *3.1.6 Blunt-Ended Corrugated Nuclei: Smooth Muscle Neoplasms*

Smooth muscle cells, and benign and well-differentiated tumours derived from them, have an elongated nucleus aligned with the long axis of the cell. In benign neoplasms the nuclei often appear palisaded, reflecting the parallel orientation of the adjacent cells. In addition, these nuclei often appear to have corrugated or serrated profiles, best seen in thin sections. Presumably this allows the nucleus to concertina when the normal cell contracts.

## 3.2 Nuclear Inclusions in Tumours

### *3.2.1 Nucleoli*

The nuclei of malignant neoplastic cells often contain multiple nucleoli. These are usually more prominent in smears and imprints than in histological sections, and they are a useful criterion in diagnostic cytology. The nucleoli in malignant cells are also commonly irregular is size and shape; these features are criteria included in some grading schemes.

### *3.2.2 Cytoplasmic Inclusions*

Cytoplasmic inclusions within tumour cell nuclei are evident in routine hsitological sections as vacuolations. To be certain of the true nature of the vacuolations it is necessary to perform electron microscopy. They are particularly conspicuous in papillary carcinoma of the thyroid (OYAMA 1989).

### *3.2.3 Dutcher-Fahey Inclusions*

Dutcher-Fahey inclusions are eosinophilic hyaline inclusions within the nuclei of the lymphoplasmacytoid type of non-Hodgkin's lymphoma. They are diastase-resistant PAS-positive and immunohistology has revealed that they consist of immunoglobulin.

## 4 Mitoses in Tumours

The presence of mitoses in *epithelial* cells is not helpful in distinguishing between normal and benign and malignant neoplasms, because almost all epithelial cells are labile in the sense that in many instances they are normally replicating. Notable exceptions include the epithelial cells of solid organs (e.g. hepatocytes), in which mitoses are infrequent except during regeneration after injury. In epithelial neoplasms, therefore, invasion outside the epithelial compartment is a more important criterion of malignancy. In connective tissue neoplasms, however, recognition of the presence of invasion requires that the cells have entered the vascular compartment; because connective tissues lack conspicuous basement membrane boundaries between cellular compartments, this same criterion cannot be used to diagnose malignancy. It is therefore in connective tissue neoplasms that the recognition and assessment of mitotic activity plays a more important role in tumour diagnosis.

The frequency of mitoses does not necessarily denote a commensurate tumour growth rate. Numerous mitoses may be counteracted by a concomitant degree of ischaemic necrosis (which may be obvious) or apoptosis (which may be inconspicuous). Also, the administration of cytostatic drugs, such as vincristine, causes arrest of cells in metaphase because synthesis of tubulin for the mitotic spindle is blocked; in such tissue the mitotic index will be artificially high because a large number of metaphase arrests have accumulated.

## 5 Nuclear Immunohistochemistry in Tumours

### 5.1 Steroid-Hormone Receptors

The nucleus is now regarded as the intracellular site of unoccupied steroid-hormone receptors. Formerly they were thought to reside in the cytoplasmic cytosol because of the ease with which they appear in the soluble fraction of tissue homogenates. However, immunohistology using monoclonal antibodies to steroid-hormone receptors such as those for oestrogens and progestogens invariably reveals that the receptors are intranuclear, as in breast carcinomas for example (GIRI et al. 1987, 1988). Although no morphological differences may be evident on conventional light microscopy, immunohistology often reveals heterogeneity of nuclear staining and this may have therapeutic and prognostic significance.

### 5.2 Proliferation Antigens

Certain nucleoproteins appear only during certain phases of the cell cycle. Of these the most widely studied is the protein detected by the Ki 67 monoclonal antibody (GERDES et al. 1984). This protein is expressed in $G_1$, S, $G_2$ and M phases of the cell cycle and is undetectable in the nuclei of non-cycling cells. It can be detected by this antibody only in frozen sections of unfixed tissue, thus limiting its applicability to archival material or conventionally fixed biopsies. As an index of proliferative activity it is prognostically useful in certain categories of tumours, such as non-Hodgkin's lymphomas.

Alternatively, cellular proliferation can be assessed by incubating fresh tumour biopsies in medium containing bromodeoxyuridine (BrdU) (MORSTYN et al. 1986). This is incorporated into the DNA in the nuclei of cells in S phase at the time of incubation. The BrdU can be detected by immunohistochemistry with a specific antibody after the tissue has been fixed and processed.

### 5.3 Nuclear Oncoproteins

Most protein products of oncogenes are to be found in the cytoplasm, on the cell membrane, or are secreted. Examples include growth factors and their receptors. The protein products of other oncogenes reside in the nucleus of cells in which they have been synthesised. Examples include the c-*myc* oncoprotein which is thought to act directly within the nucleus and is associated with DNA synthesis (EVAN and HANCOCK 1985), in particular the transition from $G_0$ to $G_1$ phases of the cell cycle (BLANCHARD et al. 1985). The study of nuclear oncoproteins is likely to have increasing impact on our understanding of the molecular biology of neoplasia and may have prognostic implications in certain types of tumour.

# 6 Nuclear Orientation

## 6.1 Nuclear Stratification in Glandular Epithelial Dysplasia

In most glandular epithelia the nuclei are basally situated, that is they reside in that part of the cell adjacent to the basement membrane and furthest from the surface or lumen. In the epithelial lesions commonly referred to as "dysplasia" there is progressive migration of the nuclei away from their basal location. In severe dysplasia the nuclei appear to reside at all levels in the epithelial surface; this gives the appearance of nuclear stratification.

## 6.2 Nuclear Streaming in Epithelial Proliferations in Mammary Ducts

Elongated nuclei may be indicators of cellular polarity if they consistently orientate parallel to the long axis of the cell. This is most useful diagnostically in the interpretation of proliferative epithelial lesions in mammary ducts, where it assists in the distinction between epitheliosis and intraduct carcinoma. In epitheliosis the nuclei frequently appear to "stream"; the nuclei of adjacent cells appear to lie in the same orientation, an appearance which has been likened to that of a field of corn being blown by the wind. In intraduct carcinoma the nuclei of adjacent cells are orientated haphazardly, a feature which, in addition to nuclear hyperchromaticism and architectural pattern, usually denotes malignancy.

## 6.3 Nuclear Palisading in Spindle Cell Tumours

Nuclei in smooth muscle tumours and in neurilemmomas are often seen to be orientated in parallel arrays or palisades. Presumably this reflects cellular orientation within these lesions, but the intercellular borders are often indistinct so it is the nuclear pattern which is the visible expression of this.

# 7 Nuclear Morphology in Tumour Grading

BRODERS (1926) pioneered the introduction of grading in the histopathological appraisal of tumours and as a prognostic index. Numerous grading systems have been proposed for different tumours with varying emphasis on the nuclear component. A few grading systems utilise nuclear features exclusively; others rely solely on tumour architecture; most incorporate both parameters. Features such as hyperchromaticism, size, mitotic activity and pleomorphism are often recorded for compilation of nuclear

**Table 2.** Examples of tumour grading systems incorporating an assessment of nuclear features

| Grading system | Tumour | Nuclear criteria | Other criteria |
|---|---|---|---|
| BLACK and SPEER (1957) | Breast ca. | Frequency of abnormal nuclei | |
| BLOOM and RICHARDSON (1957) | Breast ca. | Hyperchromaticism<br>Pleomorphism<br>Mitoses | Tubule formation |
| KEMPSON and BARI (1970) | Uterine smooth muscle tumours | Mitotic counts | |
| HARTVEIT (1971) | Breast ca. | Crowding<br>Lobulation<br>Diameter | |
| JAKOBSSON et al. (1973) | Larynx ca. | Pleomorphism<br>Mitoses | Growth pattern<br>Differentiation<br>Invasion<br>Cellular response |
| EVANS et al. (1977) | Chondrosarcoma | Multinucleate lacunae<br>Size<br>Mitoses | Cellularity |
| GAETA et al. (1980) | Prostatic ca. | Nucleoli<br>Pleomorphism<br>Vesicularity<br>Mitoses | Architecture |
| KREICBERGS et al. (1981) | Chondrosarcoma | Size<br>Mitoses | Cellularity |
| FUHRMAN et al. (1982) | Renal cell ca. | Size<br>Pleomorphism<br>Chromatin | |
| CHRISTOPHERSON et al. (1983) | Endometrial ca. | Chromatin<br>Shape<br>Nucleoli | |
| KERN (1984) | Bladder ca. | Atypia<br>Mitoses | Architecture |
| KLOPPEL et al. (1985) | Pancreatic ca. | Mitoses<br>Size<br>Pleomorphism<br>Chromatin | Architecture<br>Mucus content |
| MITTAL et al. (1988) | Endometrial ca. | Size<br>Shape | |

grades (Table 2). In general such grading schemes correlate successfully these nuclear parameters with prognosis. However, flow or static nuclear cytometry for DNA ploidy (see Chap. 5 and 4 respectively) and other parameters (see Chap. 4) may provide a much more accurate measure because a larger number of nuclei are analysed and the subjective element of microscopy is eliminated. Nevertheless, direct visual assessment of nuclear parameters by microscopy removes the possibility that the results may be biassed by contamination with stromal elements, as occurs with flow cytometry.

Grossly bizarre or pleomorphic nuclei tend to imply a high-grade malignant neoplasm, but the cells harbouring such nuclei are probably effete and incapable of replication. Nevertheless, bizarre and pleomorphic nuclear morphology may denote an unstable genome favouring the emergence of highly malignant clones.

All grading systems are prone to a significant degree of observer error because the appraisal of nuclear features like size (without morphometry) and staining density (without microdensitometry) are subjective. Some investigators have argued that formal grading systems are so prone to observer variation that the pathologist need do no more than assign individual tumours to well, moderately or poorly differentiated categories (DELIDES et al. 1982).

A recent survey by the Surveillance, Epidemiology, and End Results Program (SEER) of the National Cancer Institute, Bethesda, Maryland, in the USA has revealed that only 52.6% of cancers were graded following histological examination (HENSON 1988). Reasons given for absence of a recorded grade for tumours include the suspicion that grading is arbitrary and subjective, and so prone to observer error that it is unreliable in practice. Also, it is assumed that tumour heterogeneity vitiates grading because the histological sample is unlikely to be representative. It is also time consuming and competes with other activities which carry a more immediate or tangible reward for the histopathologist. Prognosis can also be predicted from tumour stage. This information is usually available more frequently than grade, probably because it is simpler and quicker to derive and the individual criteria contributing to the overall grade are less subjective (e.g. extent of local invasion, nodal status). But stage and grade are not mutually exclusive prognostic indices. Indeed, one enhances the usefulness of the other. The prognostic usefulness of the grade of an individual tumour is very limited unless the stage is also determined.

## 8 Conclusions

Nuclear morphology in tumours is important for two reasons. First, subtle nuclear features can be diagnostically useful and can assist in the identification of specific tumour types. Second, abnormalities of nuclear morphology are major or sole components of grading systems for tumours. Although flow and static DNA cytometry may eventually supersede the rather subjective assessment of nuclear grade, it is unlikely that, on a similar timescale, machines will replace histopathologists in the recognition of subtle but diagnostically useful nuclear features.

Current concepts of the molecular biology of cancer situate the fundamental biochemical lesion within the nucleus. Lesions such as point mutations, gene amplification, and translocations are beyond the scope of conventional histology used in most diagnostic laboratories. Although the new techniques of molecular biology, in particular in situ hybridisation, can now be applied to tumours, it is likely that a relatively simple assessment of morphological nuclear parameters will continue to hold its importance in the surgical pathology of tumours for some time to come.

## References

Alberts B, Bray D, Lewis J, Raff M, Roberts K, Watson JD (1983) Molecular biology of the cell. Garland, New York

Arey LB (1943) Nature and significance of grooved nuclei of Brenner tumours and Walthard cell islands. Am J Obstet Gynecol 45:614

Black MM, Speer FD (1957) Nuclear structure in cancer tissues. Surg Gynecol Obstet 105:97–102

Blanchard J-M, Piechaczyk M, Dani C, Chambard C, Franchi A, Pouyssegur J, Jeanteur P (1985) c-*myc* gene is transcribed at high rate in $G_0$-arrested fibroblasts and is post-transcriptionally regulated in response to growth factors. Nature 317:443–445

Bloom HJG, Richardson WW (1957) Histologic grading and prognosis in breast cancer: a study of 1409 cases of which 359 have been followed for 15 years. Br J Cancer 11:359–377

Broders AC (1926) Practical points on the microscopic grading of carcinoma. NY State J Med 32:667–671

Chan JKW, Saw D (1986) The grooved nucleus: a useful diagnostic criterion of papillary carcinoma of the thyroid. Am J Surg Pathol 10:672–679

Christopherson WM, Connelly PJ, Alberhasky RC (1983) Carcinoma of the endometrium. V. An analysis of prognosticators in patients with favourable subtypes and stage I disease. Cancer 51:1705–1709

Delides GS et al. (1982) Intralaboratory variations in the grading of breast carcinoma. Arch Pathol Lab Med 106:126–128

Evan GI, Hancock DC (1985) Studies on the interaction of the human c-*myc* protein with the cell nucleus: $p62^{c\text{-}myc}$ as a member of a discrete subset of nuclear proteins. Cell 43:253–261

Evans HL, Ayala AG, Romsdahl MM (1977) Prognostic factors in chondrosarcoma of bone: a clinicopathologic analysis with emphasis on histologic grading. Cancer 40:818–831

Fuhrman SA, Lasky LC, Limas C (1982) Prognostic significance of morphologic parameters in renal cell carcinomas. Am J Surg Pathol 6:655–663

Gaeta FJ, Asirwatham JE, Miller GE et al. (1980) Histologic grading of primary prostatic cancer: a new approach to an old problem. J Urol 123:689–693

Gerdes J, Lemke H, Baisch H, Wacker H, Schwab U, Stein H (1984) Cell cycle analysis of a cell-proliferation associated human nuclear antigen defined by the monoclonal antibody Ki-67. J Immunol 133:1710–1715

Giri DD, Dangerfield VJM, Lonsdale R, Rogers K, Underwood JCE (1987) Immunohistology of oestrogen receptor and D5 antigen in breast cancer: correlation with oestrogen receptor content of adjacent cryostat sections assayed by radioligand binding and enzyme immunoassay. J Clin Pathol 40:734–740

Giri DD, Goepel JR, Rogers K, Underwood JCE (1988) Immunohistological demonstration of progesterone receptor in breast carcinomas: correlation with radioligand binding assays and oestrogen receptor immunohistology. J Clin Pathol 41:444–447

Gray A, Doniach I (1969) Morphology of the nuclei of papillary carcinoma of the thyroid. Br J Cancer 23:49–51

Hartveit F (1971) Prognostic typing in breast cancer. Br Med J 4:253–257
Henson DE (1988) The histological grading of neoplasms. Arch Pathol Lab Med 112:1091–1096
Jakobsson P, Eneroth CM, Killander D, Moberger G, Martensson B (1973) Histological classification and grading of malignancy in carcinoma of the larynx. Acta Radiol Ther Biol 12:1–8
Kempson RL, Bari W (1970) Uterine sarcomas: classification, diagnosis and prognosis. Hum Pathol 1:331–249
Kern WH (1984) The grade and pathologic stage of bladder cancer. Cancer 53:1185–1189
Kloppel G, Lingenthal G, von Bulow M, Kern HF (1985) Histological and fine structural features of pancreatic ductal adenocarcinomas in relation to growth and prognosis: studies in xenografted tumours and clinicopathological correlation in a series of 75 cases. Histopathology 9:841–856
Kreicbergs A, Mezak E, Soderberg G (1981) The prognostic significance of different histomorphologic features in chondrosarcoma. Virchows Arch [A]390:1–10
Lutzner MA, Hobbs JW, Horvath (1971) Ultrastructure of abnormal cells in Sézary's syndrome, mycosis fungoides and parapsoriasis en plaque. Arch Dermatol 103:375–386
Mittal KR, Schwartz PE, Barwick KW (1988) Architectural (FIGO) grading, nuclear grading and other prognostic indicators in stage 1 endometrial adenocarcinoma with identification of high-risk and low-risk groups. Cancer 61:538–545
Morstyn G, Pyke K, Gardner J, Ashcroft R, de Fazio A, Bhathal P (1986) Immunohistochemical identification of proliferating cells in organ culture using bromodeoxyuridine and a monoclonal antibody. J Histochem Cytochem 34:697–701
Oyama T (1989) A histopathological, immunohistochemical and ultrastructural study of intranuclear cytoplasmic inclusions in thyroid papillary carcinoma. Virchows Arch [A]414:91–104
von Hansemann D (1890) Über asymmetrische Zellteilung in Epithelkrebsen und deren biologische Bedeutung. Virchows Arch [A]119:299–326
von Hansemann D (1892) Über die Anaplasie der Geschwulstzellen und die asymmetrische Mitose. Virchows Arch [A]129:436–449

# Ultrastructural Pathology of the Nucleus

H. M. H. KAMEL, J. KIRK and P. G. TONER

## 1 Introduction

This chapter deals with the ultrastructure of the pathological interphase nucleus. Starting with a concise account of normal structure, it aims at providing an illustrated outline of the range of alterations in ultrastructural morphology which may result from or accompany pathological conditions. Where appropriate, the mechanisms of such alterations will be described and their functional and diagnostic significance discussed. It is not possible in a chapter of this size to encompass the whole of this rapidly expanding field, so the emphasis throughout is on those aspects which are of significance in establishing pathogenetic mechanisms and in diagnosing disease.

## 1.1 Normal Nuclear Structure

Because the interphase nucleus in eukaryotic cells has a relatively constant ultrastructure, the basic structural components can be demonstrated in the nucleus of any cell, irrespective of its tissue origin, state of differentiation or functional specialisation. The same components may also readily be seen after the cell has suffered a variety of chemical, physical or biological insults, some of which may affect the shape, size and internal arrangement of the nucleus. Despite this underlying uniformity, different cell types can have distinctive nuclei, characterised by a combination of size, shape, chromatin disposition and type, number and size of nucleoli, and other features. It follows that no consideration of altered morphology should be embarked upon without an awareness of what is normal for the cell and tissue in question (WEISS 1988).

The nucleus has tended to be the Cinderella of ultrastructural studies, because it lacks the membrane-bound specialised organelles, the prominent filaments and tubules and the other formed structures which give such interest and diversity to the cytoplasm. Its relative homogeneity belies its importance as the control centre of the cell.

The four basic nuclear structural components are: the nuclear envelope, which consists of the inner and outer nuclear membranes and the perinuclear cisterna; the chromatin, which is present as dense patches under the nuclear envelope, dispersed throughout the nucleus and in and around the nucleolus; the nucleolus itself; and the nuclear matrix, including the nuclear pore-lamina complex. Much of our knowledge of these structures and of their alterations in disease comes from electron microscopic observations.

From the beginning, however, it must be clearly understood that most isolated nuclear alterations cannot be regarded as pathognomonic of any particular underlying condition. Nuclear and nucleolar enlargement, segregation of nucleolar components, irregularities of the nuclear membrane, chromatin abnormalities, some kinds of nuclear inclusions and an increased nuclear-cytoplasmic ratio can all be observed in many different circumstances, both physiological and pathological.

## 1.2 Techniques in Electron Microscopy

Although for most practical purposes, ultrastructural pathologists rely on conventional, glutaraldehyde-fixed, osmicated, heavy metal-stained, thin section transmission electron microscopy (CTEM), there is increasing use of specialised procedures on tissues which have not undergone this processing sequence, to supplement the available information. Thus, chromatin spreading (SCHEER and ZENTGRAF 1982; TSANEV and TSANEV 1986), histochemical staining (BOUTEILLE et al. 1975; DROZ et al. 1976; FAKAN 1976; GAUTIER 1976; MOYNE 1980), specific labelled molecular probes such as enzyme-gold complexes (BENDAYAN 1982, 1984), immunolabelling (GERACE et al. 1978; KISTLER et al. 1984; SPECTOR et al. 1984; SCHEER and ROSE 1984; HERMANDEZ-VENDUN 1986; SCHMIDT-ZACHMANN et al. 1984) and in situ hybridisation (HAASE 1986; MANUELIDIS 1985; NUOVA and RICHART 1989) are now available as research tools, to give an insight into the chemistry of the nucleus, and in particular its protein and nucleic acid composition. Computer-assisted three-dimensional reconstruction of serial sections and stereological techniques on random sections, to quantify the volume or identify the three-dimensional organisation of nuclear components, are now well developed (BOUVIER et al. 1980; PEBUSQUE et al. 1981; DUPUY-COIN et al. 1986; HERNANDEZ-VERDUN 1986). Combined time-lapse microcinematography and three-dimensional photographic reconstruction have been exploited by DUPUY-COIN et al. (1986) to study nuclear changes and movements in living cells during the cell cycle. Valuable information can also be derived from examination of "thick sections" by high voltage TEM, often combined with tilting stage, stereoscopic observations. Epoxy-free embedding and chromatin extraction procedures enable the examination of the nuclear matrix stereoframework (PENMAN 1985; WEISS 1988).

## 1.3 The Final Electron Microscopic Image

The appearance of the nucleus as revealed by CTEM is an image of the electron-staining properties of a single thin section, comprising only a small fraction of its total volume. The patterns of dark and light on the micrograph reflect the differential staining of its components, modified by the processing to which these structures are exposed prior to electron microscopic examination. It is important to recognise the factors which can modify the final image, some of which are of particular importance to the practising ultrastructural pathologist.

### *1.3.1 Tissue-Processing Effects*

Delay in fixation after tissue excision, postmortem delay, the resulting fall in the pH of the unfixed tissue, and high room temperature are the best known of the factors that can induce morphological alterations in the nucleus, as in other parts of the cell. Others include drying out of the specimen, mechanical trauma through rough handling and diathermy during surgery. Because such factors are very common, though variable, in pathology practice, are often uncontrollable and are rarely properly documented, the electron microscopist must constantly be on guard against misinterpreting their effects. These morphological effects mimic and can easily be confused with those due to several pathological conditions. Condensation of chromatin into heterochromatin clumps, nuclear membrane splitting due to swelling or blebbing of the perinuclear cisterna, and nucleolar condensation are examples of such effects. Comparison of the appearance of perfusion-fixed animal tissues with those excised and fixed immediately following delay (YAGISHITA et al. 1979; COLLAN and SALMENPERA 1976), with and without handling, gives some idea of the nature and scale of the problem. This exercise is strongly recommended for the novice ultrastructural pathologist.

The type and concentration of primary and secondary fixative and the duration and temperature of fixation are additional factors which not only affect the nature and quality of the final image, but can also produce artefactual structural formations which can be misinterpreted as pathological change. In the routine diagnostic laboratory, the pathologist commonly encounters tissue which was initially fixed in buffered formalin and sometimes stored for a prolonged period. Such tissues are often refixed in glutaraldehyde prior to routine CTEM processing. In the worst cases, tissues are rescued from paraffin or frozen blocks before reprocessing. The effects of suboptimal fixation and paraffin processing on nuclear morphology, such as loss of membrane definition in the nuclear envolope, condensation and degradation of the fine structure of chromatin and nucleoli and the appearance of extensive "empty" areas, should perhaps be better known, so that interpretation can be more soundly based. Autopsy tissues, which may often have been stored for prolonged periods in formalin, may also have suffered a delay of many hours before fixation, compounding the problems of ultrastructural interpretation.

Although double staining with uranium and lead is now standard practice, it should be remembered that these and other staining procedures alter the presentation of fixed components and can selectively enhance or obscure detail.

### *1.3.2 Problems of Image Interpretation*

Although with experience, many structures, such as nuclear pores and nucleoli, are recognisable without a conscious awareness of the magnification, the precise identification of several structures in nuclei is dependent on an accurate estimate of their size. This is particularly true of the granular and filamentous components, such as perichromatin and interchromatin granules and the chromatin itself. Accurate identification of virus particles in suspected infections often requires their measurement (Sect. 7.7.2).

The precise determination of magnification in the electron microscope depends on the specimen height being correctly zeroed and the instrument being accurately calibrated. Irrespective of these precautions, the routine achievement of absolute dimensional measurements in biological specimens is an illusion, since the size of a structure in section is not its size in life. Routine electron microscopic processing causes shrinkage of up to 30%.

It must also be borne in mind that focal conditions in the electron microscope can influence the final image to a significant extent at high magnification. Because of the fine scale of many nuclear structures and because out-of-focus images can show artefactual granularity and spurious enhancement of detail, it is important to use well focused images, particularly for the finer structures.

Finally, a thin section can show only a small part of the whole picture which constitutes reality and may display misleading images of various structures, depending upon the angle at which they are intersected by the plane of section. Oblique and tangential sectioning effects are particularly common in relation to the nuclear envelope. Nuclear number and nuclear inclusions are also open to misinterpretation, through a failure to appreciate the effects of plane of section. It is not possible, from a single thin section to determine with accuracy the number of nuclei in any given cell. The light microscopic appearances are more reliable in this respect.

## 2 Changes in Nuclear Number

With experience and with due consideration of the cellular and pathological context, multinucleation, or polykaryocytosis, may often be distinguished from multilobation, with which it can be confused (Figs. 1–4). Certain cytoplasmic features may accompany multinucleation. The occurrence of multiple centriole pairs, to be distinguished from multiple ciliary basal bodies, or of viral nucleocapsids, either intranuclear or intracytoplasmic, may provide useful confirmatory evidence of multinucleation in circumstances where the thin section appearances are equivocal.

### 2.1 Normal Multinucleation

Some normal tissues, such as liver, pancreas and transitional epithelium, routinely display binucleation. Only a few normal tissues, such as skeletal muscle and syncytiotrophoblast, are truly syncytial.

### 2.2 Multinucleation in Pathological Conditions

True binucleation or multinucleation occurs most often in reactive or neoplastic situations and in certain viral infections. For example, binucleate

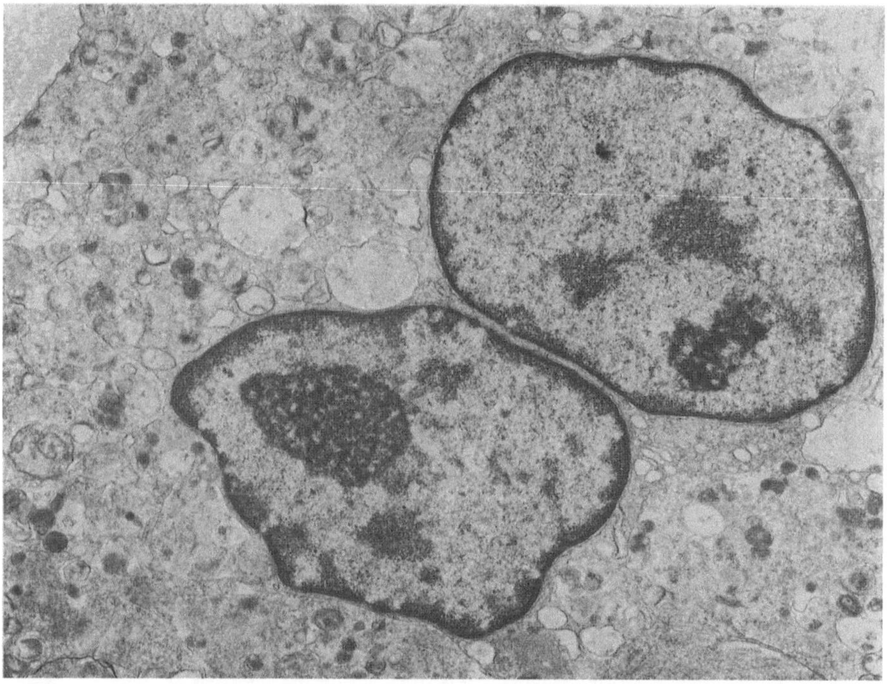

**Fig. 1.** A bilobed nucleus in a granular cell myoblastoma. A prominent fibrous lamina is present and the nucleoli are reticulated. Note the abundance of secondary lysosomes in the cytoplasm. × 9000

plasma cells may be seen in a variety of pathological conditions, ranging from neoplasms such as multiple and solitary myeloma, to reactive conditions such as chronic periodontitis, periapical granuloma, pemphigus vulgaris and other situations (JINN-FEI and EL-LABBAN 1986; YEO 1986). The Reed-Sternberg cell of Hodgkin's disease typically appears binucleate or bilobed. The binucleate Reed-Sternberg cell sometimes displays further nuclear irregularity and segmentation, giving a complex multilobated binucleated appearance (Fig. 3). The normally multinucleated muscle cells can also acquire a further increase or alternatively decrease in their nuclear number in several muscular disorders and in hypertrophic and regenerative conditions. This and other nuclear charges associated with muscle disorders have been recently reviewed (TOME and FARDEAU 1986).

### *2.2.1 Multinucleate Syncytia in Viral Infection and Neoplasia*

Multinucleation can result from nuclear division without accompanying cytokinesis, or alternatively, from the fusion of mononuclear cells.

The formation of syncytial giant cells, with many nuclei, late in the infectious cycle is a pathological feature of infections with certain enveloped

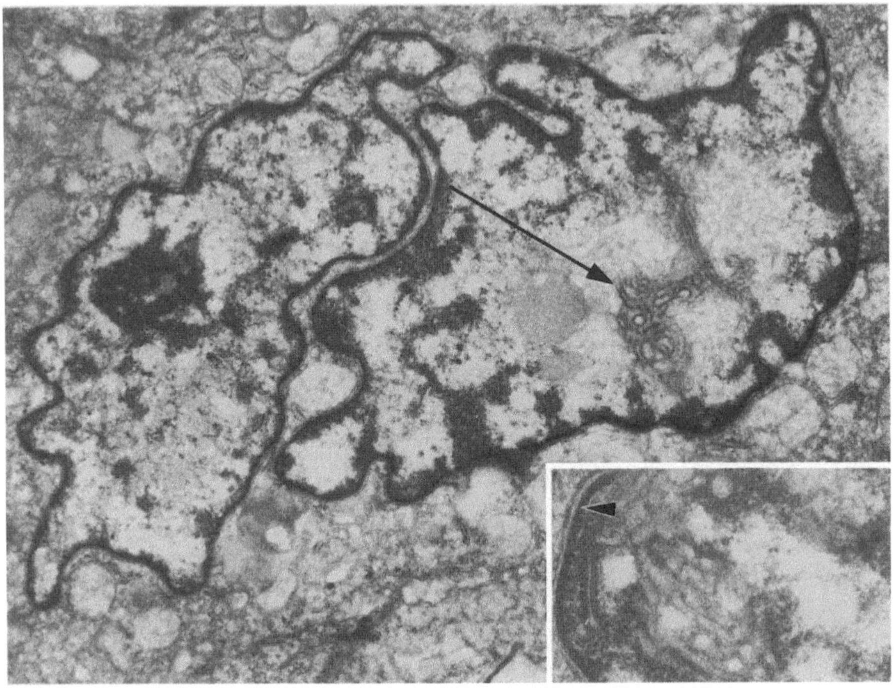

**Fig. 2.** Nuclear irregularity causes a bilobed appearance in this case of bronchiolo-alveolar cell carcinoma. Note the tubular inclusion (*arrow*). *Inset*: Tubular inclusions in bronchiolo-alveolar cell carcinoma. Note the prominent fibrous lamina (*arrowhead*). Autopsy tissue, × 9500; inset, × 28 000

viruses, including measles and herpes simplex (KINGSBURY 1985; DUPUY-COIN et al. 1982; WOLINSKY 1979). This is also a prominent feature in the brain in encephalopathy which accompanies the acquired immune deficiency syndrome (AIDS) (SHARER et al. 1985; BUDKA 1986). In a recent demonstration of giant multinucleated cells in AIDS encephalopathy, in situ hybridisation has detected genomic nucleic acid of the human immunodeficiency virus (HIV) and viral particles have been seen by TEM in the giant cells (KOENIG et al. 1986). Other studies have demonstrated their haematogenous nature (BUDKA 1986). While most authors accept that these giant cells form by fusion, this may not necessarily be the case. A recent report has demonstrated nuclear bridges in giant cells in AIDS (MIZUSAWA et al. 1987). Such nuclear bridges are much more commonly seen in the giant tumour cells of glioblastoma multiforme, where amitotic division is more likely to be responsible for giant cell formation (KROH and CERVOS-NAVARRO 1988).

The process by which monocytes transform into macrophages and under certain circumstances, including anoxia, low pH and lactic acidosis, may fuse to form epithelioid cells and giant cells, has been studied in detail by SUTTON (1967), who describes a progressive reduction in marginated

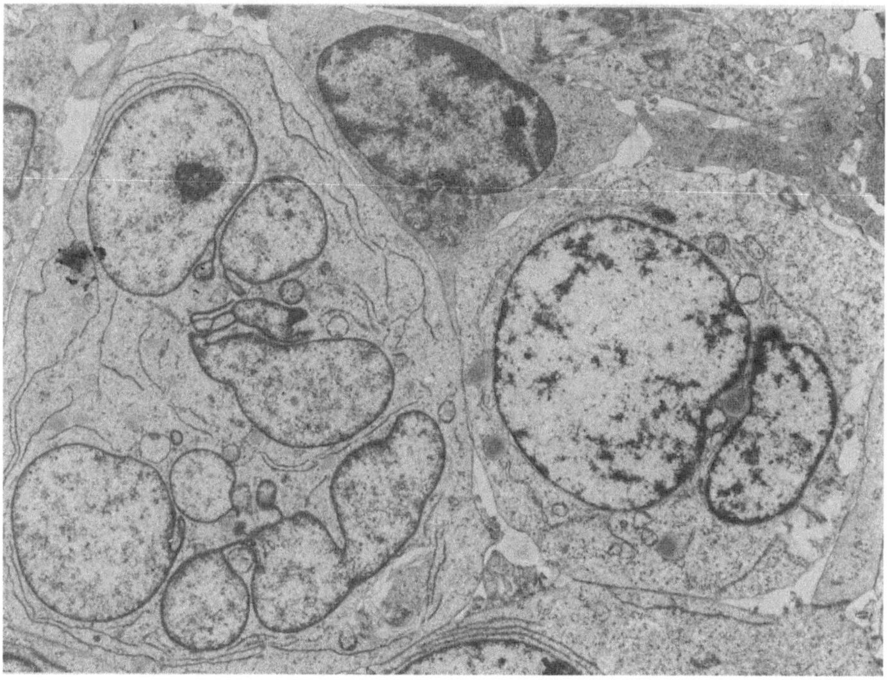

**Fig. 3.** Two cells of Reed-Sternberg type in a case of Hodgkin's disease. One appears binucleate or bilobed, the other multinucleate or multilobed; a firm distinction cannot be made on thin sections. × 4000

heterochromatin, the appearance of multiple nucleoli and the eventual fusion of contiguous cells.

Giant syncytial cell formation during measles and other paramyxovirus infection results from fusion of adjacent cells, which follows the incorporation of large amounts of a specific fusion glycoprotein, the F protein, into the plasma membrane and its activation of trypsin-like proteases in the environment (GALLAHER and BRATT 1974; KINGSBURY 1985). In neoplasia the reason for multinucleation may be lack of regulation of nuclear division, or alternatively lack of co-ordinated nuclear and cytoplasmic division. The reason for fusion of macrophages is not known, but it could involve enzymatic modification of the plasma membrane proteins and glycoproteins by latent enzymes, active only at low pH (SUTTON 1967).

### 2.2.2 *In Vitro Induced Multinucleation*

Induced multinucleation, either by fusion or by nuclear division without cytokinesis, has some value in the laboratory. Hybridoma cells, which are functional heterokaryons, are produced using heat-inactivated Sendai virus to fuse antibody-producing B-lymphocytes with myeloma cells, as a pre-

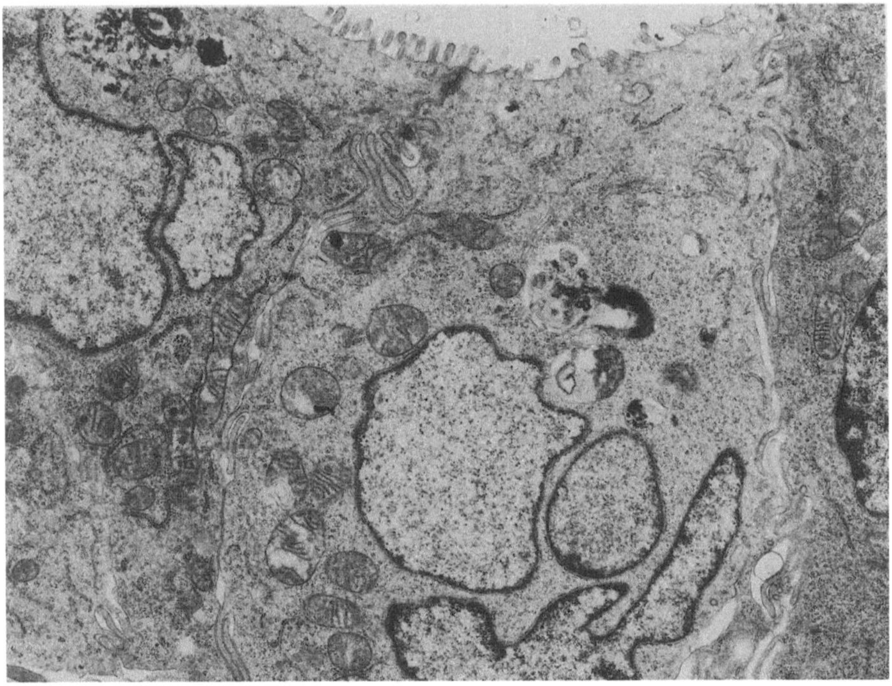

**Fig. 4.** Marked irregularity in the nucleus of a mouse small intestinal crypt epithelial cell after X-irradiation. The appearances in thin section resemble multinucleation. × 11 200

liminary to producing monoclonal antibodies (KOHLER and MILSTEIN 1975). A valuable means of distinguishing tumour from non-tumour cell lines in vitro is to apply cytochalasin B, which acts to disrupt the cytoplasmic microtubules of the mitotic spindle. Following such treatment, the nuclei of non-tumour cells divide once, giving rise to binucleate cells, while unregulated tumour cell nuclei divide repeatedly, creating distinctive polykaryons (SOMERS and MURPHEY 1980). Multinucleation can also be induced by agents that inhibit the mitotic spindle such as Colcemid. The resulting micronuclei are each formed of chromatin material of a single chromosome surrounded by nuclear envelope. This phenomenon has been exploited in investigating the nuclear envelope-chromatin interrelationship (CHAI et al. 1978).

## 3 Changes in Nuclear Chromatin

In conventional ultrathin sections, chromatin is present in two forms, the clumped granules of densely aggregated, functionally inactive heterochromatin and the finely dispersed granules of functionally active euchromatin. The distribution and relative proportions of these two forms of chromatin in different cells show considerable normal biological variation and are influenced by a variety of pathological processes. Within the nuclear substance, there occur discrete dense aggregates known as perichromatin and interchromatin granules.

### 3.1 Euchromatin and Heterochromatin

Euchromatin, an essential component in active DNA and RNA synthesis, is abundant in functionally active and secreting cells, as well as in proliferating blast and stem cells. Nuclei of resting cells or cells exposed to agents which inhibit RNA synthesis contain more heterochromatin. Conditions of accelerated growth, as in regeneration, hyperplasia and neoplasia, are, therefore, characterised by a decrease in the heterochromatin-euchromatin ratio (Fig. 3). An increase is seen in sublethally and lethally damaged cells (Figs. 5–7) and following treatment with cytotoxic agents that inhibit RNA synthesis, such as actinomycin D, ethionine and amanitine.

Exceptions to the above generalisations cannot easily be overlooked. Highly anaplastic tumours sometimes display a significantly increased

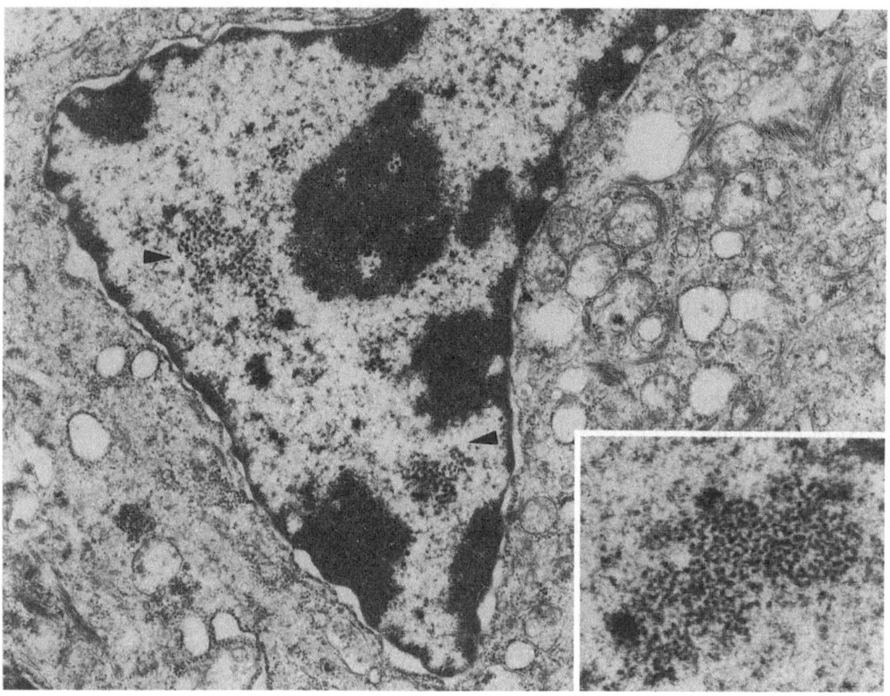

**Fig. 5.** Aggregates of interchromatin granules (*arrowheads* and *inset*) in actinomycin D-treated Ridgeway osteogenic sarcoma. Note the dilatation of the nuclear envelope and the prominence of the nuclear pores. × 18 200; inset, × 27 000

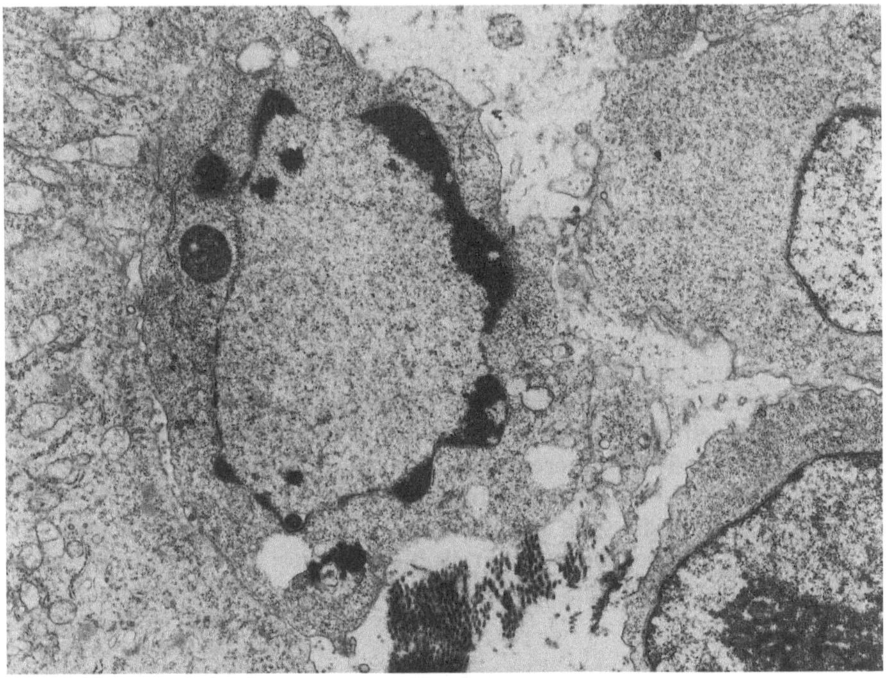

**Fig. 6.** Peripheral clumping and margination of chromatin in methotrexate-treated mouse sarcoma 180. Note also the dense inclusions in the cytoplasm. × 11 200

heterochromatin-euchromatin ratio and actively secreting plasma cells are characterised by their large peripheral dense heterochromatin clumps. It may be that in undifferentiated tumours, prominent heterochromatin aggregation reflects a state of hyperploidy, and that a relatively small euchromatin fraction is adequate to sustain proliferative activity. Alternatively, it may be that the anaplastic state is sometimes associated with loss of various functional capabilities, with the corresponding genetic material being maintained in the inactive heterochromatin form. In the case of the plasma cell, FAWCETT (1981) suggested that as these cells are committed to a particular narrowly defined function, the secretion of a single pattern of immunoglobulin molecule, only the relevant small fraction of the genomic DNA is required in the active form.

Dispersed aggregates of chromosomal material are sometimes seen in the cytoplasm of neoplastic metaphase cells. These have been attributed to asynchrony between the rate of nuclear condensation and nuclear envelope formation in the dividing population of neoplastic cells (HENDERSON et al. 1986).

Margination of the chromatin, leading to the pavementing of dense chromatin alongside the inner nuclear membrane, is a morphological phenomenon observed to a variable extent in irreversibly damaged or dying cells (Figs. 6, 7). It is thought to represent a final stage prior to cell

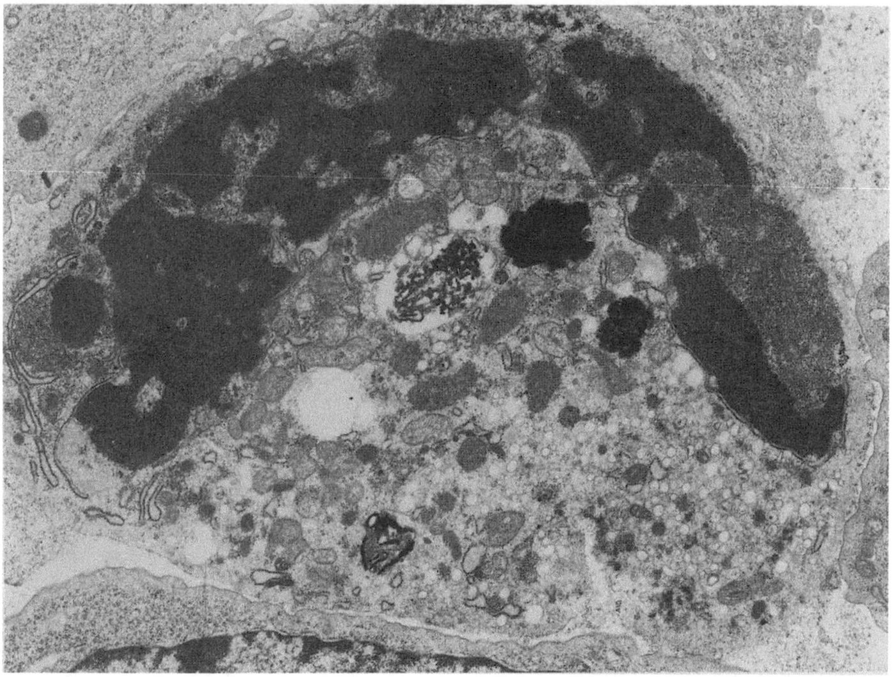

**Fig. 7.** An apoptotic body in methotrexate-treated mouse sarcoma 180, showing the condensed nuclear structure. × 11200

death and pyknosis. This change is more common after some insults than others. In the cell death by apoptosis (Fig. 7) the morphological hallmark is that of cellular shrinkage, darkening and condensation including the various nuclear components. It is seen after ischaemia and X-ray irradiation and during certain viral infections and after certain chemotherapeutic agents such as actinomycin D and methotrexate (Figs. 5–7). Peripheral margination and clumping of the chromatin are hallmarks of dying and dead cells.

The staining and distribution of chromatin can be subject to artefacts due to factors related to tissue preservation, fixative type, pH and storage temperature. These factors can considerably alter the appearance of the chromatin. The resulting common changes of dense chromatin clumps, condensation along the nuclear periphery and blebbing of the nuclear membranes mimic those seen after some agents in direct damage.

### 3.2 Perichromatin Granules

Perichromatin granules are electron-dense RNP granules, 30–35 nm in diameter (Watson 1962), substantially larger than the individual granular components of chromatin. They are scattered often singly in the peri-

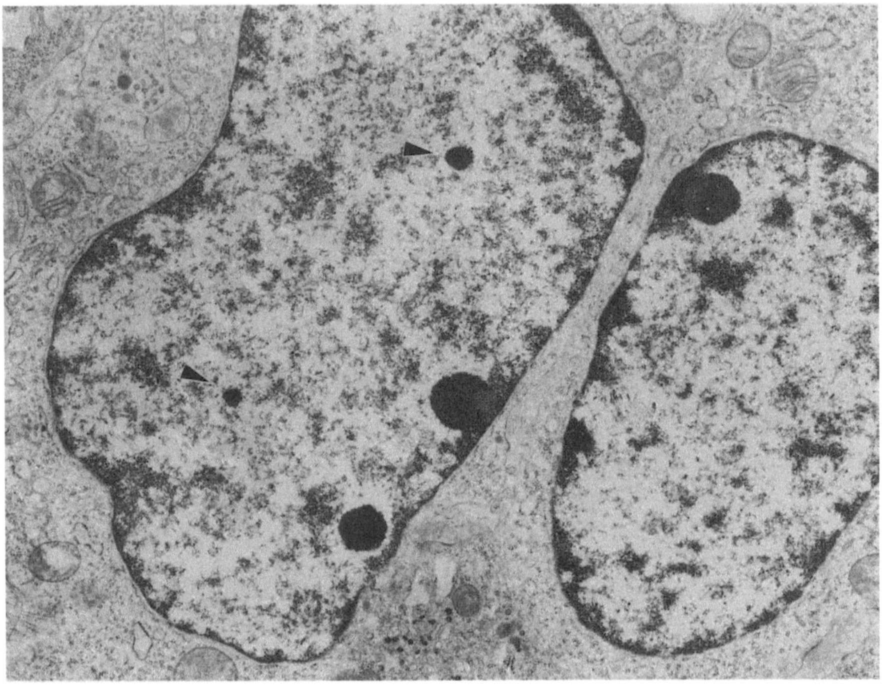

**Fig. 8.** Actinomycin D-treated mouse Ridgeway osteogenic sarcoma. Two giant perichromatin granules are seen (*arrowheads*). Note the condensed nucleoli, one of which is surrounded by a clear halo. × 7100

chromatin regions. They are surrounded by areas of electron translucency and are often localised at or near the margins of heterochromatin masses. They are frequently seen in normal hepatocytes, pancreatic cells, and neuronal and glial cells (BOUTEILLE 1972). SMETANA (1977) demonstrated the presence of chromatin filaments connecting the perichromatin granules to the nearby chromatin.

The presence of increased numbers of perichromatin granules appears to be suggestive of protein synthesis inhibition or impairment (GHADIALLY 1988). They are frequently observed after treatments known to have this effect, such as actinomycin D (KAMEL et al. 1988), aflatoxin B, (DERENZINI and MOYNE 1978), cycloheximide (DASKAL et al. 1975) and after thermic shock (HEINE et al. 1971) and hyperthermic treatment (KAMEL 1985). Increased perichromatin granules have also been noticed in several neoplastic conditions and in experimental animal tumours (GHADIALLY 1982, 1985; MURAD and SCARPELLI 1967; KAMEL 1985). The intranuclear spheroidal bodies seen in cases of nasopharyngeal fibroma are said to represent giant perichromatin granules (GHADIALLY 1985). Spheroidal bodies of similar appearance are also observed after treatment of mouse Ridgeway osteogenic sarcoma with actinomycin D (Fig. 8) (KAMEL et al.

1988). Here again, these observations may reflect aberrations in protein synthesis. Conversely, perichromatin granules are lost early in the course of autolysis (KARASEK 1975).

### 3.3 Interchromatin Granules

Interchromatin granules are 10–20 nm in diameter, but otherwise have a similar appearance to the larger perichromatin granules. They often occur in clusters interconnected by fine filamentous material (Fig. 5) (MONNERON and BERNHARD 1969). They can sometimes be mistaken for virus particles. Like the perichromatin granules, increased numbers have been described in various neoplastic conditions. The association of increased interchromatin granules aggregates with nucleolar alterations due to toxic agents or virus infections suggests a nucleolar origin for these granules (SINGER 1975; RECHER et al. 1976). The authors also found increased interchromatin granules away from the morphologically unaltered nulceoli in the mouse S 180 tumour cells treated with methotrexate (unpublished observations) (Fig. 5). This may indicate that interchromatin granules can be derived from nuclear component other than nucleolar RNP.

### 3.4 Nuclear Pockets

Nuclear pockets represent localised invaginations of nuclear or cytoplasmic material into a saccular irregularity of the nuclear envelope, with an underlying rim of chromatin material. GHADIALLY et al. (1985 b) pointed out that nuclear pockets are distinguished from nuclear pseudoinclusions by the presence of this chromatin band beneath the nuclear envelope. Nuclear pockets were classified into type 1, containing cytoplasmic material, and the less common type 2, containing nuclear material.

Nuclear pockets have been observed in malignant tumour cells such as retinoblastoma, leukaemic and lymphoma cells, in the nuclei of Leydig cells in cases of Klinefelter's syndrome (NISTAL et al. 1985), after treatment by chemotherapeutic agents such as cytosine arabinoside, methotrexate and fluorouracil, and in the granulocytes of patients with pernicious anaemia (KAMEL 1985; GHADIALLY 1988).

Current views suggest that nuclear pockets may reflect a state of chromosomal abnormality (GHADIALLY 1988). AHEARN et al. (1974) showed that there was direct correlation between the occurrence of nuclear pockets and aneuploidy in leukaemic cells, and that following treatment their frequency declined. These observations and the nature of the conditions with which nuclear pockets are associated appear to support the view that they may reflect an underlying chromosomal defect.

# 4 Changes in Nuclear Size

## 4.1 The Relationship of Size to Transcriptional Activity

The nuclei of normal interphase cells vary greatly in size. In most tissues, this is characteristic of the cell type and its metabolic state. This dimensional variation occurs despite the fact that each somatic cell contains the same amount of DNA, corresponding to the full genomic complement. When this DNA, together with its associated proteins, is condensed into the heterochomatic form, it takes up less space and therefore should need a smaller nucleus than if it were in its euchromatic form. Indeed there is often a roughly inverse relationship between the quantity of heterochromatin and nuclear volume. The size of a nucleus, however, must normally depend not only on the amount of heterochromatin but also to a large degree on the transcriptional activity of its euchromatin component. Thus transcriptionally inactive euchromatin is more condensed than active euchromatin, though not so tightly packed as to be heterochromatic (WEISS 1988). In most cells only a small proportion of euchromatin is active at any one time, but this proportion may change with the functional state of the cell and hence the volume may change in parallel, even though the amount of heterochromatin remains the same. An activated cell with a larger nucleus may thus appear to have less heterochromatin, simply because it is spread more thinly in the greater volume.

## 4.2 Estimation of Nuclear Size

One of the common features of neoplasia is increased nuclear diameter and hence volume. The incidence of such enlargement is reviewed by GHADIALLY (1985). Increased size is usually based on a subjective or at best semi-quantitative impression. Given the restrictions imposed by ultrathin sections and given the inevitable variations in plane of section, accurate measurements are very difficult without resort to serial sectioning and stereological techniques. Such methods are too tedious and time-consuming for most diagnostic purposes. Reliance must instead be placed on impressions derived from a subjective assessment of the cell populations involved, or from the direct measurement of a small sample. The most common error in assessing nuclear size is underestimation, due to measurements being carried out on the profile of the tip of an oval or lobed nucleus or any part of a rod-shaped nucleus. Such profiles, alternating with larger cross-sections on a micrograph, may erroneously suggest major variations in nuclear size within a field, which may be interpreted as indicating the presence of more than one cell type. Assessment of the amount of heterochromatin, of the appearance of the nuclear pores, whether sectioned en face or not, and of cytoplasmic features associated with nuclei

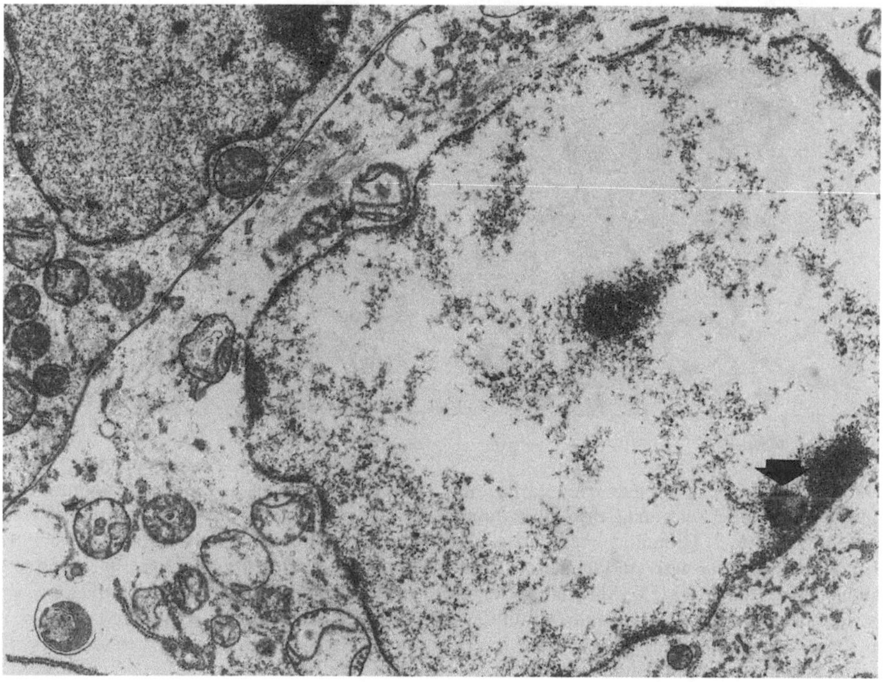

**Fig. 9.** Nuclear swelling in mouse small intestinal villous epithelium after hyperthermic treatment. The nucleus in the nearby cell ceems less severely affected. Note the ring-shaped nucleolus (*arrowheads*). × 8200

of apparently different size, may help in the avoidance of such mistakes. For relatively uniform spherical nuclei, randomly sectioned, the best measure of nuclear diameter is obtained from examination of the largest profiles in the field.

## 4.3 Pathological Swelling and Shrinkage

In some degenerating cells, swelling of the nucleus can occur, resulting in a large nucleus with abundant heterochromatin set in a watery milieu which may or may not contain the remains of the euchromatin (Fig. 9). The nucleus, with its envelope remaining intact, can expand to a maximum of four times its original volume (BRANSTETTER and GOLDBLATT 1983). This limitation in the extent of swelling may be imposed by the nuclear matrix component rather than by the nuclear membranes (RILEY and KELLER 1978). More commonly, however, degeneration of cells is accompanied by shrinkage and increased condensation. This condensation should not be regarded as the same process which results in the formation of heterochromatin from euchromatin during differentiation. Certainly the euchromatin volume decreases, indicating a cessation of transcription. This

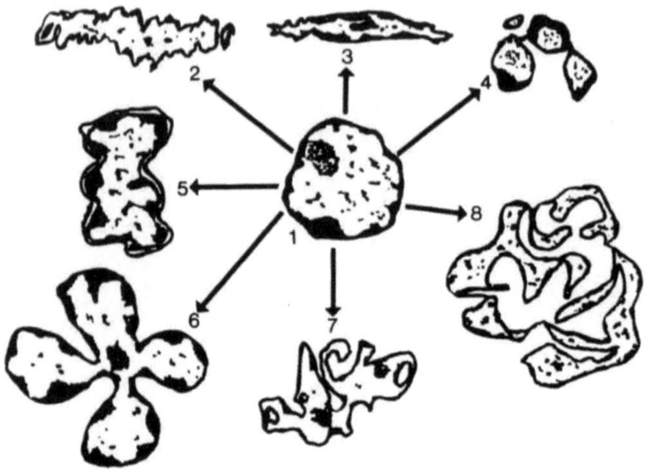


**Fig. 10.** Spectrum of nuclear irregularity: note that these diagrams assume that the nucleus in question has been sectioned through its major diameter. *1*, Round or oval nucleus with minimal irregularity. *2*, Crenated irregularity of spindle-shaped nucleus, as in smooth muscle cell. *3*, Polar irregularity with spindle shape, as in fibrocytes. *4*, Convoluted nucleus, as in polymorphonuclear leucocytes. *5*, Shrunken irregularity with separation of nuclear envelope and peripherally condensed chromatin, as in degenerating cells. *6*, Raidally segmented nucleus, as in some leukaemic cells. *7*, Haphazard irregularity, as in some histiocytes and neoplastic cells. *8*, Cerebriform, deeply convoluted nucleus as in Sezary's syndrome

is accompanied by an increase in the proportion of dense material. It is not known, however, if this material is identical to the heterochromatin found in the normal cell. The underlying mechanisms for such events in degenerating or dying cells may be disturbances in the ionic concentration balance and in membrane permeability (BLACKBURN 1971).

## 5 Changes in Nuclear Outline and Location

As a result of the higher magnification and greater resolution of electron microscopy, many normal nuclei may show an unexpected degree of surface irregularity in ultrathin sections, by comparison with their light microscopic appearance in routine paraffin sections. There is no clear understanding of the mechanisms by which nuclear irregularity is generated, whether as a routine feature of a particular cell type or as a reflection of some physiological or pathological change.

### 5.1 Normal Nuclear Irregularity

The occurrence of irregularity of contour is a normal morphological characteristic of the nucleus in many cell types, such as smooth and

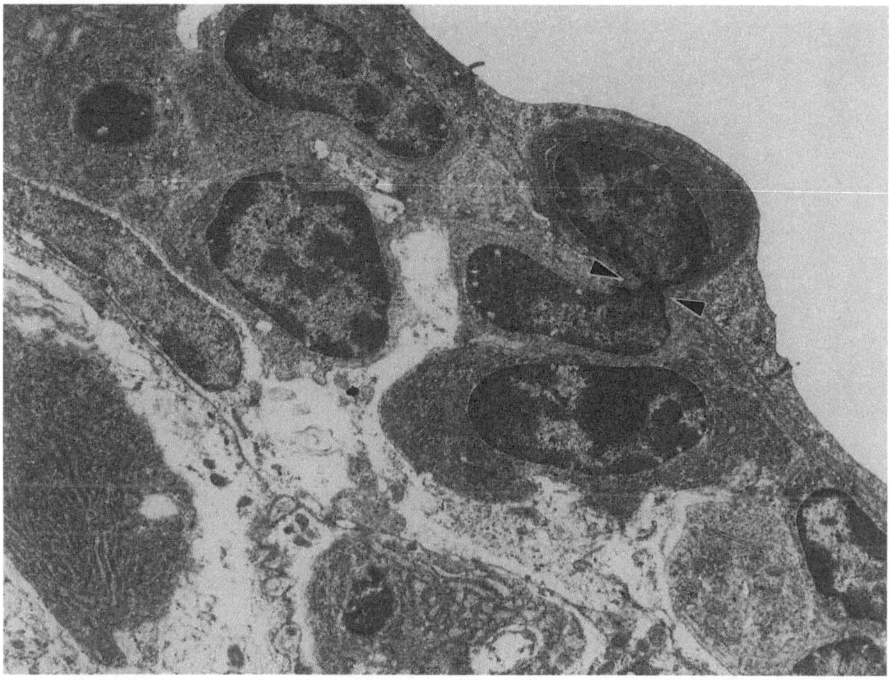

**Fig. 11.** Nuclear plasticity. An inflammatory cell traversing a vessel wall becomes temporarily bilobed, the two parts connected by a narrow waist or bridge (*arrowheads*). × 3900

striated muscle and endothelium (Fig. 10) (WEISS 1988). This can be seen as an adaptive mechanism in response to substantial changes in the overall size and contour of these cells, related to their mechanical functions and physical state. Inflammatory cells crossing vessel walls often show extreme bilobing and bridge formation (Fig. 11), apparently as a result of the enforced narrowing and constriction which is required to enable the cell to pass through very small apertures (MIGLIORISI et al. 1987). Nuclear irregularity is also a feature of the ageing process in liver, pituitary and adrenal gland cells (KLEINFELD and KOULISH 1957; SOBEL et al. 1969). Increasing segmentation and lobulation are characteristic features of the maturing neutrophil polymorphonuclear leucocyte.

## 5.2 Acquired Nuclear Irregularity

Elaborate nuclear irregularity and segmentation, nuclear invagination and deep clefting of the nuclear envelope are frequently observed in benign and malignant neoplasms and in secondary tumour deposits (Fig. 12) (GHADIALLY 1985, 1988; HENDERSON et al. 1986). These irregularities are prominent in certain types of leukaemia and lymphoma and can be of diag-

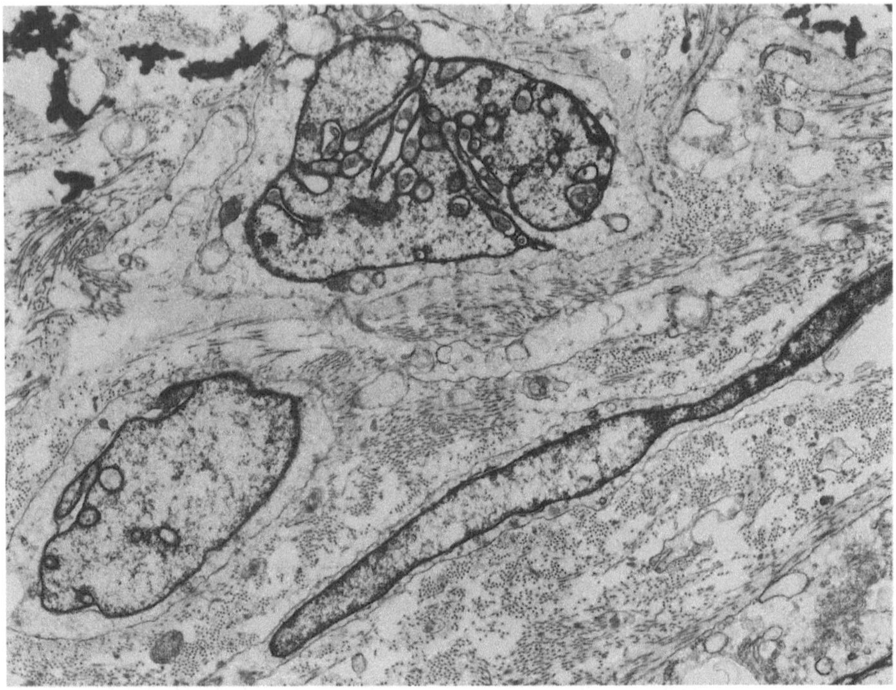

**Fig. 12.** Malignant fibrous histiocytoma showing variations in nuclear shape: extreme elongation of a spindle-shaped fibroblastic type nucleus is seen. The other two nuclei show elaborate irregularity with lobulation, segmentation and invaginations. × 6300

nostic significance in cases of Sézary's and Reider's syndromes (GHADIALLY 1985, 1988; HENDERSON et al. 1986 for references).

The pattern of nuclear irregularity, as much as its mere presence, can be of diagnostic value (Fig. 10). The radially segmented nuclei of Reider's cells characterise a range of lymphomas and leukaemias, including B-cell lymphomas, acute and chronic lymphoblastic leucaemia, acute myeloid leukaemia, chronic myelomonocytic leukaemia and lymphocytic leukaemia (GHADIALLY 1985, 1988; HENDERSON et al. 1986). The cerebriform nuclei of Sézary's syndrome display a peculiarly distinctive segmentation, with deep identations. Nuclear irregularity in most types of neoplasia, however, has no one characteristic pattern, but takes the form of apparently random surface disturbances, widely variable in nature and extent. No particular pattern is completely exclusive to any one pathological entity, although its presence and frequency of occurrence in a neoplastic cell population can sometimes be of assistance in favouring one diagnosis against another.

## 5.3 Pathogenesis of Nuclear Irregularity

An acquired or increased irregularity of the nucleus provides an expanded nuclear surface area, which in turn allows for increased nucleocytoplasmic exchange. This may explain the prominent irregularity sometimes observed in conditions characterised by increased cellular activity, including various forms of malignancy. Increased irregularity, however, is sometimes also seen in the nuclei of slowly growing and metabolically quiescent benign tumour cells. This phenomenon, as well as the occurrence of similar nuclear changes with age, would seen to argue against heightened metabolic activity as the sole correlate of nuclear irregularity in neoplasia.

An increased irregularity of nuclear contour may be seen in "insulted" or "degenerating" cells, particularly in certain viral infections and after exposure to irradiation and to some chemical agents and chemotherapeutic drugs such as methotrexate (KAMEL 1985). Increased nucleocytoplasmic exchange can again be hypothesised as a compensatory change, reflecting the enhanced cellular metabolism required to combat the effects of the injurious agent. Alternatively, the observed irregularity may simply be a non-specific secondary manifestation of cell injury induced by such changes as the membrane permeability and ionic concentration. Normal irregularity in the nuclei of muscle cells has been attributed more to their function-related changes in the ionic concentration than to the mechanical effects (FRANKE and SCHINKO 1969).

Other hypotheses have been advanced to explain the pathogenesis of acquired nuclear irregularity. One such hypothesis proposes that nuclear contour is affected by changes in the microtubular or filamentous components of the cytoplasm (BELTRAN and STUKEY 1972; ZUCKER-FRANKLIN et al. 1974; NORBERG 1971) and that contraction of these structures in different pathological conditions is responsible for the altered morphology (BESSIS 1961). Inhibition of formation of the radially segmented nuclei of Reider's leukaemic cells occurs after treatment with agents that depolymerise their microtubular component such as colchicine, low temperature and vinblastine (NORBERG and SODERSTROM 1967; NORBERG 1969). Prominent irregularity, however, can occur in the absence of significant microtubular or filamentous components and vice versa. Another possibility is that such changes are induced by rapid alterations in nuclear or cytoplasmic volume and may be temporal events. There is no doubt that the detailed configuration of the nucleus in any cell is constantly changing in life, although apparently within certain typical limits.

## 5.4 Nuclear Location

The limitations of ultrathin sectioning make it difficult to define the true location of the nucleus from electron micrographs, since random sections will show widely variable appearances. Time lapse cinematography has shown that the nuclei of various cell types undergo cycles of regular or irregular rotation (DUPUY-COIN et al. 1986). Some cells do show a preferential location for the nucleus, which may be evident on elecron micrographs, although others do not.

The location may change during development, as in muscle cells, where the central nuclei of developing myotubes give way to a peripheral situation in the mature cell. Re-internalisation or centralisation of the muscle nuclei may be observed as a pathological change in regenerative disease and in the centronuclear type of congenital myopathy (TOME and FARDEAU 1986). The reasons and mechanisms for such changes are not well understood, although involvement of the intermediate filaments has been suggested in the case of muscle cells.

## 6 The Nuclear Envelope and Matrix

The nuclear envelope and matrix are intimately related structures. The outer and inner nuclear membranes and the nuclear pore-lamina complex, together with the proteinaceous part of the nuclear matrix, share a common function in nucleocytoplasmic exchange. The nuclear envelope represents the outer nuclear boundary, which separates and protects it from the cytoplasm and delineates its morphology. The matrix is the pool in which lie the nuclear chromatin and the nucleolus.

### 6.1 The Nuclear Envelope

The nucleus is surrounded by a double-membraned envelope. The outer nuclear membrane is in continuity with the rough endoplasmic reticulum. The perinuclear cistern is a 20–30 nm wide space that separates the outer from the inner nuclear membranes (both are 5–7 nm thick), except where they fuse to form the nuclear pores. The nuclear envelope with its pores is strategically situated, both bridging and separating the discrete but interdependent structures of the nucleus and the cytoplasm (FRANKE 1977; FRANKE et al. 1981; GERACE and BLOBEL 1982; GERACE and BURKE 1988).

In functional terms, the nuclear envelope plays a regulatory role, permitting controlled nucleocytoplasmic exchange, perhaps mainly through the nuclear pores (CLAWSON et al. 1980; DE ROBERTIS and DE ROBERTIS 1987). The anchorage of the nuclear chromatin at its inner aspect might indicate a role in DNA transcription and processing the genetic material (FRANKE and SCHEER 1984; COMINGS and OKADA 1973; ZBARSKY 1978), while the continuity of its outer membrane with the rough endoplasmic reticulum could suggest a link with protein and membrane synthesis (FRANKE and SCHEER 1974; FRANKE et al. 1981) while providing a mechanism for the dissolution and reconstitution of the nuclear envelope during mitosis. Its strategic location raises the possibility that the nuclear membranes have a protective or barrier role through, for example, their associated drug metabolising enzymes and might be capable of modulating the effects of agents such as chemicals and hormones that can gain access to the nucleus (BRANSTETTER and GOLDBLATT 1983).

### *6.1.1 Spectrum of Nuclear Envelope Alterations*

The common changes caused by various pathological conditions or injurious agents can be broadly, if rather arbitrarily categorised as either proliferative or non-proliferative (Fig. 13). It should be emphasized that either type of alteration can be produced by the same agent. The nature of the induced changes, however, will depend on the extent of the pathological event or the dose of the cytotoxic agent, as well as the effect on other cytoplasmic or nuclear structures. A particular agent at different dosage can thus produce either proliferative or non-proliferative membrane changes. Irradiation in non-lethal doses tends to produce largely proliferative changes, while damaged radiosensitive cells often show non-proliferative alterations. The changes can involve one membrane, often the inner, or both. In general, envelope changes are seen mainly in association with viral infections, described later in more detail, in neoplasia and following exposure to irradiation and certain chemical agents (BLACKBURN 1971; BRANSTETTER and GOLDBLATT 1983; GHADIALLY 1988).

### *6.1.2 Non-proliferative Membrane Alterations*

Non-proliferative membrane alterations include membrane separation or fusion, evaginations or blebbing, sloughing or fragmentation, with or without the formation of annulate lamellae, temporary defects (FLICKINGER 1974, 1978) and nuclear envelope rupture or lysis (Fig. 13). Generally, such changes are seen in degenerating cells and are associated with cytoplasmic damage or decreased RNP synthesis, whether mediated by starvation, thermal insult, viral infection, toxic compounds or irradiation (BLACKBURN 1971; GHADIALLY 1988). Some of these changes, however, also accompany maturation and ageing.

Nuclear membrane separation or "blebbing" is seen in degenerating cells, while nuclear membrane rupture occurs in necrotic cells. However, temporary defects in the nuclear envelope are not lethal events in themselves and may be repaired by the overlying endoplasmic reticulum (FLICKINGER 1974, 1978). Some nuclear inclusions which arise from the cytoplasm may gain access to the nucleus through these defects, which are later sealed by the endoplasmic reticulum.

Dense particulate inclusions are sometimes seen within the perinuclear cistern (BLACKBURN 1971). They can be of viral origin or induced by toxic

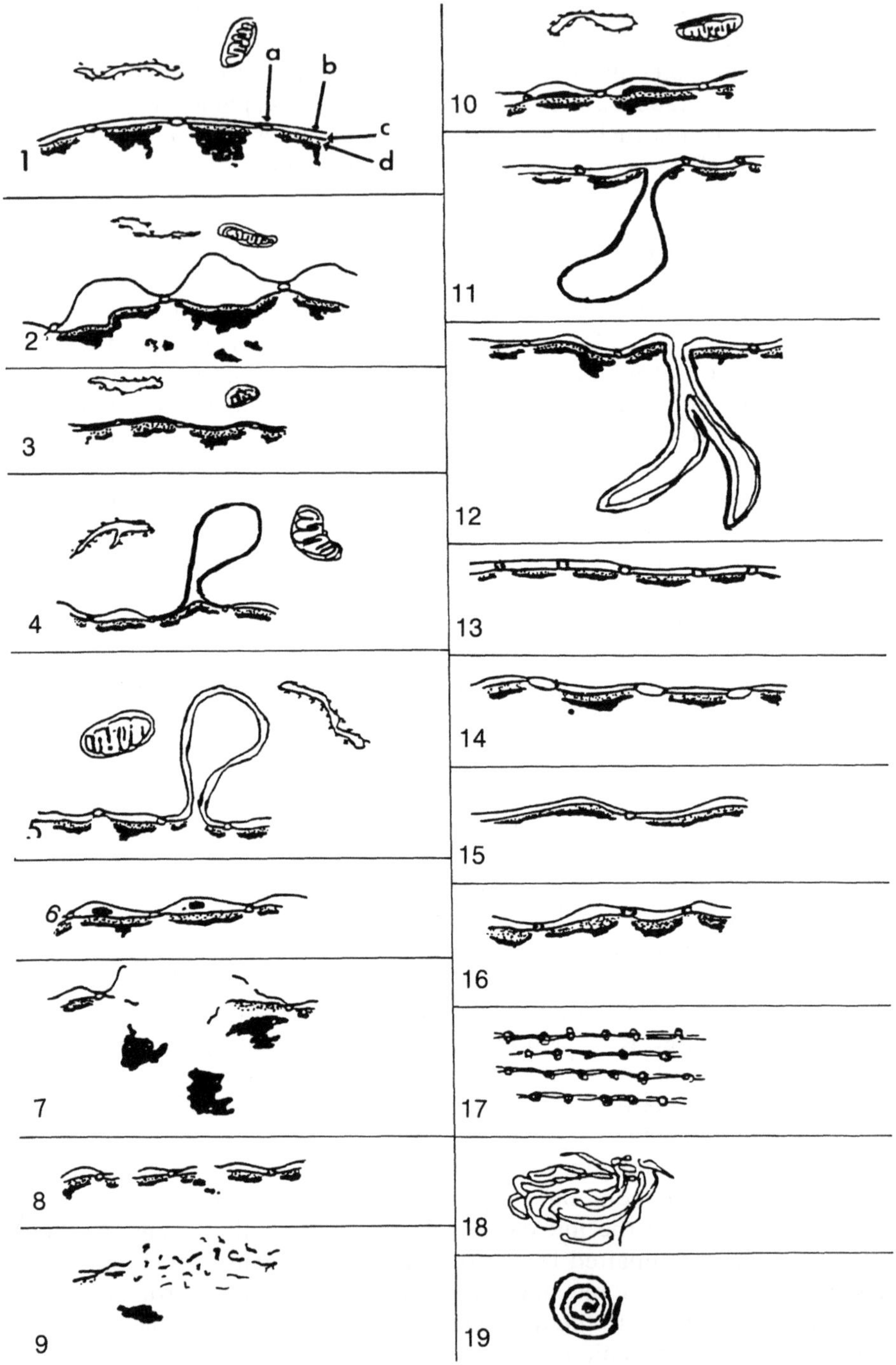
a
b
c
d
1
2
3
4
5
6
7
8
9
10
11
12
13
14
15
16
17
18
19

agents. Accumulation of immunoglobulin material in the perinuclear space is described in cases of multiple myeloma (DJALDETTI and LEWINSKI 1978). In effect, any material which may be found in the cavities of the endoplasmic reticulum may also gain access to this compartment.

Patchy or macular thickening of the inner nuclear membrane characteristically occurs in certain viral infections and is thought to be stimulated by viral antigens or replicative activity. It is also sometimes seen in senile cells or cells with low metabolic activity (BLACKBURN 1971).

### *6.1.3 Proliferative Membrane Alterations*

The nuclear envelope may form finger-like projections into the cytoplasm or the nucleus, or may produce elaborate tortuous branching invaginations with or without reduplication of the nuclear membranes (Fig. 13). Though these are described mainly in neoplasia and viral infections (BLACKBURN 1971; Henderson et al. 1986; GHADIALLY 1988), they are also seen in less common pathological situations such as storage, genetic muscular dystrophy diseases and following exposure to certain chemical agents, irradiation or hypothermia (BLACKBURN 1971; SZEKELY et al. 1980).

Nuclear envelope proliferation may also result in the formation of concentric lamellar membranous arrays, nuclear pockets, annulate lamellae and tubular inclusions (Figs. 2, 13; see also Sect. 7.6.3).

Intranuclear tubular formations derived from the inner nuclear membrane are seen in association with various pathological conditions. Well known examples include idiopathic pulmonary fibrosis, various forms of fibrotic lung disease, sarcoidosis, collagen vascular disease and hypertrophy of human cardiac muscle (TERZAKIS 1965; FERRANS et al. 1975; ENGEDAL et al. 1977; KAWANAMI et al. 1979). They have also been described in some human and experimental animal tumours (KARASAKI 1970, 1973; FERRANS et al. 1975; HENDERSON et al. 1986; GHADIALLY et al. 1985a; TORIKATA and ISHIWATA 1977).

Hormone-dependent post-ovulatory tubular formations, arising from the inner nuclear membrane and connected with or nearly with the nucleolus, are a well recognised feature of the human endometrium, thought to be related to the acyl group in the 17-beta position of the D ring of

◀ **Fig. 13.** Diagrammatic presentation of various alterations of the nuclear envelope (NE) and pore-lamina complex. *1*, Normal NE and pore-lamina complex: *a*, nuclear pore; *b*, outer nuclear membrane; *c*, inner nuclear membrane; *d*, nuclear fibrous lamina. *2*, Separation of component membranes of NE resulting in dilatation. *3*, Fusion of membranes of NE. *4*,Evagination of outer nuclear membrane. *5*, Evagination of both nuclear membranes. *6*, Dense inclusions in perinuclear cistern. *7*, Rupture of NE. *8*, Temporary defects in NE. *9*, Lysis and fragmentation of membranes. *10*, Macular thickening of inner nuclear membrane. *11*, Invagination of inner nuclear membrane. *12*, Invagination and reduplication of both nuclear membranes. *13*, Increased number of nuclear pores. *14*, Enlargement of nuclear pores. *15*, Decreased number of nuclear pores. *16*, Thickening of fibrous lamina. *17*, Formation of annulate lamellae. *18*, Formation of tubular inclusions. *19*, Formation of concentric laminated inclusions

progesterone. They can, therefore, be induced by several synthetic progestational agents with similar chemical structures. The functional significance of these tubular formations is not known, although it has been suggested that they may represent a pathway for transport of materials such as RNA and lipids through the perinuclear cistern, to and from the nucleus and nucleolus (KARASAKI 1973).

Tubular formations inside the nucleolus are also amongst the commonest of intranucleolar inclusions. They are described in normal cells and in serveral human and animal tumours (BOURGEOIS et al. 1979). The frequent contact of the nucleolus with the nuclear envelope may facilitate the entrapment of membrane-derived inclusions. Alternatively, these tubular inclusions may originate from the membranes of the nucleolar-envelope junction.

### *6.1.4 The Nuclear Envelope in Viral Infections*

Infections by the adenovirus and the herpes-cytomegalic virus groups are associated with profound morphological alterations in the nuclear envelope (GRIMLEY and HENSON 1983). This association can perhaps be attributed to the strategic location of the nuclear envelope and its contribution to virus assembly. Firstly, the virus has to traverse the nuclear membrane on its way to and from the nucleus and in the case of Herpesviridae, may become envoloped during this journey (GINSBERG 1980a, b; WOLINSKY 1979). Sometimes the virus particles accumulate in vesicular structures derived from the proliferating inner nuclear membrane (BIBOR-HARDY et al. 1982a, b).

The catalogue of virus-induced alterations in the nuclear envelope includes fusion, proliferation and reduplication of its two membranes, inner membrane macular thickening and the formation of dense inclusion bodies in the perinuclear space (BIBOR-HARDY et al. 1982a, b). Virus-induced proliferation can cause nuclear envelope invagination, the formation of small vesicles along the inner nuclear membrane. Formation of nuclear or cytoplasmic concentric lamellae may also be due to virus-induced nuclear membrane proliferation (SIMARD et al. 1986). These changes are thought to be induced by contact of the virus antigen with the nuclear envelope.

### *6.1.5 The Nuclear Envelope in Neoplasia*

Changes in the nuclear envelope in neoplasia are largely proliferative, involving either or both of the nuclear membranes, but mainly the inner one. Nuclear envelope proliferation may produce finger-like projections into either the nucleus or the cytoplasm (Figs. 2, 12). Alternatively, it may result in the formation of concentric lamellar membranous arrays, tubular inclusions or intracytoplasmic and intranuclear lamellae and nuclear pockets (HENDERSON et al. 1986; BRANSTETTER and GOLDBLATT 1983;

GHADIALLY et al. 1985a; GHADIALLY 1988). In some experimental animal tumours, intranuclear and intranucleolar tubular inclusions occur due to membrane prolieferation. It should, however, be stressed that noeplasia-associated changes in the nuclear envelope are infrequently, if not rarely seen, and that they do occur in other non-neoplastic lesions. They should not be relied upon solely as markers of neoplasia.

### 6.2 The Nuclear Fibrous Lamina

This is a fine band of faintly to moderately electron-dense material, marginating or lining the inner or nuclear aspect of the inner nuclear membrane (Fig. 13). It is considered part of the nuclear matrix (BEREZNEY 1984; VERHEIJEN et al. 1988), playing a part in nucleocytoplasmic exchange in conjunction with the nuclear envelope, the nuclear pores and the proteinaceous component (SCHATTEN and THOMAN 1978). The fibrous lamina is also assumed to participate in maintaining nuclear shape, by acting as scaffolding beneath the nuclear envelope. It is very difficult to image in many normal cell types, but it is frequently seen in many mesenchymal cells such as smooth muscle cells, fibroblasts, endothelial and Schwann cells, osteocytes and osteoblasts (COHEN and SUNDEEN 1976; RIFKIN and HEIJL 1979). It is also prominent in the interstitial cells of the testis and in the epithelial cells of Brunner's glands (LEESON and LEESON 1968).

In pathological conditions, the fibrous lamina is not a marker of particularly active or rapidly proliferating cells, being seen rarely in malignant tumours (Fig. 2). It is more described in slowly growing benign tumours (GHADIALLY 1988), such as pleomorphic salivary adenoma, chondromyxoid fibroma, elastofibroma, nasopharyngeal angiofibroma, chondroblastoma and granular cell myoblastoma (Fig. 1). It has been described in the plasma cells of Hodgkin's disease, in myofibroblasts of repaired tissue and in the synovial cells of rheumatoid arthritis. The significance of its prominence in pathological conditions remains largely obscure.

### 6.3 The Nuclear Pores

The nuclear pores, randomly distributed across the nuclear envelope, are considered to play a major role in nucleocytoplasmic exchanges (CLAWSON et al. 1980; DE ROBERTIS and DE ROBERTIS 1987; SCHEER et al. 1988). The number and diameter of the nuclear pores (Fig. 13) correlate with DNA content, nuclear surface area, nuclear volume and, in paricular, transcriptional activity (FRANKE and SCHEER 1974; MAUL 1977; MAUL et al. 1980). It is, therefore, not surprising that they are found to increase in size and number in conditions requiring increased RNA synthesis or heightened metabolic acitivity (BLACKBURN 1971; SCHEL et al. 1978), such as neoplastic disease (MAUL 1977), and in regenerative or reparative activity. They also decrease in conditions associated with lowered transcriptional activity and decreased protein synthesis (LODIN et al. 1978), such as starvation and in mature and aged cells. Their number was noticed to decrease in the pancreatic acinar cells of children with kwashiorkor.

On the contrary, it has been suggested that in certain neoplastic disorders, such as thyroid papillary carcinoma, nuclear pore deficiency is itself responsible for increased nuclear activity, in order to compensate for any resulting decrease in the rate of nucleocytoplasmic exchange (JOHANNESSEN et al. 1982). MAUL (1977), on the other hand, attributed the increased frequency of nuclear pores in neoplastic cells either to increased synthetic activity or to the increased cell and nuclear surface areas.

Some experimental studies link increases or decreases in nuclear pore diameter to the metabolic state of the cell. Enlargement of the nuclear pores is observed in yeast cells exposed to nitrogen starvation (WILLISON and JOHNSTON 1978). Other studies, however, reported insignificant variation in pore size due to metabolic activity (SEVERS and JORDAN 1978).

The annulate lamellae, seen in the cytoplasm and to a lesser extent in the nuclei, are currently regarded as stores of nuclear pore complexes regularly studded on parallel stacks of membranes. They are formed in actively dividing embryonic and neoplastic cells, actively secreting cells and cells treated by tubulin-binding compounds (MAUL 1977). They are also seen in virally transformed cells. In order to explain their origin, this author proposed that they are formed in the endoplasmic reticulum from nuclear complex precursors that aggregate to form more complexes. The latter become attached to membranes of the endoplasmic reticulum to form the annulate lamellae.

The fact that the nuclear envelope and its associated structures, the pores and the fibrous lamina, are often neglected or overlooked during examination of pathological material may be attributed to several factors. First, proper assessment of these structures requires proper fixation and preservation, not easy to guarantee in routine diagnostic material. Second, the range of magnification necessary for proper examination of these structures often exceeds the level necessary for most routine diagnostic pruposes. Third, the observations documented are often of non-specific or incidental findings. Careful study of the nuclear envelope is required in a wide range of pathological conditions, combined with a better understanding of the functions of its various components.

## 6.4 The Nuclear Matrix

The nuclear matrix or skeleton is the residual non-chromatin structural framework that is resistant to successive treatment of purified nuclei by detergents, salts and nucleases (BEREZNEY 1984; BEREZNEY and COFFEY 1977; KAUFMANN et al. 1986; SIMARD et al. 1986). It can be looked upon as the nuclear counterpart of the cytoplasmic cytoskeleton.

The nuclear matrix is formed of a non-chromatin, non-nucleolar intranuclear fibrogranular network, along with the nuclear pore-fibrous lamina complex and a residual component of the nucleolus (BEREZNEY 1984; SIMARD et al. 1986; VERHEIJEN et al. 1988). Functionally, it provides a scaffolding upon which are built the various nuclear morphological features, including size and shape. There is much evidence to suggest that the nuclear matrix may be a dynamic structure (BEREZNEY 1979, 1984) involved in the regulation of DNA synthesis and gene expression (KAUFMANN et al. 1986; JACKSON et al. 1984; VAN DER VELDEN and WANKA 1987), RNA nucleocytoplasmic transport (HERLAN et al. 1979), hormone binding (BARRACK and COOFEY 1982), carcinomagenesis and virogenesis (BIBOR-HARDY et al. 1982a; COVEY et al. 1984; EISENMAN et al. 1985; SIMARD et al. 1986). However, the extent and significance of such involvement are not completely understood (KAUFMANN et al. 1986).

It is difficult to ascertain whether the presence of physiological and pathological process-related products in the nuclear matrix is due to incidental entrapment during the isolation procedure, or the the acitve engagement of the nuclear matrix in the assembly and storage of these components (SIMARD et al. 1986; KAUFMANN et al. 1986). However, with regard at least to HSV virogenesis, it has been shown that the nuclear matrix participates in viral capsid assembly and possibly also in its gene regulation and DNA replication. Using immunogold labelling studies, it has been shown that the HSV capsid proteins accumulate in the nuclear matrix after being synthesized in the polyribosomes and conveyed through the nuclear pores. There, the nuclear matrix provides an anchoring site on which the assembly of the viral capsid takes place (SIMARD et al. 1986).

Other studies have demonstrated the presence of an intimate relationship between changes in the structural organisation of the nuclear matrix components and alterations in the nucleolar structure (BRANSTETTER and GOLDBLATT 1983; HERLAN et al. 1978). Actinomycin D-induced alterations in nucleolar shape were found to be associated with corresponding changes in the isolated nuclear matrix of the same cells (HERLAN et al. 1978). It may be speculated that an induced dynamic or structural event in the volume of the nuclear matrix will have its echo on the entangled nucleolus and vice versa. The possibility that this morphological interdependence is due to a functional interrelationship, however, cannot be ruled out.

## 7 Intranuclear Inclusions

### 7.1 Introduction

Intranuclear inclusions, as described here, cannot be too rigidly defined in our present state of knowledge. They include most non-chromatin, non-histone, non-nucleolar, non-matrical bodies or substances contained within the nucleus and not in continuity with cytoplasm. Some structures such as perichromatin and interchromatin granules and normal nuclear bodies are traditionally not regarded as "inclusions". Furthermore the term inclusion has sometimes been applied to altered chromatin, nucleoli or matrix, or to their component parts. Attempts to limit the designation of an inclusion to a structure formed inside the nucleus are unnecessarily restrictive, as they would exclude some definite inclusions like the visible core of an invading adenovirus which has just passed through a nuclear pore. In any case, the morphologist must deal with structures as they are, since their origins are often not determinable. While our chief concern here is with inclusions found in pathologically altered nuclei, the reader must also be aware that nuclear inclusions of various kinds may be found in some normal cells.

## 7.2 Inclusion or Pseudo-inclusion?

Fingerlike projections of cytoplasm into a nucleus can, if the connection is not visible in the plane of section, resemble true nuclear inclusions (Fig. 14). In this case such structures are correctly called pseudoinclusions. Another rare form of pseudoinclusions can arise from an inpushing of the inner membrane of the nuclear envelope into the nucleoplasm. In this case, single-membrane-bound pseudoinclusions can be formed with contents derived not from the cytoplasms but from the cistern of the nuclear envelope. To insist on the term pseudoinclusion for totally detached cytoplasmic fragments, surrounded completely by nuclear envelope or by the inner nuclear membrane and contained within the nucleus, is illogical for it implies that they are not truly inclusions within the nucleus, which they patently are.

The electron microscopist can usually distinguish invaginations of both types from true inclusions by determining whether or not they are membranebound and contain cytoplasmic organelles, in which case they are more likely, though not certain, to be simply cytoplasmic invaginations. However, as indicated above, cytoplasmic invaginations can become pinched off to form true inclusions. Ultimately, the acid test remains serial sectioning of the whole nucleus, which is usually impractical. Thus are the limitations of single ultrathin sections exposed.

In summary then, it may not always be possible to distinguish true inclusions from pseudo-inclusions. Thorough sampling, experience and knowledge of the range of morphology of those inclusions which have been described will aid the distinction.

## 7.3 Origins and Mechanisms of Formation

With the exception of inclusions associated with viral infection and a few other specific examples, the origin and significance of most intranuclear inclusions has remained obscure. They occur in such a wide variety of situations, both normal and pathological, that they can rarely, if ever, be described as pathognomonic. However, they may act as pointers to unusual features of cellular activity, such as hypermetabolism. Several theories have been offered to explain the origins and mechanisms of formation of the various types of inclusion. Only a few of these have been substantiated by morphological evidence and experimental studies. The most commonly offered explanations include:

1. Cytoplasmic material has been entrapped within the forming nuclear envelope at the end of mitosis. This may occur due to rapidity or frequency of mitosis or due to some defect or abnormality in its control, as in neoplasia (GHADIALLY 1988).
2. Cytoplasmic material has gained access to the nuclear compartment by way of agent-induced temporary defects in the nuclear envelope. Such defects are later sealed by the overlying endoplasmic reticulum (FLICKINGER 1974, 1978). Inclusions formed in this way are non-membrane-bound true inclusions.
3. Cytoplasmic products or material, such as lipids or immunoglobulins, have gained access to the nucleus through the endoplasmic reticulum-nuclear envelope channels (BLOM et al. 1976).
4. Cytoplasmic invaginations have been formed by either cytoplasmic swelling and expansion or increased nuclear surface irregularity (SOBEL et al. 1969). As such they may appear as pseudoinclusions if sectioned in certain ways. Some may be pinched off and separated from the main nuclear envelope and from the exterior cytoplasm, at which point they become nuclear envelope-bound true inclusions. Consequent dissolution of their surrounding envelope (GHADIALLY 1988), most likely as a result of lack of maintenance from the rough endoplasmic reticulum, leads to a non-membrane-bound true inclusion.
5. They are derived from basic cellular components made in the cytoplasm, but imported into the nucleus before assembly, as for example some filamentous, microtubular and crystalline inclusions.
6. They are products of viral infection and replication either in the nucleus or in the cytoplasm, being either virions, parts of virions or by-products of viral replication. In the latter case they may be produced in the cytoplasm and move to the nucleus for assembly, as above.
7. They are formed in situ from nuclear components, as in the case of tubular structures derived from the nuclear envelope.
8. They are invaginated and trapped cytoplasmic structures that undergo subsequent degenerative changes into, for example, myelin figures (LEDUC and WILSON 1959).

## 7.4 Classification

With experience and meticulous searching, many nuclear inclusions can often be detected and identified using the light microscope. They appear as optically clear bodies, or as dense eosinophilic or basophilic structures of various shapes. The application of CTEM has allowed a more sensitive and specific identification and recognition of components. Terms such as intranuclear concentric laminated inclusions, intranuclear lamellae, tubules, vesicles and filaments were all introduced following the application of ultrastructural methods. Beyond the major subdivision into true inclusions on the one hand and pseudo-inclusions derived from and still in continuity with the cytoplasm on the other, nuclear inclusions can be further described and classified on the basis of their structure and perceived composition. Specified structural components of inclusions include glycogen, lipid, viruses and haemoglobin. Unspecified descriptive components include tubular, filamentous, crystalline and amorphous structures.

## 7.5 Specified Intranuclear Inclusions

The term "specified intranuclear inclusions" describes the occurrence within the nucleus of cytoplasmic components or metabolites which can be recognised by their specific ultrastructure. Intranuclear viral inclusions will be dealt with separately (Sect. 7.7).

### *7.5.1 Cytoplasmic Inclusions*

Cytoplasmic inclusions are often designated pseudo-inclusions, because they represent cytoplasmic invaginations which may be separated from the nuclear matrix by a double-membrane boundary derived from the nuclear envelope. They vary widely in size and number and may contain almost any cytoplasmic structure (Figs. 14, 15a). They are often seen in ageing cells and after exposure to certain drugs and toxic compounds. They are present in cardiac muscle cells in cases of hypertrophic cardiomyopathy (FERRANS et al. 1975). They are described as characteristic of several neoplasms and can be incidental findings in almost any tumour (GHADIALLY 1985, 1988; TRUMP et al. 1978; HENDERSON et al. 1986). Their presence in bronchiolo-alveolar cell carcinoma and in thyroid papillary carcinoma is responsible for the optically clear nuclei or nuclear vacuoles seen by light microscopy.

### *7.5.2 Intranuclear Glycogen*

Irregularly shaped compact masses of glycogen, either monoparticulate or aggregated, have been described in hepatocytes in many diseases, includ-

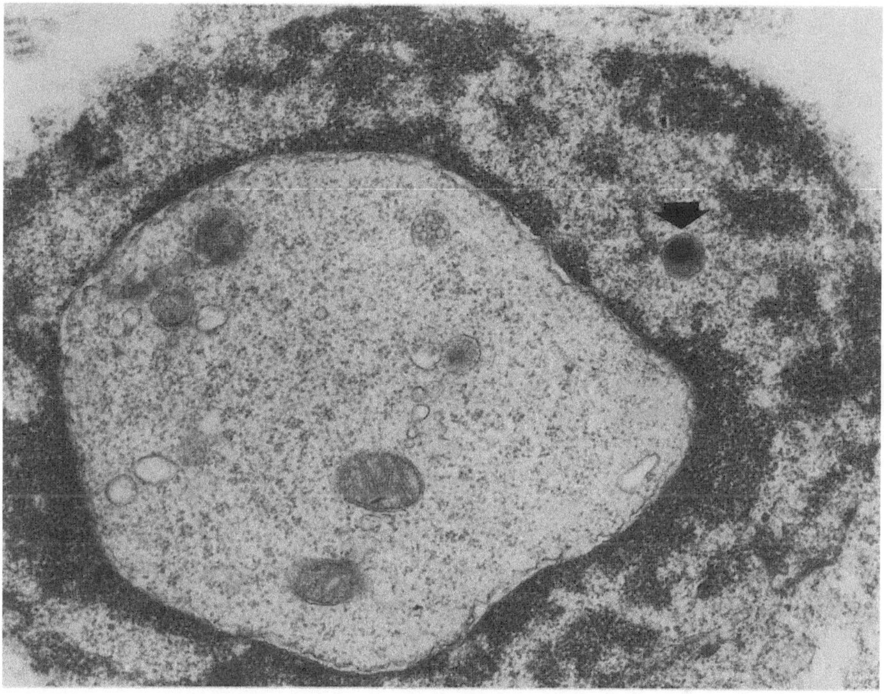

**Fig. 14.** A large cytoplasmic pseudoinclusion results from indentation of the nuclear surface. A small dense inclusion (*arrow*) is also present. × 18 800

ing diabetes mellitus, viral hepatitis, lupus erythematosus, Graves' disease, and Wilson's disease. They are also seen in the liver cells in association with several neoplastic disorders such as Hodgkin's disease, hepatocellular carcinoma and carcinomas of the stomach and the pancreas (for references to occurrence and origin of intranuclear glycogen, see GHADIALLY 1988). Glycogen particles can sometimes be mistaken for viruses. It should also be noted here that the CTEM appearance of glycogen depends very much upon the fixation and staining procedures involved (GHADIALLY 1988).

### *7.5.3 Intranuclear Lipid*

Beside being ubiquitous in some normal cells such as the liver cells, lipid inclusions also occur in a spectrum of neoplastic and non-neoplastic conditions, ranging from chromosomal disorders such as the Leydig cells in Klinefelter's syndrome (NISTAL et al. 1985), through reactive conditions such as the synovial cells in lipohaemarthrosis (GHADIALLY and ROY 1969), to a wide variety of neoplastic disorders in different tissues.

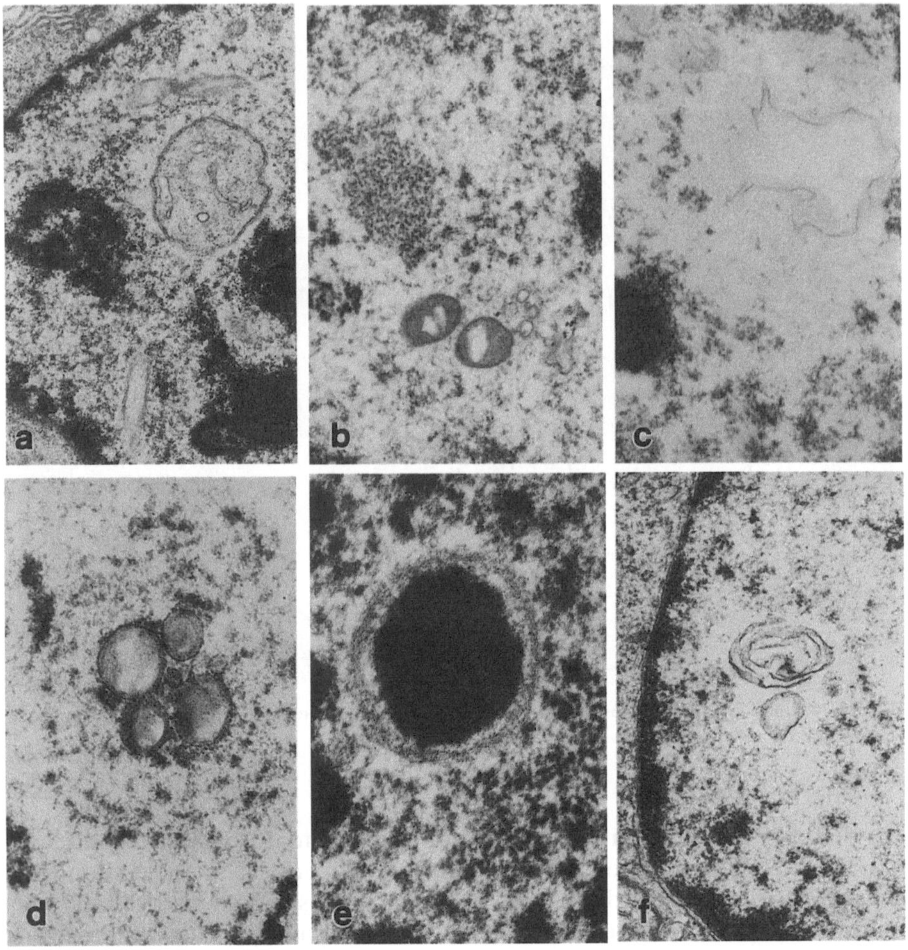


**Fig. 15a–f.** Intranuclear inclusions. **a** Inclusion resembling a mitochondrion, apparently unenveloped within the nucleus in an experimental animal tumour. × 14250. **b** Two inclusions resembling dense pyknotic mitochondria; note also the vesicular structures and the aggregation of interchromatin granules in a methotrexate-treated mouse sarcoma 180. × 18800. **c** Membrane-bound intranuclear vacuolation of uncertain cause. Similar appearances can arise artefactually; actinomycin D-treated Ridgeway osteogenic sarcoma. × 11500. **d** Vesicular nuclear body with fibrillar exterior; astrocyte in multiple sclerosis autopsy tissue. × 39000. **e** Nuclear body with fibrillar exterior and dense core; astrocyte in Alzheimer's disease biopsy tissue. × 25500. **f** Concentric laminated inclusion in an experimental animal tumour. × 14250

### *7.5.4 Lead and Bismuth*

These two heavy metals can induce characteristic intranuclear dense elemental inclusions, with a distinctive ultrastructural morphology. Lead inclusions are often characterised by a markedly dense core surrounded by a less dense network of radiating fibrillar material. Bismuth inclusions are often well delineated, oval to rounded homogeneous dense structures, which sometimes show some granularity. Both types of inclusion are typically detected in the renal tubular cells, but lead inclusions can also be seen in hepatocytes. Electron-probe X-ray micoranalysis offers a definitive means for their precise identification. The mere presence of these inclusions does not necessarily indicate acute intoxication, but can point to previous exposure to either of these elements.

## 7.6 Unspecified Intranuclear Inclusions

Unspecified nuclear inclusions (Fig. 15 b, c) are seen in normal cells and in several pathological conditions. These include mainly viral infections and neoplastic disorders, but also some other miscellaneous conditions, including muscular and neuronal disorders and toxic or hormone-induced disturbances. In pathological situations they may be more numerous or more prominent than normal, or they may appear in cells which do not normally contain them.

### *7.6.1 Nuclear Bodies*

Nuclear bodies are among the most commonly seen but most poorly understood forms of nuclear inclusion, as documented in the earlier literature (BOUTEILLE et al. 1967; KUHN 1967; HENRY and PETTS 1969). Numerous studies and reviews of their morphology are available and the reader is referred to these for comprehensive information on their structure and incidence in various situations. Attempts to determine the mechanisms of formation and the function, if any, of nuclear bodies, have been unrewarding. These non-membrane-bound, usually spherical bodies vary widely in detailed morphology, but tend most commonly to be composed of various combinations of fibrillar, granular and vesicular components (Fig. 15 d, e).

Nuclear bodies have been observed in a wide variety of tumours including, for example, malignant fibrous histiocytoma (GOTZOS et al. 1986) and parotid acinar cell carcinoma. They occur in virus infections such as measles, after treatment with certain drugs and after hormonal stimulation. TSH-stimulated cultured thyroid cells show frequent granular nuclear bodies (VÁGNER-CAPODANO et al. 1980). Nuclear bodies can also sometimes be seen in apparently normal nuclei, such as those of astrocytes. Examination of the list of conditions in which they occur shows that they tend to be present in metabolically active or activated cells (GHADIALLY 1985). In some cases they may represent cell proteins (i.e., cytoplasmic-, nuclear- or nucleolar-derived proteins) which have polymerised inside the nucleus. The formed morphological component will therefore depend on the nature of the protein product, which in turn will depend on the differentiated function of the cell.

### *7.6.2 Intranuclear Concentric Laminated Inclusions*

These are single-membrane-bound and consist of concentric alternating electron lucent and electron opaque non-membranous layers (Fig. 15 f). They are often seen in serous and mucin-secreting cells, especially those of salivary gland origin (GHADIALLY 1988). They are described in cells of adenoma of the breast (GHADIALLY 1985), acinar cell carcinoma of the parotid gland (ERLANDSON and TANDLER 1972) and adrenal cortical adenoma associated with Conn's syndrome (PROPST 1970). They have also been observed in methotrexate-treated mouse sarcoma S180 (unpublished observation) and in the pancreatic acinar cells in cases of Reye's syndrome (ITABASHI et al. 1976). These inclusions may be derived from intranuclear lipid, serous or mucoid pseudo-inclusions that have undergone secondary degenerative changes (LEDUC and WILSON 1959).

### *7.6.3 Filamentous, Fibrillary, Microtubular and Crystalline (FFMC) Inclusions*

This heterogeneous group of inclusions has been described under a variety of names in an extensive literature and includes the so-called intranuclear rodlets, vermicellar bodies, and paracrystalline inclusions. They have been grouped here since (a) there is confusing overlapping in their nomenclature in the literature, (b) intermediate forms between the different variants are common, and (c) they are mostly seen in the same wide range of pathological conditions.

It is probably more confusing than helpful in this limited space to list the full range of normal and pathologically altered cells and the various situations in which these inclusions have been described, for they are of diverse origin and differing significance. The reader is referred to the review literature for such information (GHADIALLY 1988; TONER et al. 1980; PAYNE and NAGLE 1983). However, there are some circumstances in which particular forms are unusually common, such as the distinctively different forms in normal neurons and in a number of viral infections.

Morphological subcategorisation is difficult in view of the overlap between different forms, such as filamentous rodlets and paracrystalline arrays. A few, such as the tubular nucleocapsids of measles virus in subacute sclerosing panencephalitis or the filamentous forms of papovavirus, are clearly distinctive when seen in good quality micrographs, preferably en masse and with other indications of viral infection. Even these, however, are likely to be missed if present in small quantities in average or poor quality micrography, particularly when their occurrence is not anticipated. Both of these viral inclusions are formed from viral-encoded subunits or capsomeres, but the origin of the materials which form most FFMC inclusions is unknown, although most are presumed to be host cytoplasmic or nuclear protein.

A general difficulty in describing FFMC inclusions lies in the accuracy or otherwise of the measurement of their component filaments and tubules. It is doubtful whether too much reliance can be placed on all published fine size distinctions at the level of microfilaments (4–7 nm) and intermediate filaments (9–10 nm). Most electron microscopes, if correctly set up and calibrated, would approach 95% accuracy. However, because of specimen height and hysteresis effects, not to mention uncontrollable dimensional variations due to tissue processing and sectioning, they are undoubtedly unreliable in practice. Accurate measurements on enlarged prints also depend on well focused negatives, a known enlargement factor, good printing, proper control of paper shrinkage on drying and an adequate sample size. Conditions are rarely so favourable in a diagnostic pathology laboratory.

Any consideration of how the intranuclear rodlets and fibrillar lattices (Fig. 16) are formed within nuclei must take into account the fact that virtually identical structures have been reported from cytoplasmic locations, where they may form very large bodies (Hirano bodies) in some pathological conditions and smaller ones artefactually as a result of post-mortem autolysis (YAGISHITA et al. 1979). It may be inferred that at least for this class of FFMC, the constituent structural proteins may be present at higher concentrations in the cytoplasm than in the nucleus.

In this context, it is noteworthy that there are endogenous nuclear proteins, the lamins, which are normal components of the nuclear lamina and which share biochemical and structural characteristics with the components of all known cytoplasmic intermediate filaments (GERACE 1985). Furthermore it has been shown that the lamina of Xenopus oocytes consists of 8- to 10-nm filaments arranged in a regular, near tetragonal lattice (GERACE 1985; WEISS 1988). Such lamin proteins are therefore prime candidates for involvement in the formation of filamentous lattices in the nucleus, while cytoplasmic intermediate filament proteins may be involved in formation of analogous structures in the cytoplasm.

In an effort to understand the significance of the occurrence of FFMC inclusions in the central nervous system, where they are common, attempts have been made to induce their formation experimentally (PAYNE and NAGLE 1983; GHADIALLY 1988).

## 7.7 Intranuclear Inclusions in Virus-Infected Cells[1]

### *7.7.1 Introduction*

By light microscopy, intranuclear viral inclusions have been categorized either by their morphology, such as Cowdry type A or Cowdry type B inclusions, or by their staining reaction, such as basophilic, acidophilic and amphophilic inclusions. The ultrastructural counterpart of such descriptions has not been fully determined in all cases.

---

[1] Reviews and similar texts are cited extensively in this section as they provide the maximum assistance to the ultrastructural pathologist newly entering the field.

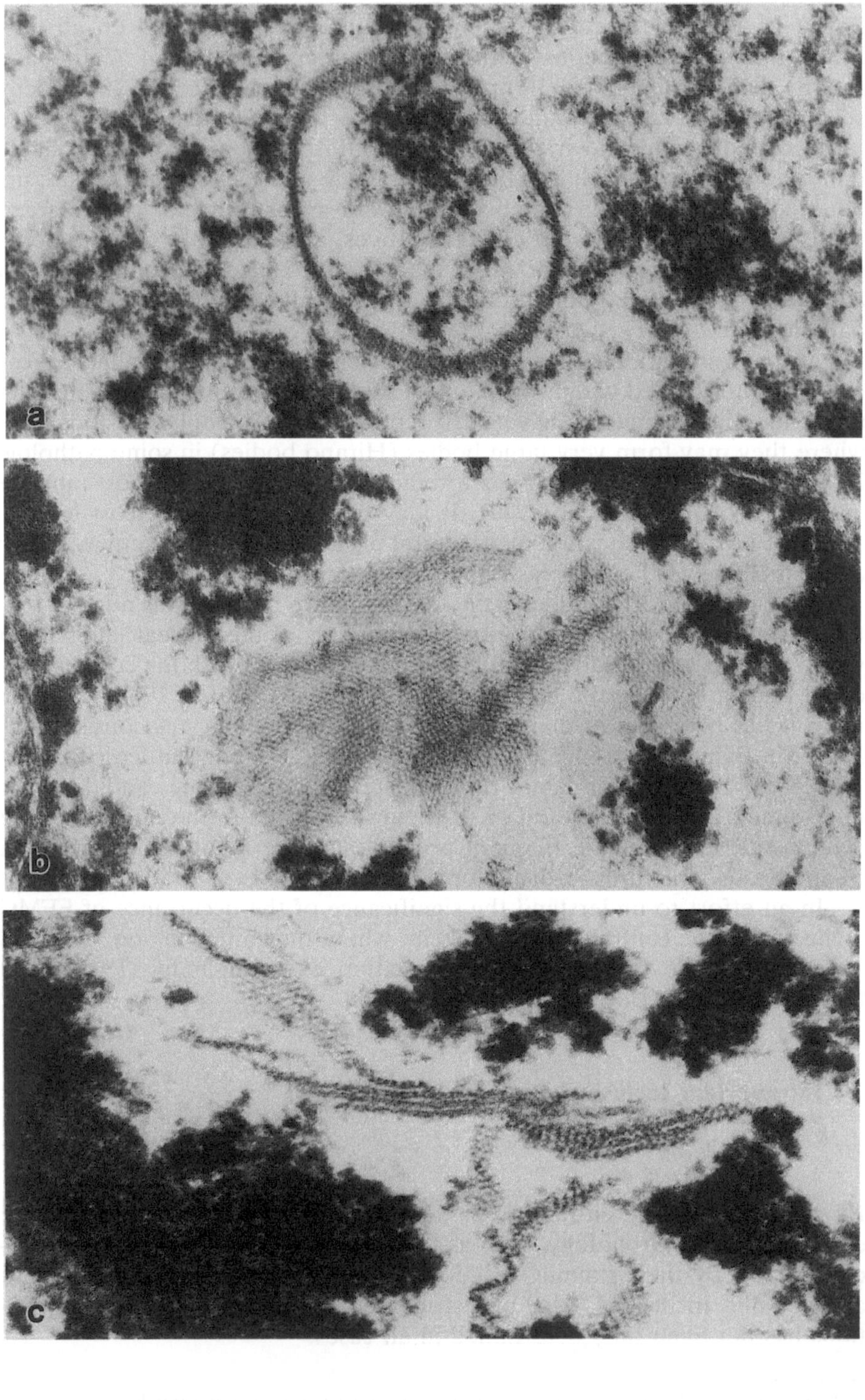
a
b
c

Amongst the many roles which have been identified for the electron microscope in virology (GRIMLEY and HENSON 1983; BURNS 1980; BOOS and ESIRI 1986), two stand out as being of fundamental importance in viral diagnosis using pathological tissue sections. First, it is used to supplement the non-specific light microscopic finding of inclusion bodies in paraffin- or epoxy-embedded sections. The ease with which this can be accomplished depends, for paraffin-embedded material, on whether epoxy resin-embedded tissue can be prepared from the appropriate region of the light microscopic section itself (YUNIS et al. 1977) or is available from adjacent sections. The routine use of large area toluidine blue-stained epoxy sections for light microscopic screening for viral inclusions as a prelude to more precisely targeted CTEM study has been described (BURNS 1978). It should be noted, however, that the light microscopic appearances of some nuclear inclusions are less distinctive in plastic sections than in routine H & E-stained paraffin sections (WILLS 1983).

Examination of such light microscopic inclusion-positive areas by CTEM can often reveal the viral particles which constitute the inclusion and permit their identification. Alternatively, electron microscopy may show that the "inclusion" was actually a giant nucleolus or cytoplasmic pseudoinclusion, unrelated to viral infection.

The second clearly defined role of CTEM is to search for evidence of virus where the index of suspicion is high but where no inclusions are seen by light microscopy. This method has proven successful, for example, in the case of herpes encephalitis (WILLS 1983), but may be extremely time-consuming unless pathological markers such as inflammation, necrosis or other focal lesions are used to narrow the area for intensive search.

The nucleocapsids of many viruses which form intranuclear inclusions, including members of measles, papova-, and herpesvirus families, are surprisingly resistant to post-mortem autolysis, formalin fixation, paraffin embedding and even freezing, so it is often possible to identify them in the most unpromising tissue specimens (WILLS 1983). However, the membranes of enveloped viruses are less well preserved in such circumstances and poor preservation may preclude identification of the infected cells. Such tissues are, however, more readily amenable to immunocytochemical or in situ hybridisation, or polymerase chain reaction studies for viral antigen or genome at both light and electron microscopic levels.

Even when viral particles cannot be detected in the nucleus by CTEM, the genome of certain viruses – herpesviruses, adenoviruses, parvoviruses, papovaviruses, retroviruses and hepadnaviruses – may still be present, either as intergrated or unintegrated viral DNA (GREEN 1985;

◀ **Fig. 16 a–c.** Filamentous lattice structures. **a** In normal well fixed neurons, a ring-shaped structure, possibly representing a saucer or hollow sphere, shows a lattice substructure. × 48 800. **b** and **c** Inclusions in autopsy brain cells (possible artefacts) show a variety of appearances which are all explicable as different orientations of multiple layers of near-orthogonal filamentous lattices. **b** × 25 500; **c** × 31 800

BLACKLOW and CUKOR 1985; FAUCI 1988). The development of sensitive in situ hybridisation methods offers the opportunity to detect and visualise such viral passengers at both light and electron microcopic levels (HAASE 1986; MANUELIDIS 1985; AKSAMIT et al. 1985).

### *7.7.2 Recognition, Measurement and Interpretation*

Intranuclear inclusions found in virus-infected cells comprise a heterogeneous collection of both host and virus-derived structures (Figs. 17–19; GHADIALLY 1988; GRIMLEY and HENSON 1983; WILLS 1983). They include not only complete virus (virions), but also empty viral capsids (the protein shells which surround the viral nucleic acid) and other subcomponents or precursors, surplus pools of viral nucleic acids, budding virions and modified or disordered host proteins and nucleic acid structures. They may also be composed of the massive accumulations of abnormal-appearing, nonfunctional nucleocapsids. Recognition and identification of intranuclear viral inclusions requires knowledge, therefore, not only of the classic structure of the typical mature virus particle but also of immature and aberrant viral forms and associated structures. All of these viral forms must also be distinguished from normal and modified host nuclear components, such as nuclear pores, nucleolar components and chromatin, with which they have been confused in the past (GRIMLEY and HENSON 1983; WILLS 1983; GHADIALLY 1988).

Furthermore, even if an inclusion contains only perfectly formed icosahedral virions of uniform size, these will vary in appearance in ultrathin sections, depending upon how much of the particle is included within the thickness of the tissue examined. Such variation is more apparent in the case of virions with a diameter significantly greater than the section thickness, such as the Herpesviridae, and less so in those of small diameter such as the Parvoviridae. Measurements of icosahedral, near spherical, virus particles should be carried out on the largest profiles, because a grazing section will give a low estimate of diameter. These measurements of viruses in ultrathin tissue sections cannot be compared directly with the dimensions cited in most virological texts, which are based almost inevitably on measurements of negatively stained particles. WILLS (1983) has estimated that the manipulations involved in the preparation of ultrathin sections reduces viruses to about 80% of their negatively stained size.

**Fig. 17a–c.** Icosahedral (near spherical) nuclear inclusions in viral infections. **a** Nucleocapsids of the Herpesviridae family are approximately 100 nm in diameter and may be hollow or have ring-shaped or dense cores. Lung cell in disseminated cytomegalovirus infection. Autopsy tissue. **b** Virions of the Adenoviridae are similar but smaller (60–70 nm diameter) and typically from close-packed crystalline arrays. Avian adenovirus in cultured cells (courtesy of Dr. B. M. Adair, Veterinary Research Labs, Stormont, Belfast). **c** Virions of the Papovaviridae are smaller (30–45 nm) and typically show no substructure. Elongated forms, not illustrated here, commonly coexist with spherical forms. Experimental papilloma virus infection. All × 50000 ▶

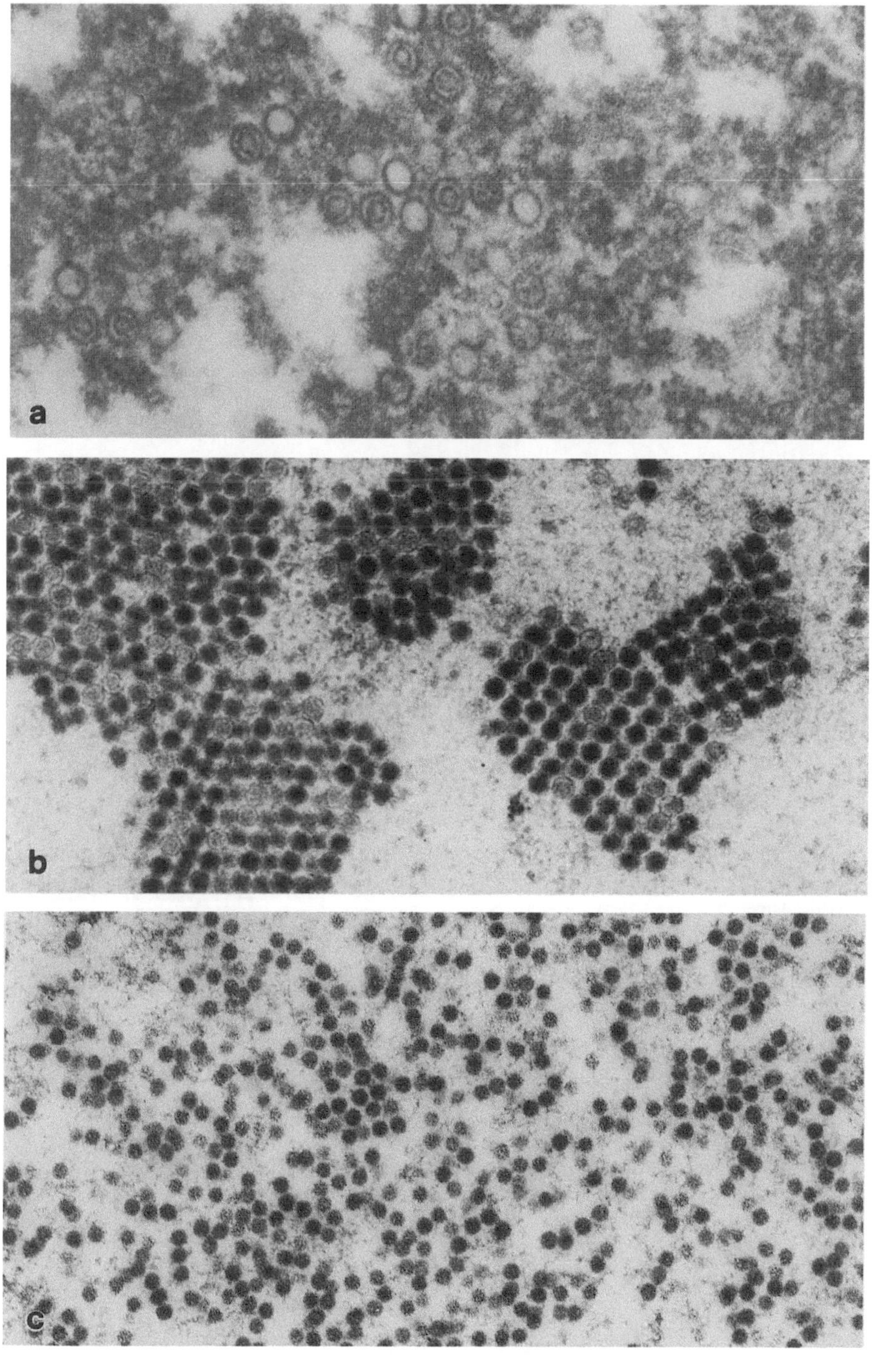
a
b
c

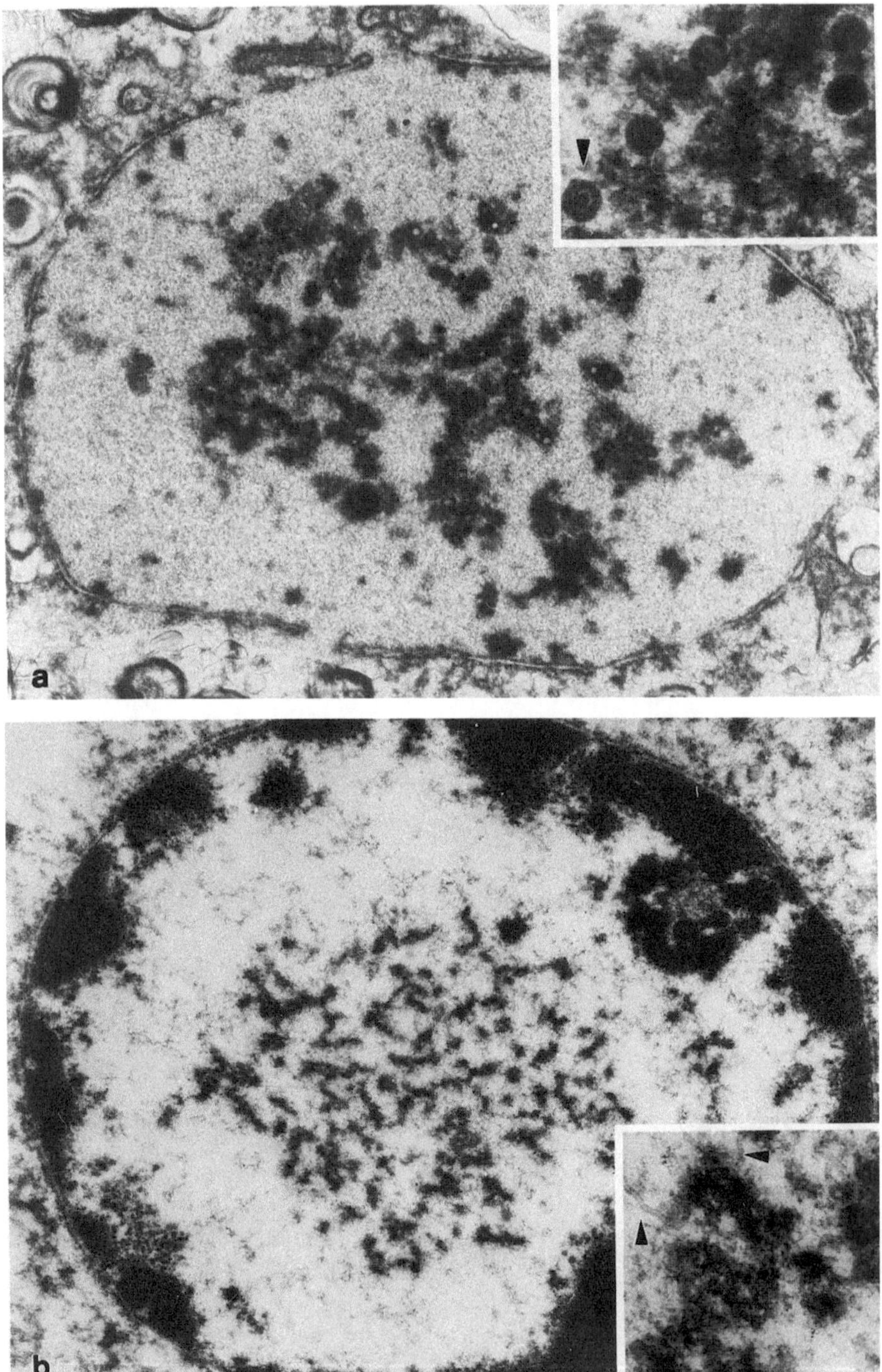
a
b

Most published illustrations of intranuclear viral inclusions are selected to show considerable numbers of particles in cells at an advanced stage of the viral replicative cycle. As the study of such accumulations reveals the main features of virus structure, we will concentrate on such inclusions in the following paragraphs. Pathologists examining tissue sections, however, will often be unable to find such advanced-stage infected cells and should remember that there are also earlier stages, from the entry of infecting virus into the nucleus (GINSBERG 1980a), through the induction of changes in the host structures and the formation of viral precursors, to the assembly of the first viral particles. Between the two extremes of viral entry on the one hand and the stage of massive inclusions on the other, there are changes which in some cases may be characteristic or at least indicative of viral infection and which the ultrastructural pathologist may learn to recognise (GRIMLEY and HENSON 1983). We will draw attention to such changes in the descriptions of the main virus families. Needless to say, the nucleus should not be examined in isolation, for in many infections the occurrence of viral inclusions in the nuclei is accompanied by virus-associated changes in the cytoplasm (KINGSBURY 1985; WILLS 1983; GRIMLEY and HENSON 1983).

### *7.7.3 Mature Virus Particles (Virions)*

In some cases, such as adenoviruses, papovaviruses and parvoviruses, mature infectious virus particles (virions) are assembled and accumulate in the nucleus. In the first two of these, the virions, which may be accompanied by empty capsids and viral nucleic acid pools, may be so numerous that they form large inclusions detectable by light microscopy (GINSBERG 1980a; WILLS 1983). In all three, the ultrastructure is sufficiently distinctive to permit recognition of the virus family (Fig. 17b, c). With the possible exception of the polyoma and papilloma subgroups of Papovaviridae, which differ significantly in size (28–40 nm and 43–46 nm respectively) (WILLS 1983; GHADIALLY 1988), subtyping or distinction of genus cannot yet be accomplished routinely in pathological tissues by ultrastructural examination of individual nuclear virions in thin sections.

### *7.7.4 Immature Virus*

In other viral infections, most notably in the members of the Herpesviridae (cytomegalovirus, HSV-I, HSV-II, Epstein-Barr virus, varicella zoster virus) but also in the Hepadnaviridae (hepatitis B virus), only imma-

◀ **Fig. 18a, b.** Irregular inclusions visible at low magnification in autopsy tissue from viral infections, with *insets* showing diagnostic detail (*arrowheads*) at higher magnifications. **a** Herpes encephalitis (courtesy of Dr. C. H. S. Cameron, Dept. of Pathology, Queens University Belfast). × 14300; inset, × 54500. **b** Subacute sclerosing panencephalitis. × 18300; inset, × 70000

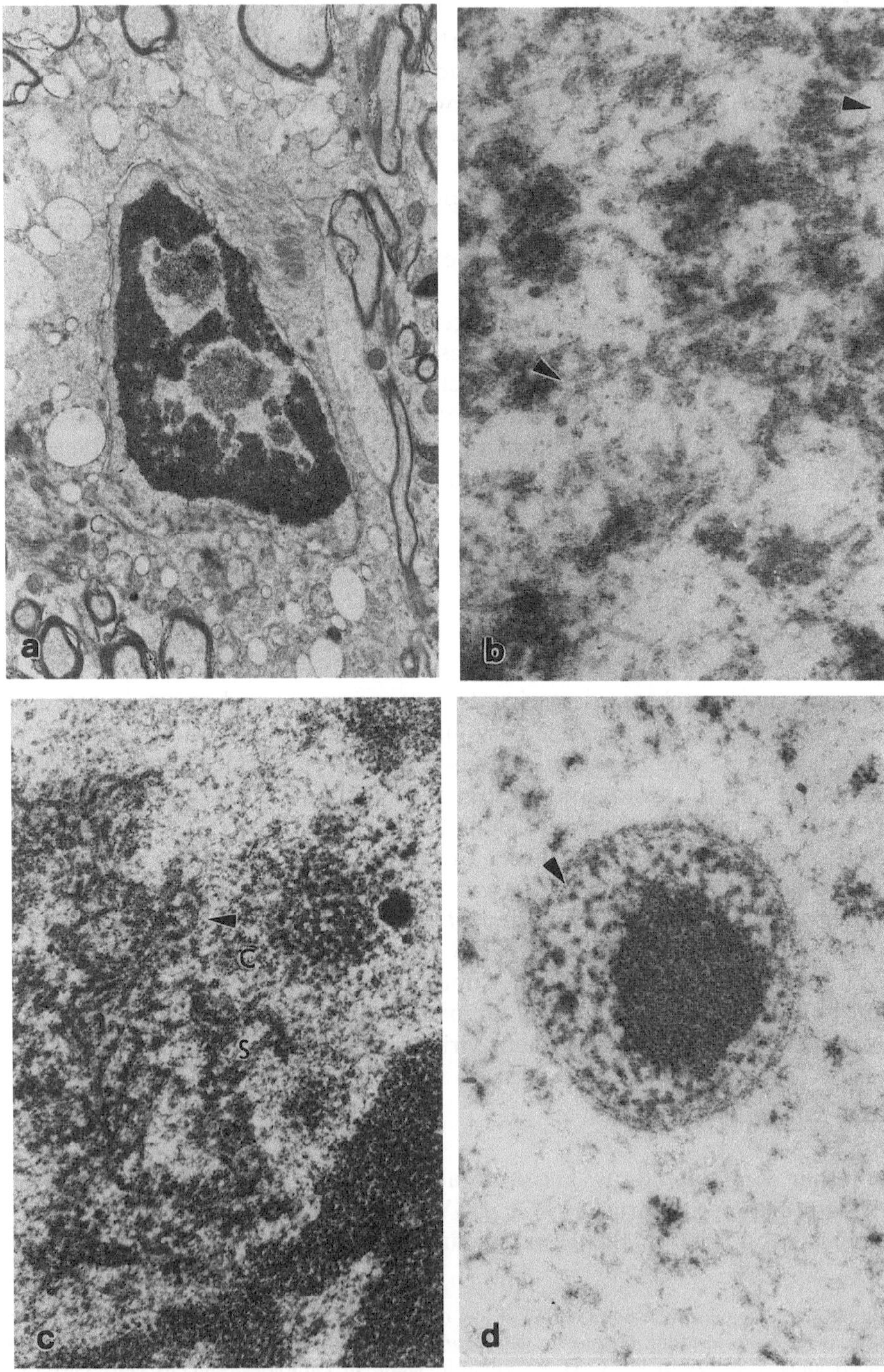
a
b
c
C
S
d

ture virions (nucleocapsids) are assembled in the nucleus (Figs. 17a, 18a). These must leave and complete their development elsewhere. Their maturation involves addition of a lipoprotein membranous envelope at the nuclear envelope, endoplasmic reticulum, Golgi apparatus or plasma membrane. However, the nuclear inclusions, composed both of immature virus particles and of empty and malformed capsids and other viral components, can be very large and may be visible as inclusions by light microscopy. The ultrastructure of the intranuclear, incomplete herpes virus is distinctive for the family but not for the genera within the family, at least in routine pathological specimens (WILLS 1983). Immunostaining, in situ hybridisation and correlative clinical and pathological information must be used for these finer distinctions.

### *7.7.5 Non-functional (Dead-end) Nucleocapsids*

In certain circumstances in which there is a block to maturation, measles virus infection results in the accumulation of large numbers of nucleocapsids within the nucleus (GRIMLEY and HENSON 1983). These nucleocapsids are assembled from nucleic acid and capsid components which have earlier been synthesised in the cytoplasm, the normal site of nucleocapsid assembly. In a productive measles infection, the cytoplasmic fuzzy-coated nucleocapsids mature by budding through the plasma membrane (KINGSBURY 1985). Massive nuclear inclusions composed of relatively smooth nucleocapsids in neurons and glia are a distinctive feature of the late stages of subacute sclerosing panencephalitis (Figs. 18b, 19; DUBOIS-DALCQ 1979). The nucleocapsids which form the nuclear inclusions in poliovirus-infected neurons (described in detail in Sect 7.7.6.8) are another example of this category of inclusion.

### *7.7.6 Inclusions of Medically Important Viruses*

#### 7.7.6.1 Adenoviridae (e.g. human adenoviruses)

DNA genome; un-enveloped icosahedral capsid ca. 57–72 nm diameter in thin sections (Fig. 17b). Large, dense granular-fibrillar patches or ring-

◀ **Fig. 19a–d.** Further examples show the heterogeneity of the measles virus nucleocapsid inclusions in subacute sclerosing panencephalitis. **a** Nucleocapsid clumps in an apoptotic nucleus. Note the highly condensed chromatin masses which have separated from the nuclear envelope; biopsy tissue, × 7000. **b** High magnification of a nuclear inclusion reveals a mass of sinuous, striated, hollow, tubular nucleocapsids (*arrowheads*); snap frozen autopsy tissue, refixed for electron microscopy. Measles virus-specific RNA was extracted from the remains of the frozen block (M. Taylor, personal communication). × 87500. **c** In this biopsy tissue both curved (*c*) and straight (*s*) nucleocapsids are seen. × 46700. **d** In an infected brain cell in another biopsy, nuclear body-like inclusions show a fibrillar exterior, tubular intermediate layer and dense core. Some authors have stained similar nuclear bodies with measles antibodies and suggest that the tubules (*arrowheads*) are nucleocapsids. × 28800

shaped filamentous structures may precede and accompany the appearance of the very distinct and regular dense-cored or hollow capsids in the nucleus (YAMAMOTO and SHAHRABADI 1971; GRIMLEY and HENSON 1983; SHELBURNE et al. 1983). Capsids often appear in paracrystalline groups within the patchwork of granular and fibrillar material (GRIMLEY and HENSON 1983). Nuclear envelope thickening and reduplication have also been described (GRIMLEY and HENSON 1983). Following infection, viral entry to the nucleus involves alignment of the capsid with the outer face of a nuclear pore and injection of the DNA genome through the pore into the nucleoplasm (GINSBERG 1980 a).

### 7.7.6.2 Herpesviridae (e.g. herpes simplex and cytomegaloviruses)

DNA genome; icosahedral capsid (ca. 100 nm diameter in thin sections). Chromatin tends to marginate, leaving the centre of the nucleus to the virus factory. Viral capsids at various stages of development may be seen around the margin of globular masses or along chain-like strands within a 'viroplasmic matrix' (Fig. 18 a; GRIMLEY and HENSON 1983). Herpesvirus nucleocapsids assemble while attached to the fibres of the nuclear matrix, as shown by whole-mount electron microscopy of chromatin-depleted cells and other specialised techniques (BEN-ZE'EV 1983; PENMAN 1985). Although hollow capsids are common, the complete nucleocapsid has a ring-like (hollow spherical) or dense core (Fig. 17 a). The nuclear envelope may proliferate and budding through the nuclear envelope may be seen (SHELBURNE et al. 1983). A wide variety of cytoplasmic forms are seen.

### 7.7.6.3 Papovaviridae

*Genus 1. Polymavirus genus [e.g. JC virus, cause of progressive multifocal leukoencephalopathy (PML)]*
DNA genome; icosahedral capsid. The reported range of diameters in thin sections (28–40 nm) is unexpectedly wide and may include some inadvertent measurements of thin, transversely sectioned, filamentous nucleocapsids, which occur commonly in association with the typical spherical particles. In PML, a large number of virions are found in the enlarged (2–3 times) nuclei of oligodendrocytes within small demyelinating lesions or at the advancing edge of larger ones (WALKER and PADGETT 1983; ZU RHEIN 1969). Virus is also seen in the cytoplasm of glia and in association with myelin sheaths. Smaller numbers of virions have also been described in the nuclei of bizarrely shaped (? transformed) astrocytes (MAZLO and TARISKA 1982).

*Genus 2. Papillomavirus genus (e.g. common wart virus)*
DNA genome; icosahedral capsid ca. 43–46 nm in thin sections (Fig. 17 c; GHADIALLY 1988; WILLS 1983). Virus particles are most common in plantar warts, where they form massive intranuclear arrays which displace the

chromatin to the margin, but they may also be found in the cytoplasm (WILLS 1983). SHAH (1985) suggests that cytoplasmic virus only occurs following damage to the nuclear membrane. Mature papilloma virions are attached to nuclear matrix fibres within the nucleus (PENMAN 1985).

7.7.6.4 Hepadnaviridae (e.g. hepatitis B virus)

DNA genome; probably icosahedral capsid ca. 27 nm diameter. The nuclei of hepatocytes show no distinct inclusions by light microscopy but may assume a ground-glass appearance. TEM reveals numerous distinctive small hollow and naked capsids, which resemble parvovirus capsids and which contain the hepatitis B core antigen (YUNIS et al. 1977; GRIMLEY and HENSON 1983; BIANCHI and GUDAT 1983).

7.7.6.5 Parvoviridae (e.g. adeno-associated virus and human parvovirus B 19, cause of aplastic crisis in sickle cell anaemia)

DNA genome; icosahedral capsid (ca. 17–24 nm diameter in thin sections). Nuclear inclusions consist of small hollow-appearing capsids. Although not of particular clinical importance, capsids of adeno-associated parvovirus may sometimes be seen in small numbers adjacent to the intranuclear adenovirus crystals (GRIMLEY and HENSON 1983; BLACKLOW and CUKOR 1985).

7.7.6.6 Paramyxoviridae (e.g. measles, mumps, respiratory syncytial virus)

RNA genome; the helically coiled nucleocapsids appear as striated sinuous tubules, with reported outer diameters of 14–17 nm in the morbillivirus and paramyxovirus genera and around 12.5 nm in the pneumovirus genus HOWATSON and FORNASIER 1982; DUBOIS-DALCQ 1979; GHADIALLY 1988). Allowing for the shrinkage expected in ultrathin preparations (see Sect. 7.7.1), these figures accord well with the accepted sizes (18 nm and 12–15 nm diameter respectively) of negative-stained nucleocapsids (MELNICK 1984). Although nuclear accumulation is not an essential step in the life cycle, inclusions composed of large numbers of tubular, usually smooth, nucleocapsids have been described, particularly in subacute sclerosing panencephalitis (Figs. 18 b, 19; GHADIALLY 1988; DUBOIS-DALCQ 1979). Additionally in subacute sclerosing panencephalitis, complex nuclear bodies, some enclosing tubular nucleocapsid-like structures, are a characteristic feature (Fig. 19 d) and have been shown to contain measles antigen (HADFIELD et al. 1972; MARTINEZ et al. 1974).

7.7.6.7 Orthomyxoviridae (e.g. influenza)

RNA genome; Single viral genes, made in the nucleus, appear as nucleocapsids in the cytoplasm, and are enveloped at the plasma membrane (KINGSBURY 1985; GINSBERG 1980 b). However although virus-specified RNA and proteins, including nucleocapsid protein, have been detected immunocytochemically in the nuclei of influenza-infected cells (JACKSON et al. 1982; KLENK and ROTT 1988), distinct helical nucleocapsids (ca. 7–11 nm diameter) are not normally demonstrable by electron microscopy in this site, but only in the cytoplasm (COMPANS and CHOPPIN 1973). Ultrastructural abnormalities of influenza virus-infected cell nuclei centre on the nucleoli in which there is accumulation of dense, sometimes branching material and, in some reports, 40 nm diamter 'tubules' (APOSTOLOV, FLEWETT and KENDAL 1970; COMPANS and CHOPPIN 1973). Eventually the normal nuceolar structure is lost and the nuclear membrane may break down.

#### 7.7.6.8 Picornaviridae (e.g. poliovirus, coxsackievirus, human enteroviruses including hepatitis A)

RNA genome; icosahedral nucleocapsids 18 to 26 nm diameter in thin sections. Picornavirus replication requires no nuclear functions as it has been shown to multiply in the cytoplasm of enucleated cells (RUECKERT 1985). However Cowdry type B intranuclear inclusions have been described in poliovirus-infected neurons. Ultrastructural study by ANZAI and OZAKI (1969) has revealed that they are composed of crystalline accumulations of icosahedral viral nucleocapsids. Poliovirus replication is also known to be accompanied by nuclear shrinkage, pyknosis and chromosomal fragmentation (BELSHE 1984).

#### 7.7.6.9 Flaviviridae – flavivirus genus (e.g. yellow fever virus, Japanese encephalitis virus and dengue)

RNA genome; icosahedral nucleocapsids approximately 24 nm in diameter in thin sections, found in the cytoplasm and enveloped by an altered host cell membrane during maturation. The replication of these viruses is incompletely understood and the widely held belief that there is no specific nuclear involvement has been challenged (TADANO et al. 1989). For example enucleation inhibits the replication of Japanese encephalitis virus and virus-specified proteins have been detected immunocytochemically in the nuclei of cells infected by a number of different flaviviruses. However these are not necessarily structural proteins and old CTEM reports of particles in the nucleus of Japanese encephalitis-infected cells should be interpreted with caution (YASUZIMA et al. 1964; MURPHY et al. 1968).

### *7.7.7 Virus-like Particles and Inclusions*

As mentioned above, normal or pathologically altered host structures can resemble virus particles or virus-associated structures. With experience, it is possible to recognise the more common of these structures for what they are, i.e. nuclear pores, glycogen granules, nuclear bodies and perichromatin and interchromatin granules. Nevertheless, mistaken interpretations are still published in the literature (COUVREUR et al. 1984). Difficul-

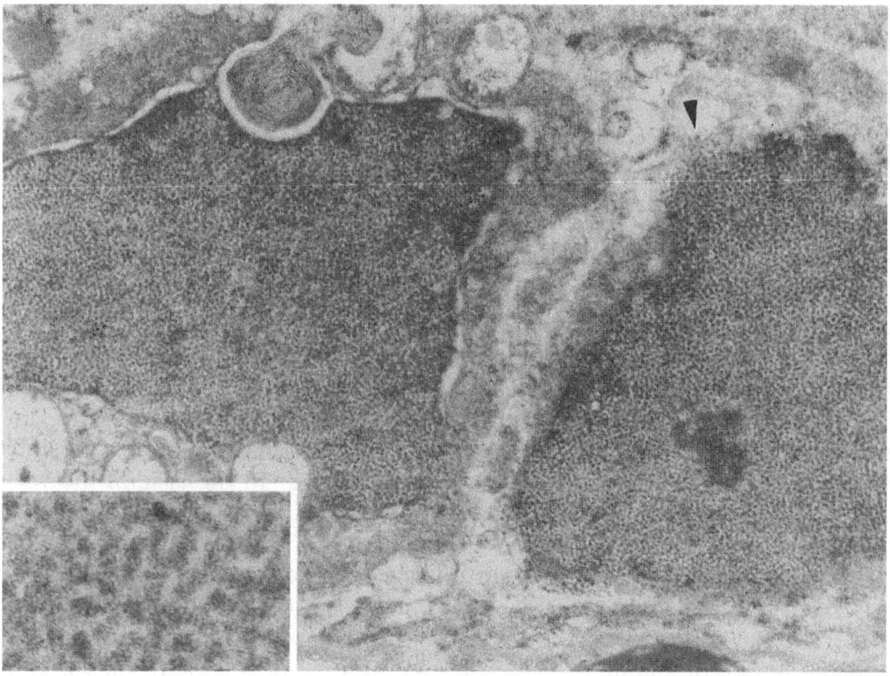

**Fig. 20.** "Paramyxovirus-like" non-viral inclusions in human autopsy brain tissue. Two apparently abnormal perivascular nuclei contain numerous fuzzy filaments, some apparently spilling into the cytoplasm (*arrowhead*). × 15400. At high magnification (*inset*) the dense fuzzy filaments which may be derived from chromatin do not show the distinct hollow tubular structure typical of Paramyxoviridae. × 75000

ties arise with the less common intranuclear structures such as filaments, tubules and vesicles, which may frequently resemble viral or virus-associated structures. The presence of similar structures in normal and pathological control tissues, which must be examined, should suggest a non-viral interpretation.

The ultrastructural pathologist daily encounters structures and appearances which cannot be fully explained. This seems to be particularly true in searches of tissue for evidence of viral infection in disease of uncertain aetiology.

A classic example of this problem is the description by PRINEAS (1972) of nuclei packed with paramyxovirus-like filaments in cases of multiple sclerosis (Fig. 20). After some initial excitement, it became apparent that these filaments were actually composed of altered chromatin (LAMPERT and LAMPERT 1975; GHADIALLY 1988). Three features of these "inclusions" made it unlikely that they were paramyxovirus. First, they could also be found in perivascular cells in control cases of non-infectious neurological disease. Second, the inclusions filled and indeed appeared to spill from the nucleus, whereas paramyxovirus inclusions are always accompanied by a rim of chromatin. Finally, it was not possible to demonstrate convincingly the expected hollow tubular nature of the supposed paramyxovirus nucleocapsids. Incidentally, this artefactual explanation does not account for all of the paramyxovirus-like intranuclear inclusions which have been described in multiple sclerosis (reviewed in KIRK and HUTCHINSON 1978).

The problem becomes even more difficult when the unusual appearances are not demonstrable in control tissues, yet lack the specific features of the suspected virus. Such inclusions must be put into the category of the unexplaned, until some further evidence becomes available, perhaps from the use of immunocytochemistry or in situ hybridisation. Examples of such inclusions are those found in osteoclasts in Paget's disease of bone (HOWATSON and FORNASIER 1982) and in endothelial cells and myocytes in patients with endocardial fibroelastosis. The hope that viral immunocytochemistry would quickly establish the identity of such inclusions where CTEM was equivocal has not been fulfilled. In the case of the very distinct 12.5 nm microtubules found in the nuclei of osteoclasts in Paget's disease of bone (HOWATSON and FORNASIER 1982), immunocytochemical demonstration of antigens of both measles and respiratory syncytial virus has been claimed (MILLS et al. 1982). The size of the tubules is clearly closer to the latter virus though their disposition is apparently different and the apparent staining for measles antigen is unexplained. Similarly, in inclusion body myositis the reported immunocytochemical staining of nuclear "filamentous" inclusions with antibodies against mumps antigen (CHOU 1986) has not been confirmed (NISHINO et al. 1989).

In Creutzfeldt-Jakob disease (CJD), a transmissible dementia analogous to ovine scrapie and bovine spongiform encephalopathy, peculiar membranous vacuoles have been described in affected neurones (KIM et al. 1988). The author's speculation that they might be diagnostic in CJD is undermined by reports of their occurrence both in normal animals (JEFFREY, personal communication, 1989) and also as a non-specific degenerative feature (Fig. 15c).

## 8 Changes in the Nucleolus

The nucleolus has a key role in intracellular synthetic processes. It is the initial site for ribosomal RNA formation and it is actively involved in the synthesis and distribution of both cytoplasmic and nuclear RNA. Depending on its functional differentiation, the number of nucleoli in a particular cell can vary widely, but there appears to be some uniformity between cells of a given type. In general, metabolically active and proliferating cells have more numerous and prominent nucleoli.

Although apparently a simple structure by light microscopy, the nucleolus displays great ultrastructural diversity. It has four main components, the fibrillar centres, the fibrillar components, the granular component and the nucleolar-associated chromatin (BOUTEILLE et al. 1982; GOESSENS 1984; SMETANA and BUSCH 1974; JORDAN 1984). The organisation and proportions of these components, as well as the position, number, shape and size of the nucleoli, can vary with cell and tissue type, the stage of the cell cycle, the state of cellular metabolism, the degree of cellular differentiation and circadian rhythm (HERNANDEZ-VERDUN 1986; GOESSENS 1984; BOUTEILLE et al. 1982).

## 8.1 Morphological Variants

The compact variant of the nucleolus displays fibrillar centres of variable size, surrounded by a dense fibrillar component. The granular component surrounds and permeates these structures. The pattern is circumscribed, rather than spread out in the more typical meshwork configuration (Fig. 21) (DERENZINI et al. 1983).

The reticulated variant of the nucleolus comprises a network, or nucleolonema, consisting of thread-like fibrillar and granular components incorporating the fibrillar centres into its structure and demarcating small nucleolar vacuoles. This variant is often associated with an active metabolic state or with rapid cellular proliferation as in neoplasia (PLOTON et al. 1983).

The segregated variant of the nucleolus contains the standard components, but these coalesce into discrete structures which become separated into contiguous territories, rather than being intimately interspersed with one another (Fig. 21; KAMEL et al. 1988).

The nesting or ring-shaped nucleolar variant has a single fibrillar centre, a thin surrounding fibrillar component forming a shell around it, and peripheral to this, an often inconspicuous granular component, which may even be absent (SMETANA et al. 1970, 1973; SMETANA and BUSCH 1974, RASKA et al. 1983).

Intermediate forms or patterns of nucleolus are not infrequently encountered, particularly in pathological conditions. The segregated and ring patterns have various characteristic associations, both normal and pathological, which will be discussed below in greater detail. Nucleolar shape is, in fact, highly labile and is influenced by numerous, often temporary or reversible factors.

## 8.2 Location and Connections Within the Nucleus

Three-dimensional reconstructions of ultrathin serial sections, combined with morphometric studies in computerised image analysis systems, have contributed to our understanding of the positional organisation of the nucloeolus within the nucleus (DUPUY-COIN et al. 1986; HERNANDEZ-VER-

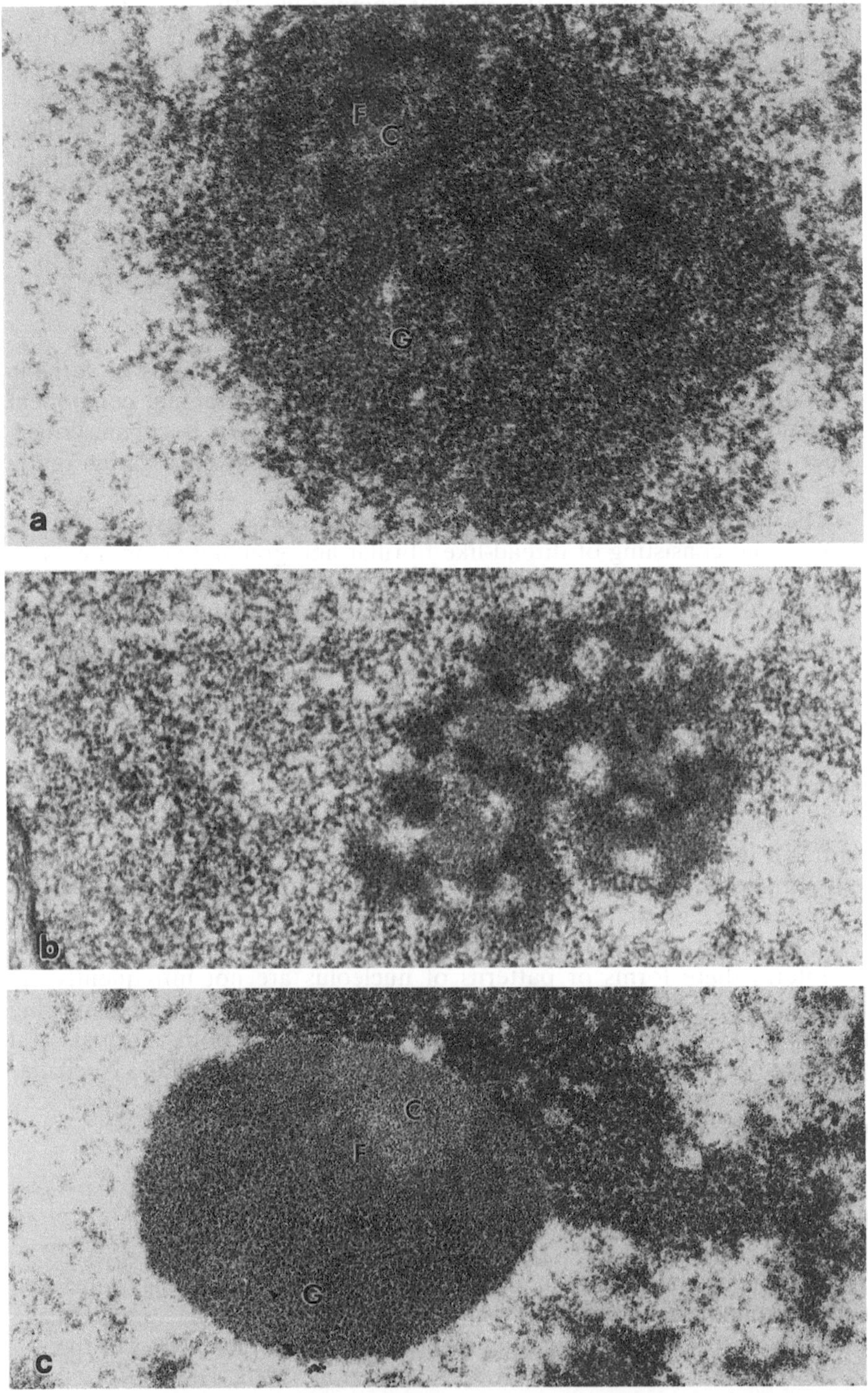
F
C
G
a
b
C
F
G
c

DUN 1986). Nucleoli are arranged in a polarised manner inside the nucleus; they are connected to the nuclear envelope through a nucleolus-envelope junction derived from the nuclear matrix (BOURGEOIS et al. 1979, 1982; BOUTEILLE et al. 1983), or they lie in direct contact with the nuclear envelope (BOURGEOIS et al. 1979) and apparently do not change location as they change their structure (HERNANDEZ-VERDUN 1986). There appears, however, to be a cell cycle-dependent temporal variation in the intranuclear location of the nucleoli (DUPUY-COIN et al. 1986).

## 8.3 Systematic Studies of Nucleolar Morphology

The recognition of the key role of the nucleolus in cellular function, coupled with its morphological variability, has stimulated many systematic studies of this organelle. The following general themes can be identified: the determination of the role of the various nucleolar components in protein synthesis; the characterisation of the effects of various pathological states on the structural organisation of the nucleolus; the linking of particular conditions and agents to specific structural and function alterations; and the attempted identification of specific diagnostic features associated with certain diseases and experimental manipulations.

## 8.4 Common Nucleolar Changes

### *8.4.1 Enlargement*

Nucleolar enlargement occurs in various active functional and physiological states, such as in embryonic and stem cells and in cells engaged in active protein secretion. Enlargement is also a feature of repair and regeneration, hyperplasia and neoplasia. The common link is an increased demand for ribosomal protein synthesis, which is ultimately under nucleolar control.

### *8.4.2 Margination*

In this condition, the nucleolus is seen lying unusually close to, or in particularly intimate contact with the inner membrane of the nuclear en-

◀ **Fig. 21 a–c.** Segregation of the nucleolar components in an actinomycin D-treated mouse. Ridgeway osteogenic sarcoma. **a** Normal distribution of the fibrillar centres (*C*), fibrillar component (*F*) and granular component (*G*) in untreated tumour. × 55 000. **b** Early stages in segregation; the fibrillar centres and surrounding fibrillar components appear prominent and the granular component is mainly at the periphery. × 55 000. **c** Segregation of the nucleolar components; the fibrillar centre is one large mass surrounded by a band comprising the fibrillar component. The granular component is aggregated on the other side with surrounding translucency. × 55 000. (**a** and **c** from J Submicrosc Cytol Pathol 20:225–235, 1988, with permission)

velope. This is said to facilitate nucleocytoplasmic exchange in conditions requiring increased synthetic activity. It is commonly observed in malignant cells, as well as in some benign lesions, such as keratoacanthomas, presumably in their growth phase (GHADIALLY 1988). Margination is also observed in regenerative states, such as in liver cells after partial hepatectomy.

### *8.4.3 Segregation*

Segregation is characterised by migration and separation of the fibrillar centres and fibrillar component from the granular component of the nucleolus. This unique morphological pattern is observed in response to various cytotoxic and chemotherapeutic agents, including actinomycin D, aflatoxin, amsacrine, mitomycin C and adriamycin; to some carcinogens, such as 4-nitroquinoline-*N* oxide; and to physical agents such as hyperthermia and radiation (SIMARD 1966; SCHOEFL 1964; LAPIS and BERNHARD 1965; DASKAL et al. 1975; REYNOLDS and MONTGOMERY 1967; MONTGOMERY et al. 1966; HARRIS et al. 1968; SITTORI and BOSISIO 1966; DIMOVA et al. 1979; KAMEL 1985; KAMEL et al. 1988; JENSEN et al. 1985). Nucloeolar segregation is also seen after herpes simplex virus infection (BOUTEILLE et al. 1982). The underlying common mechanism is interference with or inhibition of RNA synthesis.

The ultrastructural details of nucleolar segregation were described some 25 years ago and since then have been correlated with the DNA-dependent RNA-inhibitory effect of various agents on different normal and neoplastic tissues both in vitro and in vivo. Depending on cell sensitivity, agent dose and cytotoxicity, and period of exposure, segregation can be observed as early as 30 min after the initial exposure. The following is a brief outline of the morphological changes observed at different stages of development of nucleolar segregation in mouse Ridgeway osteogenic sarcoma cells after exposure to actinomycin D (Fig. 21; KAMEL et al. 1988).

In the early stages, there is concentric peripheral aggregation of the nucleolar chromatin and granular components, with increased prominence of the fibrillar component and fibrillar centres. As segregation progresses, there is a tendency for polarisation of the granular component often at the side of the nucleolus facing away from the nuclear periphery, while the fibrillar centres, encased crescentically by the fibrillar component, gather at the side of the nucleolus near to the nuclear rim. This polarisation and aggregation of the granular component at one side of the nucleolus has been referred to in earlier studies as the formation of a "nucleolar cap".

At a later stage, the fibrillar centres seem to fuse to form a few clumps or a single larger mass, with the encasing fibrillar component forming a linear density between this and the granular component. As the process progresses further, a paragranular halo of electron translucency separates the granular component from the perinucleolar chromatin, perhaps marking the start of the gradual disappearance or "eating up" of the granular component. With larger doses, or in susceptible tumour cells, condensed ring-shaped nucleoli are also observed. The final stages of segregation are characterised by a pronounced reduction in nucleolar size that has been attributed mainly to the loss of granular and fibrillar components.

### *8.4.4 Ring-Shaped Nucleoli*

As indicated earlier, resting or ring-shaped nucleoli show a single central fibrillar centre, demarcated by a thin fibrillar component with a surrounding, sometimes inconspicuous granular component, which may even be absent. In mature lymphocytes, the fibrillar centre in the ring-shaped nucleolus can sometimes be in contact with the associated condensed chromatin, with apparent absence of both the granular and the fibrillar component.

Ring-shapoed nucleoli are found in normal mature smooth muscle cells, monocytes, lymphocytes and plasma cells (SMETANA and BUSCH 1974). They also occur in pathological states, such as after irradiation, treatment with acridine derivatives, actinomycin D or adriamycin, following hypo- or hyperthermia, and in the myeloblasts and promyelocytes of acute leukaemias. Ring-shaped nucleoli are also described after viral infection (BOUTEILLE et al. 1982). The presence of ring-shaped nucleoli may, therefore, indicate reduced ribosomal protein synthesis.

### *8.4.5 Other Nucleolar Alterations*

Other nucleolar alterations include complete degranulation with formation of fibrillar nucleoli in the absence of the granular component (GOESSENS 1978), the production of granular nucleoli (BOUTEILLE et al. 1982; SMETANA and BUSCH 1974) formed almost completely of the granular component in the absence of the fibrillar component. Fragmentation of the nucleoli is another morphological event that occurs after exposure to RNA synthesis-inhibitory toxic agents. (For reviews see BERNHARD 1971; HERNANDEZ-VERDUN 1986; FAKAN 1986; SIMARD 1970; SIMARD et al. 1974.)

## 8.5 Nucleolar-Associated Condensed Chromatin

Nucleolar-associated chromatin is aggregated eterochromatin which can sometimes be seen either surrounding the nucleolus (perinucleolar chromatin) or interspersed within the nucleolar components (internucleolar chromatin) (Fig. 22) (GRANBOULAN and GRANBOULAN 1964; SMETANA and BUSCH 1964). In conventionally stained sections, however, it can sometimes be difficult to differentiate between this and the granular component. Nucleolar-associated chromatin is very reduced or absent in undifferentiated cells and in cultured cells.

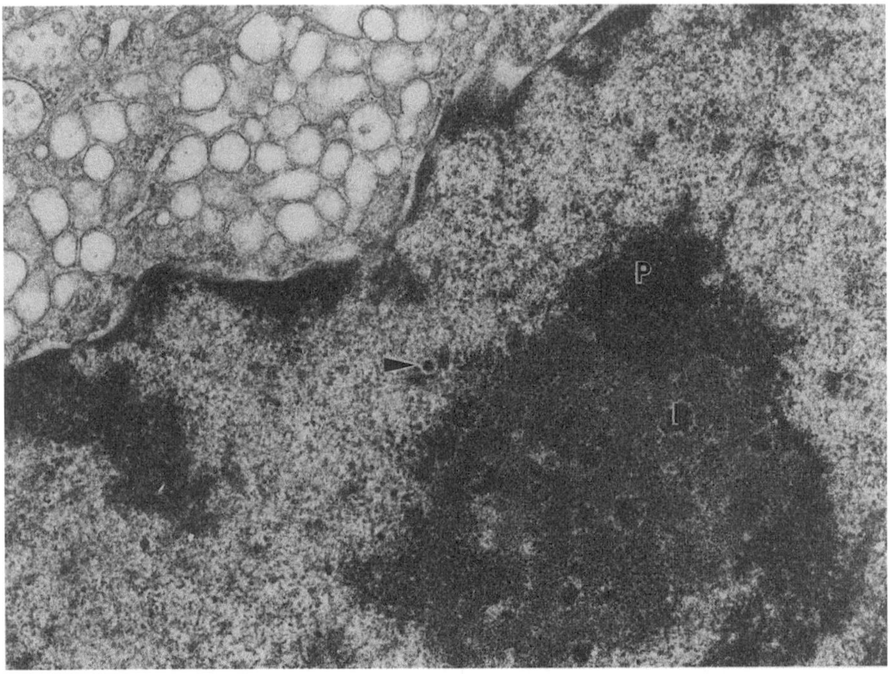


**Fig. 22.** Nucleolus showing intranuclear (*I*) and perinucleolar (*P*) chromatin in methotrexate-treated mouse sarcoma 180. A perichromatin granule (*arrowhead*) is also present. × 34 500

## 8.6 The Nucleolus in Neoplasia

Common morphological features of the nucleolus in neoplasia include enlargement, increased numbers, irregularity and margination along the nuclear perimeter. In general, the higher the grade of malignancy, the more pronounced are these morphological alterations. However, as with all generalisations, exceptions exist. Early neoplastic lesions derived from cells that do not normally have significant secretory activity or prominent nucleoli may show no significant nucleolar alterations. Nucleolar morphology in neoplasia is to some extent determined by the same various factors that operate in non-neoplastic conditions.

In highly malignant neoplasms, whatever their type, there is often a spectrum of profound morphological change including irregularity, enlargement, increased number and margination of nucleoli (GHADIALLY 1982, 1985; HAGUENAU 1969). The underlying mechanism for these changes lies in increased RNA synthesis, whether associated with rapid cellular proliferation or impaired regulation of production (MILLER et al. 1979; PLOTON et al. 1982). Impairment in the regulation of DNA synthesis is apparently another contributory factor.

Enlarged prominent nucleoli, apart from being compatible with high-grade malignancy, also occur in tumours which are characterised by active secretory or metabolic activity. The list includes adenocarcinomas of hepatic, renal, and ovarian origin, melanomas, choriocarcinomas and many other tumours.

Non-neoplastic cell populations at each particular stage of their development usually have more or less equal numbers of nucleoli and possess a distinctive pattern of nucleolar morphology. It is, however, not uncommon for neighbouring neoplastic cells to display remarkable qualitative and quantitative nucleolar variations. Such heterogeneity, if present, is undoubtedly a useful ultrastructural pointer to neoplasia.

GHADIALLY (1985) invented the term "meandering nucleolus" to describe the characteristic nucleoli often although not exclusively seen in germ cell tumours, such as seminoma, and to a lesser extent in embryonal rhabdomyosarcoma and metastatic melanoma (GHADIALLY 1988). These are large reticulated immature or open pattern nucleoli that characteristically occupy a considerable part of the nucleus.

### 8.7 Fibrillar Centres and Nucleolar Organiser Regions

The fibrillar centres of the nucleolus are the sites of the ribosomal genes in the interphase nucleus. During metaphase, hundreds of copies of these ribosomal genes are distributed at ten specific sites in five different chromosome pairs. These sites are the nucleolar organiser regions. Fusion of fibrillar centres often occurs after mitosis, with the result that teir total number, as well as the number of nucleoli, is often less than the number of nucleolar organiser regions. Specific proteins associated with the nucleolar ribosomal genetic material are stained specifically by ammoniacal silver techniques, which delineate argyrophilic structures known as AgNORs. These are amenable to qualitative and quantitative light microscopic studies. This subject is reviewed in detail in Chapter 5 of the current issue.

## 9 The Nucleus in Mitosis

No other aspect of cell structure is subject to such radical morphological change as the nucleus in mitosis. Although this chapter as a whole deals largely with the interphase nucleus, it is appropriate in this section to consider, at least briefly, the ultrastructure of the mitotic nucleus.

### 9.1 Morphological Features

The fundamental cytological changes of the nucleus in mitosis are familiar to any microscopist. Centriolar division precedes nuclear division, resulting in the formation of the spindle poles. The four principal stages of nuclear division, prophase, metaphase, anaphase and telophase, are followed by cytokinesis, or cell division, which results in the separation of two distinct cytoplasmic territories around the reconstituted daughter nuclei. The standard morphological features of this process are outlined in every textbook of biology (WEISS 1988) and are described in full elsewhere. (KUBAI 1975; INOUE 1981; BRINKLEY et al. 1984; EARNSHAW et al. 1984; ZIMMERMAN and FORER 1981).

There are several ultrastructural landmarks as the interphase nucleus moves into mitotic prophase. The earliest stages of reorganisation of the chromatin into individual chromosomal masses are accompanied by the fragmentation and disappearance of the nuclear envelope and by the disintegration and dispersal of the nucleolus. Concurrent with the disappearance of the boundary between nucleus and cytoplasm is the migration of the reduplicated centriole pairs to opposite sides of the cell, providing the polar structures between which the mitotic spindle is organised.

By now, the start of metaphase, the chromosomes have adopted their definitive structural identity and have become aligned along the equatorial plate of the mitotic spindle (Fig. 23).

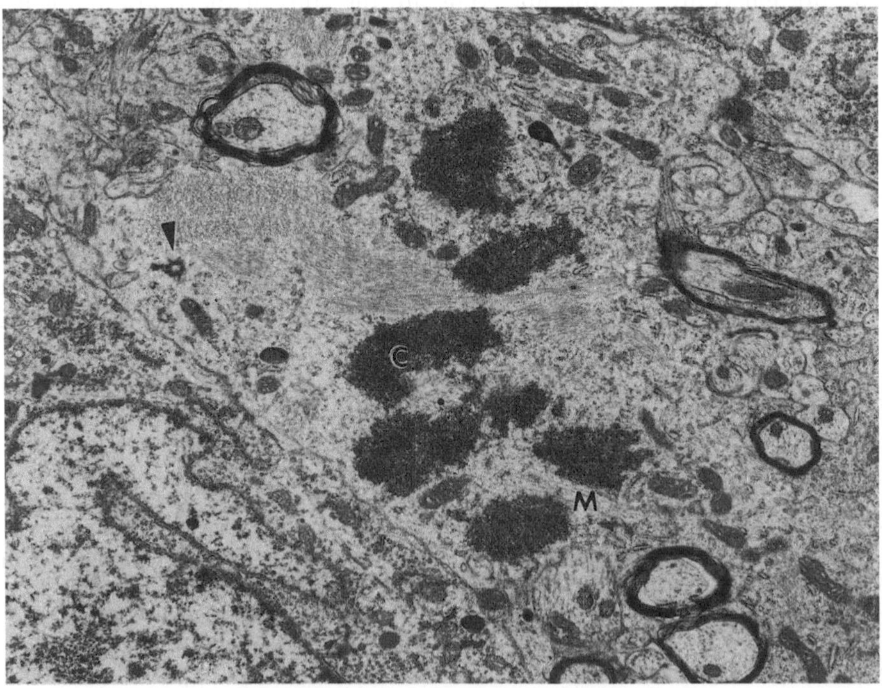


**Fig. 23.** During mitosis, the chromosomes (*C*) appear as dense clumps in the cytoplasm, unbounded by nuclear envelope. Spindle microtubules (*M*) and a centriole (*arrowhead*) can also be seen in this dividing rodent astrocyte. × 9700

The spindle structure, composed at least in part of identifiable microtubular components, is linked to each chromosome by attachments to a specific region termed the kinetochore. The tubules linking the kinetochore to the spindle pole are known as the kinetochoric elements. The other microtubular components of the spindle are divided into a polar group, which extends between the two poles of the spindle, and a free group (KUBAI 1975; INOUE 1981; BRINKLEY et al. 1984). There is reason to believe that conventional TEM techniques do not visualise all of the constituents of the mitotic spindle apparatus (HEATH 1981).

During anaphase, each of the two sets of chromosomes is drawn towards the corresponding pole of the mitotic spindle, presumably by mechanisms involving shortening of the kinetochoric microtubules and lengthening of the polar elements. When the final stage, or telophase, is reached, the chromosomes gather again into an aggregated mass of dense chromatin, initially of an irregular configuration, around which the nuclear envelope re-assembles by fusion of membrane elements derived from the endoplasmic reticulum. At this point the nucleolus also reappears.

The mitotic spindle is thus the essential structural framework for the dynamic changes of mitoses. It organises and sets in position the chromosomal components of the metaphase plate and it redistributes them actively between the daughter cells in anaphase and telophase. The spindle apparatus disperses at the end of mitosis, but a remnant of the spindle, known as the mid-body, may sometimes persist after mitosis and can be mistaken for some other form of cytoplasmic specialisation.

## 9.2 Ultrastructural Interpretation

Electron microscopy has played an important part in establishing many of the basic facts of the structural organisation of the cell in mitosis (HEATH 1981; MCINTOSH 1982). This applies not only to the fine structural features of the centrioles and of the mitotic spindle, but also to the basic nature of the chromatin during the process of mitosis and the presence and structure of the kinetochore. As with most other aspects of modern cell biology, the unravelling of complex ultrastructural details has contributed significantly to the understanding of the functional behaviour of the cell (HEATH 1981; MCINTOSH 1982).

Although the basic ultrastructural biology of mitosis is relatively well documented, the situation is less straightforward with regard to pathological alterations in the mitotic nucleus. This contrasts with the wealth of cytogenetic information on chromosomal aberrations in various categories of disease. Such information as we have tends to be derived more from studies of cells in culture conditions than from tissue biopsies or pathological specimens. Pathological alterations in the mitotic nucleus are particularly difficult to appreciate in CTEM images. There are various reasons for this.

First, pathological changes which exist at the molecular level may not be expressed clearly at the level of cellular structure. Moreover, the mitotic chromosome is in some ways even less rewarding, in ultrastructural terms, than the interphase nucleus, since it is represented by uniform darkly stained granular material, similar to heterochromatin, but without the anatomical and textural variations of euchromatin and heterochromatin in interphase.

Second, chromosomal defects, as they are typically documented, commonly involve the deletion or addition of chromosomes or gross structural defects in single chromosomes. Qualities such as these are difficult, if not impossible, to determine by routine TEM owing to the uncontrolled orientation of the plane of section and the artefacts of fixation and processing (HEATH 1981). For changes such as these, the whole-mount, metaphase-spread specimen of the cytogeneticist is greatly superior to the thin section of the electron microscopist.

Third, when seen in vivo by time-lapse cinematography on light microscopy, the process of mitosis is one of the most turbulent moments in the life of the cell. The rapid changes in the cytological and molecular organisation of the chromosomes throughout the various phases of mitosis cannot easily be related to the static morphological patterns detectable by electron microscopy. It is, for example, quite difficult, from ultrastructural criteria alone, to determine the precise phase of a mitotic figure, even leaving aside the problems of plane of section and incomplete imaging. Consequently, it is extremely difficult to determine whether a particular appearance in a given thin section represents a normal event in one of the standard phases of mitosis, or a morphological aberration associated with some specific cytogenetic abnormality.

Finally, as already indicated, it is inherent in the limitations of the ultrathin section that some structural components of a three-dimensional assembly will not appear within the image at hand. In addition, the structural interrelationships between imaged components and components not included in the plane of section must, of necessity, remain obscure. A single thin section comprises a tiny fraction of the total substance of a mitotic nucleus. The significance of any negative finding in such an image is, therefore, very limited indeed.

## 9.3 The Role of Other Technologies

For the above-stated reasons CTEM alone is ill-adapted to the study of mitosis. Our current knowledge is derived from a much broader range of techniques, including three-dimensional reconstruction, scanning electron microscopy, high voltage TEM using thick sections, X-ray diffraction, phase contrast and polarising light microscopy, immunofluorescence and time-lapse cinematography, coupled with modern molecular and genetic techniques (MCINTOSH 1982; VALDIVIA et al. 1986; DE ROBERTIS and DE

ROBERTIS 1987). In the study of mitosis, just as in the interphase nucleus, the new molecular technologies will be of great importance, particularly the use of nucleic acid probes for in situ hybridisation and the application of immunogold labelling methods at the elctron microscopic level. Several human autoantibodies are recognised that can identify chromosomal molecular components, such as DNA, histones and non-histone proteins, nucleosides and kinetochores in routine TEM sections, regardless of orientation (TAN 1979; LERNER et al. 1980; BRENNER et al. 1981; MOROI et al. 1980; VALDIVIA et al. 1986).

### 9.4 Agent Effects

Agent-induced effects on mitosis depend on the phase during which the agent gains access to the cell. Due to the relative infrequency of mitosis in the overall life span of cells, and due to the problems of interpretation of ultrastructural images of the mitotic apparatus, data regarding the morphological effects of insults during mitosis are scarce, and are mainly derived from cell culture studies.

Some cytotoxic agents have characteristic effects, a few of which are associated with distinctive morphological appearances, particularly in the mitotic spindle. Colchicine and Colcemid, hypothermia, microbeam ultraviolet irradiation and excess hydrostatic pressure are known to inhibit the formation of the microtubules of the mitotic spindle and induce their depolymerisation (BORGERS and DEBRABANDER 1975). The vinca alkaloids, vincristine and vinblastine, used as chemotherapeutic agents, also inhibit the formation of the mitotic spindle. These agents cause the disappearance of the spindle microtubules (BORGERS and DEBRABANDER 1975). The centrioles do not migrate and, as a result, remain in the centre of the cell, encircled by the double complement of chromosomes.

Following exposure to various forms of irradiation, radiosensitive cells show profound morphological changes during mitosis. These changes include abnormalities in the mitotic spindle, such as tripolarity, and abnormalities in the chromosomes, such as fragmentation, coalescence or bridging and altered forms (EVANS 1962; WOLFF 1960).

## 10 Conclusions

When considered as a tool for the pathologist, electron microscopy is both uniquely powerful and severely limited. It allows us to image structural details in tissues and cells far beyond the limits of light microscopy, down to the level of macromolecular dimensions. On the other hand, there is a biological "uncertainty principle" which dictates that the act of observing so changes the structure, through tissue fixation and processing, that its interpretation becomes increasingly difficult as the scale of magnification increases.

Nevertheless, for the last 40 years, ultrastructural techniques have played a major role in the revolution in cellular and molecular biology and pathology. Although now supplemented and in some areas superseded by increasingly sophisticated functional techniques, electron microscopy seems set to continue to make an indispensable contribution to further developments. It must be admitted, however, that the earliest and largest dividends to date have more often been paid on investigations of tissue architecture and cytoplasmic organisation than on studies of the detailed morphology of the cell nucleus.

The diagnostic pathologist is still able to make only very few practical statements about the ultrastructure of the nucleus which are meaningful, yet simple and unequivocal. Its indeterminate fine structure, when fixed in any one moment of time, contrasts with its complex, dynamic functions. The often questionable significance of its diverse inclusions and alterations is a constant source of frustration. Its contours and patterns are modulated by its histogenetic identity, its functional state and even its age. The mask of artefact, the distorting mirror of fallible interpretation and the cosmetics of fixation and staining leave us constantly uncertain as to the true face of the nucleus.

Despite these difficulties, changes in the nucleus associated with viral infection have been among the most rewarding for electron microscopists, enabling them to play a major role in establishing the viral aetiology of two serious neurological diseases, subacute sclerasing panencephalitis and progressive multifocal leukoencephalopathy. CTEM and immuno-electron microscopy have been successfully used to identify and establish the distribution of HBcAg particles and unassembled antigen in nucleus and cytoplasm of infected hepatocytes. Both techniques may continue to be useful in identifying the aetiological agents of non-A non-B hepatitis. CTEM of nuclear inclusions has an important role in the diagnosis of viral encephalitis and of viral pneumonia in the immunocompromised host. The involvement of electron microscopy in diseases of uncertain aetiology can be criticised on the grounds that it has created more heat than light. We may hope that the in situ application of both viral gene probes and antiviral antibodies at the electron microscopic level will prove more definitive in the future.

In the case of tumours, the problem again is not that we lack ultrastructural information, but that we have it in such abundance. The list of distinctive, but ultimately non-specific nuclear alterations is almost limitless, but there remains no single unique criterion or marker of the malignant state. It remains to be seen whether some combination of morphometric and functional techniques may provide the long sought-after key to this question.

Thus, despite a voluminous literature on fine structural phenomena, which can only be touched upon in a chapter such as this, our understanding of the interrelationships between nuclear morphology and disease remains incomplete and largely circumstantial. There are several obvious reasons for this. First, our historical catalogue of ultrastructural data has only recently been backed by something approaching an understanding of the molecular mechanisms of the nucleus in health and disease. Second, despite undoubted recent progress in this area, we possess only a sketchy knowledge of the linkage between various aspects of normal nuclear function and the structure of its component parts. Without a clear definition of normality, it is impossible to set clear boundaries to the abnormal, let alone to chart the detailed morpho-functional relationships of disease. Third, the inherent limitation of ultrastructural technique in the exploration of functional properties has held back progress, although as already indicated, recent advances in immunohistochemistry and in situ hybridisation promise a brighter future.

There remain, however, some difficult barriers to future advance. The nucleus is a uniquely dynamic component, changing with time, most remarkably in the process of mitosis. In the consideration of agent effects, nuclear alterations and cellular consequences are materially influenced by such changes. Moreover, the response of the nucleus to a particular insult, such as viral infection, may be unpredictable, leading to degeneration, lysis and cell death in one case and neoplastic transformation in another. The malignant nuclear phenotype itself, in ultrastructural terms, is probably a rather poor mirror for the many complex processes involved in oncogenesis.

These problems are perhaps best expressed through a simple analogy. The cytoplasmic organelles of the cells of the body, although complex and dynamic, can be simplistically considered as the machinery of an elaborate factory. There is an inherently discoverable and comprehensible linkage, in such a system, between the structure of a machine and its function. In such an analogy, however, the nucleus represents the computer system. The different machine tools in a factory may all be operated by superficially similar onboard computers, but the key to the factory's production schedule lies not in their wiring, but in their programming! For this reason, for the foreseeable future, the ultrastructural study of nuclear morphology in health and disease is likely to prove considerably more taxing than most other investigations of cellular structure.

*Acknowledgements.* The authors are grateful to Dr. C.H.S. Cameron for access to ultrastructural material and to Mr. T. McLaughlin and the staff of the EM Unit for technical support. Miss Bernadine Turley gave expert help in the printing of the illustrations and Mrs. Jacqueline Gallagher and Miss Paula Coney took endless pains in the typing and repeated revision of the manuscript.

John Kirk is indebted to the Multiple Sclerosis Society of Great Britain and Northern Ireland for continuing grant support and to Professor I.V. Allen for provision of EM facilities in the neuropathology laboratory.

## References

Ahearn MJ, Trujillo JM, Cork A, Fowler A, Hart JS (1974) The association of nuclear blebs with aneuploidy in human acute leukaemia. Cancer Res 34:2887–2896

Aksamit AJ, Mourrain P, Sever JL, Major EO (1985) Progressive multifocal leukoencephalopathy: investigations of three cases using in-situ hybridisation with JC virus biotinylated DNA probe

Anzai T, Ozaki Y (1969) Intranuclear crystal formation of poliovirus: electron microscopic observations. Exp Mol Pathol 10:176–185

Apostolov K, Flewett TH, Kendall AP (1970) In: Barry RD and Mahy BWJ (eds) The biology of large RNA viruses. Academic Press, London, pp 3–26

Barrack ER, Coffey DS (1982) Biological properties of the nuclear matrix: steroid hormone binding. Recent Prog Horm Res 38:133–195

Belshe RB (1984) Textbook of human virology. pp 407–483, PSG, Littleton, Mass

Beltran G, Stuckey WJ (1972) Nuclear lobulation and cytoplasmic fibrils in leukaemic plasma cells. Am J Clin Pathol 58:159–164

Bendayan M (1982) Ultrastructural localisation of nucleic acids by the use of enzyme-gold complexes: influence of fixation and embedding. Biol Cell 43:153–156

Bendayan M (1984) Enzyme-gold electron microscopic cytochemistry: a new affinity approach for the ultrastructural localisation of macromolecules. J Electron Microsc Technique 1:349–372

Ben-Ze'ev A, Abdulafia R, Bratosin S (1983) Herpes simplex virus assembly and protein transport are associated with the cytoskeletal framework and the nuclear matrix in infected BSC-1 cells. Virology 129:501–507

Berezney R (1979) Dynamic properties of the nuclear matrix. In: Busch H (ed) The cell nucleus, vol 7. Academic Press, New York, pp 413–456

Berezney R (1984) Organisation and functions of the nuclear matrix. In: Hnilica L.S. (ed) Chromosomal nonhistone proteins, vol IV. Structural Associations, CRC Press, Boca Raton, pp 1–119

Berezney R, Coffey D (1977) Nuclear matrix isolation and characterisation of a framework structure from rat liver nuclei. J Cell Biol 73:616–637

Bernhard W (1971) Drug-induced changes in the interphase nucleus. Adv Cytopharmacol 1:49–67

Bernhard W, Granboulan N (1963) The fine structure of the cancer cell nucleus. Exp Cell Res [Suppl 9]:19–53

Bessis MC (1961) Ultrastructure of lymphoid and plasma cells in relation to globulin and antibody formation. Lab Invest 10:1040

Bianchi L, Gudat FG (1983) Histo- and immunopathology of viral hepatitis. In: Deinhardt F, Deinhardt J (eds) Viral hepatitis: laboratory and clinical science. Marcel Dekker, New York, pp 335–382

Bibor-Hardy V, Pouchelet M, St-Pierre E, Herzherg M, Simard R (1982a) The nuclear matrix is involved in herpex simplex virogenesis. Virology 121:296–306

Bibor-Hardy V, Suh M, Pouchelet M, Simard R (1982b) Modifications of the nuclear envelope of BHK cells after infection with herpes simplex virus type 1. J Gen Virol 63:82–94

Blackburn W (1971) Pathobiology and nucleocytoplasmic exchange. Pathol Annu 1:1–31

Blacklow NR, Cukor G (1985) Parvoviruses. In: Fields BN (ed) Virology. Raven Press, New York, pp 411–432

Blom J, Mansa B, Wiik A (1976) A study of Russell bodies in human monoclonal plasma cells by means of immunofluorescence and electron microscopy Acta path microbiol scand Sect A 84:339–349

Boos J, Esiri MM (1986) Viral encephalitis: pathology, diagnosis and management. Blackwell, Oxford

Borgers M, Debrabander M (eds) (1975) Microtubules and microtubule inhibitors. Amsterdam, North Holland

Bourgeois CA, Hemon D, Boutelle M (1979) Structural relationship between the nucleolus and the nuclear envelope. J Ultrastruct Res 68:328–340

Bourgeois CA, Hemon D, Bouteille M (1982) Changes in the nucleolus-envelope region during interphase in synchronized TG cells. J Ultrastruct Res 81:257–267
Bouteille M (1972) Ultrastructural localisation of proteins and nucleoproteins in the interphase nucleus. Karolinska symposia on research methods in endocrinology. 5th Symposium, May 29–31. Gene transcription in reproductive tissue
Bouteille M, Kalifat SR, Delarue J (1967) Ultrastructural variations of nuclear bodies in human diseases. J Ultrastruct Res 19:474–486
Bouteille M, Dupuy-Coin AM, Moyne G (1975) Techniques of localisation of proteins and nucleoproteins in the cell nucleus by high resolution autoradiography and cytochemistry. Methods Enzymol 40:3–41
Bouteille M, Hernandez-Verdun D, Dupuy-Coin AM, Bourgeois CA (1982) Nucleoli and nucleolar-related structures in normal, infected and drug-treated cells. In: Jordan EG, Cullis CA (eds) The nucleolus. Cambridge University Press, Cambridge, pp 179–211
Bouteille M, Bouvier D, Seve AP (1983) Heterogeneity and territorial organisation of the nuclear matrix and related structures. Int Rev Cytol 83:135–182
Bouvier D, Dupuy-Coin AM, Bouteille M, Moens PB (1980) Three-dimensional electron microscopy of the nuclear matrix components of HeLa cells. Biol Cell 39:121–124
Branstetter D, Goldblatt P (1983) Pathobiology of the nuclear envelope. In: Trump BF, Laufer A, Jones RT (eds) Cellular pathobiology of human disease, Gustav Fischer, New York, pp 73–97
Brenner S, Pepper D, Berns MW, Brinkley BR (1981) Kinetochore structure, duplication, and distribution. in mammalian cells: analysis by human autoantibodies from scleroderma patients. J Cell Biol 91:95–102
Brinkley RB, Brenner SL, Tousson A, Valdivia MM (1984) Microtubular organising centres: evolution and biochemical characterisation of centromeres, kinetochores of mammalian chromosomes. In: Seno S, Okada Y (eds) III International Cell Biology Congress. Academic Press, New York, p 117
Budka H (1986) Multinucleated giant cells in brain: a hallmark of the acquired immune deficiency syndrome (AIDS). Acta Neuropathol 69:253–258
Burns WA (1978) Thick sections: techniques and applications. In: Trump BF, Jones RT (eds) Diagnostic electron microscopy, vol 1. John Wiley, New York, pp 141–166
Burns WA (1980) Micro-organisms. In: Trump BF, Jones RT (eds) Diagnostic electron microscopy, vol 3. John Wiley, New York, pp 37–96
Chai LS, Weinfeld H, Sandberg AA (1978) Effect of divalent-cation chelators and chloramphenicol on the spatial relationship of the nuclear envelope to chromatin in micronuclei of Chinese hamster cells. J Supramol Struct 9:459–471
Chou S-M (1986) Inclusion body myositis: a chronic persistent mumps myositis? Hum Pathol 17:765–777
Clawson GA, James J, Woo CH (1980) Pertinence of nuclear envelope nucleoside triphosphatase activity to ribonucleic acid transport. Biochemistry 19:2748–2756
Cohen A, Sundeen J (1976) The nuclear fibrous lamina in human cells: studies on its appearance and distribution. Anat Rec 186:471–476
Collan Y, Salmenpera M (1976) Electron microscopy of postmortem autolysis of rat muscle tissue. Acta Neuropathol 35:219–233
Comings DE, Okada T (1973) DNA replication and the nuclear membrane. J Mol Biol 75:609–618
Compans RW, Choppin PW (1973) Orthomyxoviruses and paramyxoviruses. In: Dalton AJ and Haguenau F (eds) Ultrastructure of animal viruses and bacteriophages: an atlas. Academic Press, New York, pp 213–237
Couvreur Y, Leeman M, Ketelbant P (1984) 'Viral' intranuclear inclusions. Ultrastruct Pathol 7:67–69
Covey L, Choi Y, Prives C (1984) Association of simian virus 40T antigen with the nuclear matrix of infected and transformed monkey cells. Mol Cell Biol 4:1384–1392
Daskal I, Merski JA, Hughes JB, Busch H (1975) The effects of cycloheximide on the ultrastructure of rat liver cells. Exp Cell Res 93:395–401
Daskal Y, Prestayko AW, Busch H (1974) Ultrastructural and biochemical studies of the iso-

lated fibrillar component of nucleoli from Novikoff hepatoma ascites cells. Exp Cell Res 88:1–14
Derenzini M, Moyne G (1978) The nucleolar origin of certain perichromatin-like granules: a study with $\alpha$-amanitine. J Ultrastruct Res 62:213–219
Derenzini M, Hernandez-Verdun D, Pession A, Novello F (1983) Structural organisation of chromatin in nucleolar organiser regions of nucleoli with a nucleolonema-like and compact ribonucleoprotein distribution. J Ultrastruct Res 84:161–172
De Robertis EDP, De Robertis EMF Jr (1987) Cell and molecular biology, 8th ed. Lee & Febiger, Philadelphia, pp 355–360
Dimova RN, Gajdardjieva KC, Dabeva MD, Hadjiolov AA (1979) Early effects of D-galactosamine on rat liver nucleolar structures. Biol Cell 35:1–10
Djaldetti M, Lewinski UH (1978) Origin of intranuclear inclusions in myeloma cells. Scand J Haematol 20:200–205
Droz B, Bouteille M, Sandoz D (1976) Techniques in radioautography. J Microsc Biol Cell 27:No. 2–3
Dubois-Dalcq M (1979) Pathology of measles virus infection of the nervous system: comparison with multiple sclerosis. Int Rev Exp Pathol 19:101–135
Dupuy-Coin AM, Bouteille M, Moens PB, Fournier JG (1982) Distribution of nuclear organelles in measles virus-induced polykaryons. Biol Cell 43:55–68
Dupuy-Coin AM, Moens P, Bouteille M (1986) Three-dimensional analysis of given structures: nucleolus, nucleoskeleton and nuclear inclusions. In: Jasmin G, Simard R (eds) Methods and achievements in experimental pathology, vol. 12. Karger, Basel, pp 1–25
Earnshaw WC, Halligan N, Cooke C, Rothfield N (1984) The kinetochore is part of the metaphase chromosome scaffold. J Cell Biol 98:352
Eisenman RN, Tachibana CY, Abrams HD, Hann SR (1985) v-*myc* and c-*myc*-encoded proteins are associated with the nuclear matrix. Mol Cell Biol 5:114–126
Engedal H, Jensen H, Saetersdal TS (1977) Ultrastructure of abnormal membrane inclusions in nuclei of human myocardial cells. Br Heart J 39:145–151
Erlandson RA, Tandler B (1972) Ultrastructure of acinic cell carcinoma of the parotid gland. Arch Pathol 93:130–140
Evans HI (1962) Chromosome aberrations induced by ionizing radiations. Int Rev Cytol 13:221
Fakan S (1976) High-resolution autoradiography as a tool for the localisation of nucleic acid synthesis and distribution in the mammalian cell nucleus. J Microsc 106:159–171
Fakan S (1986) Structural support for RNA synthesis in the cell nucleus. In: Jasmin G, Simard R (eds) Methods and achievements in experimental pathology, vol 12. Karger, Basel, pp 105–140
Fauci AS (1988) The human immunodeficiency virus: ineffectivity and methods of pathogenesis. Science 239:617–622
Fawcett DW (1981) The cell, 2nd edn. Saunders, Philadelphia
Feldherr CM, Kallenbach E, Schultz N (1984) Movement of a karyophilic protein through the nuclear pore of oocytes. J Cell Biol 99:2216–2222
Ferrans V, Jones M, Maron B et al. (1975) The nuclear membranes in hypertrophied human cardiac muscle cells. Am J Pathol 78:427–460
Flickinger C (1974) The role of endoplasmic reticulum in the repair of amoeba nuclear envelopes damaged microsurgically. J Cell Sci 14:421–437
Flickinger C (1978) Interactions between endoplasmic reticulum and nuclear membranes of different types of cells during repair of damaged amoeba nuclei. Exp Cell Res 111:427–436
Franke JH, Schinko W (1969) Nuclear shape in muscle cells. J Cell Biol 42:303–316
Franke WW (1977) Structure and function of nuclear membranes. Biochem Soc Symp 42:125–135
Franke WW, Scheer U (1974) Structures and functions of the nuclear envelope. In: Busch H (ed) The cell nucleus, vol 1. Academic Press, New York, pp 220–347
Franke WW, Scheer U, Krohne G, Jarrasch E-D (1981) The nuclear envelope and the architecture of the nuclear periphery. J Cell Biol 91:39s–50s

Gallaher WR, Bratt MA (1974) Conditional dependence of fusion from within and other cell membrane alterations by Newcastle disease virus. J Virol 14:813–820
Gautier A (1976) Ultrastructural localisation of DNA in ultrathin sections. Int Rev Cytol 44:113–191
Gerace L (1985) Traffic control and structural proteins in the eukaryotic nucleus. Nature 318:508–509
Gerace L (1986) Nuclear lamina and organisation of nuclear architecture. Trends in Biochemical Science 11:443–446
Gerace L, Blobel G (1982) Nuclear lamina and the structural organisation of the nuclear envelope. Cold Spring Harb Symp Quant Biol 46:967–978
Gerace L, Burke B (1988) Functional organisation of the nuclear envelope. Ann Rev Cell Biol 4:335–374
Gerace L, Blum A, Blobel G (1978) Immunocytochemical localisation of the major polypeptides of the nuclear pore complex-lamina fraction. J Cell Biol 79:546–566
Ghadially FN (1985) Diagnostic electron microscopy of tumours, 2nd edn. Butterworths, London
Ghadially FN (1988) Ultrastructural pathology of the cell and matrix. A test and atlas of physiological and pathological alterations in the fine structure of cellular and extracellular components, vol 1, 3rd edn. Butterworths, London, pp 1–180
Ghadially FN, Roy S (1969) Ultrastructural changes in the synovial membrane in lipohaemarthrosis. Ann Rheum Dis 28:529
Ghadially FN, Oryschak AF, Mitchell DM (1974) Nuclear fibrous lamina in pathological human synovial membrane. Virchows Arch [Cell Pathol]15:223–228
Ghalially FN, Harawi S, Khan W (1985 a) Diagnostic ultrastructural markers in alveolar cell carcinoma. J Submicrosc Cytol 17:269–278
Ghadially FN, Senoo A, Fuse Y (1985 b) A serial section study of nuclear pockets containing nuclear material. J Submicrosc Cytol 17:687–694
Ginsberg HS (1980 a) Adenoviruses. In Davis BD, Dulbecco R, Eisen HN, Ginsberg HS (eds) Microbiology. Harper & Row, Hagerstown, pp 1047–1060
Ginsberg HS (1980 b) Orthomyxoviruses. In: Davis BD, Dulbecco R, Eisen HN, Ginsberg HS (eds) Microbiology. Harper & Row, Hagerstown, pp 1120–1138
Goessens G (1978) Nucleolar ultrastructure during reversible inhibition of RNA synthesis in chick fibroblasts cultivated in vitro. J Ultrastruct Res 65:83–89
Goessens G (1984) Nucleolar structure. Int Rev Cytol 87:107–158
Gotzos V, Campelli-Gotzos B, Gross WO, Petropoulos P, Conti G (1986) A morphologic study of cells from a human malignant fibrous histiocytoma "in-situ" and cultured "in vitro". Arch Anat Cytol Pathol 34:164–171
Granboulan N, Granboulan P (1964) Cytochimie ultrastructurale du nucleole. I. Mise en evidence de chromatine a l'interieur du nucleole. Exp Cell Res 34:71–87
Green M (1985) Transformation and oncogenesis: DNA viruses, In: Fields BN (ed) Virology Raven Press, New York pp 183–234
Grimley PM, Henson DE (1983) Electron microscopy in viral infections. In: Trump BF, Jones RT (eds) Diagnostic electron microscopy, vol 4. John Wiley, New York, pp 1–73
Haase AT (1986) Analysis of viral infections by in-situ hybridisation. J Histochem Cytochem 34:27–32
Hadfield MG, David RB, Rosenblum WI (1972) Coiled nucleocapsid configuration in subacute sclerosing panencephalitis (SSPE). Acta Neuropathol 21:263–271
Haguenau F (1969) Ultrastructure of the cancer cell. In: Bittar EE, Bittar N (eds) The biological basis of medicine, vol 5. Academic Press, New York, pp 433–486
Harris C, Grady H, Svoboda D (1968) Alterations in pancreatic and hepatic ultrastructure following acute cycloheximide intoxication. J Ultrastruct Res 22:240–251
Heath JB (1981) Mitosis through the electron microscope. In: Zimmerman AM, Forer A (eds) Mitosis/cytokinesis. Academic Press, New York, pp 245–269
Heine U, Severak L, Kondratic J, Bonar RA (1971) The behaviour of HeLa-$S_3$ cells under the influence of supranormal temperatures. J Ultrastruct Res 34:375–396
Henderson DW, Papadimitriou JM, Coleman M (1986) Ultrastructural appearances of tumours, 2nd edn. Churchill Livingstone, Edingburgh, pp 17–72

Henry K, Petts V (1969) Nuclear bodies in human thymus. J Ultrastruct Res 27:330
Herlan G, Quevedo R, Wunderlich F (1978) Structural transformation of the nuclear matrix in-situ. Exp Cell Res 115:1ß3–110
Herlan G, Eckert WA, Kaffenberge W, Wunderlich F (1979) Isolation and characterisation of an RNA-containing nuclear matrix from *Tetrahymena* macronuclei. Biochemistry 18:1782–1788
Hernandez-Verdun D (1986) Structural organization of the nucleolus in mammalian cells. In: Jasmin G, Simard R (eds) Methods and achievements in experimental pathology. Karger, Basel, pp 26–62
Howatson AF, Fornasier VL (1982) Microfilaments associated with Paget's disease of bone: comparison with nucleocapsids of measles virus and respiratory syncytial virus. Intervirology 18:150–159
Inoue S (1981) Cell division and the mitotic spindle. J Cell Biol 91:3
Itabashi M, Hruban Z, Wong T-W, Chou S-F (1976) Concentric nuclear inclusions. Virchows Arch [Cell Pathol]20:103–111
Jackson DA, Caton AJ, McCready SJ, Cook PR (1982) Influenza virus RNA is synthesised at fixed sites in the nucleus. Nature 296:366–368
Jackson DA, McCready SJ, Cook PR (1984) Replication and transcription depends on attachment of DNA to the nuclear cage. J Cell Sci [Suppl 1]:59–79
Jensen CG, Wilson WR, Bleumink AR (1985) Effects of amsacrine and other DNA-intercalating drugs on nuclear and nucleolar structure in cultured V79 Chinese hamster cells and PtK2 rat kangaroo cells. Cancer Res 45:717–725
Jinn-Fei Y, El-Labban NG (1986) An ultrastructural study of binucleate plasma cells. J Oral Pathol 15:118–121
Johannessen JV, Sobrinho-Simoes M, Finseth I, Pilstrom L (1982) Papillary carcinomas of the thyroid have pore-deficient nuclei. Int J Cancer 30:409–411
Jordan EG (1984) Nucleolar nomenclature. J Cell Sci 67:217–220
Kamel HMH (1985) Ultrastructural aspects of tumours and anti-tumour therapy. PhD Thesis, Glasgow University, Glasgow
Kamel HMH, Merry S, Toner PGT (1988) Mechanism of actinomycin D-induced resistance in Ridgeway osteogenic sarcoma: an ultrastructural study. J Submicrosc Cytol Pathol 20:225–235
Karasaki S (1970) An electron microscope study of intranuclear canaliculi in Novikoff hepatoma cells. Cancer Res 30:1736–1742
Karasaki S (1973) Passage of cytoplasmic lipid into interphase nuclei in preneoplastic rat liver. J Ultrastruct Res 42:463–478
Karase J (1975) Ultrastructural nuclear change of extranucleolar ribonucleoprotein structures during autolysis of normal liver cells. Virchows Arch [Cell Pathol]18:337–346
Kaufmann SH, Fields AP, Shaper JH (1986) The nuclear matrix: current concepts and unanswered questions. In: Jasmin G, Simard R (eds) Methods and achievements in experimental pathology, vol 12. Nuclear submicroscopy. Karger, London, pp 141–171
Kawanami O, Ferrans VJ, Fulmer JD, Crystal RG (1979) Nuclear inclusions in alveolar epithelium of patients with fibrotic lung disorders. Am J Pathol 94:301–322
Kim JH, Lach B, Manuelidis EE (1988) Creutzfeld-Jacob disease with intranuclear vacuolar inclusions: a biopsy case of negative light microscopic findings and successful animal transmission. Acta Neuropathol 76:422–426
Kingsbury DW (1985) Orthomyxo- and paramyxoviruses and their replication. In: Fields BN (ed) Virology. Raven Press, New York, pp 1157–1178
Kirk J, Hutchinson WM (1978) The fine structure of the CNS in multiple sclerosis. I. Interpretation of cytoplasmic papovavirus-like and paramyxovirus-like cytoplasmic inclusions. Neuropath App Neurobiol 4:343–356
Kistler J, Duncombe Y, Laemmli UK (1984) Mapping nucleolar proteins with monoclonal antibodies. J Cell Biol 99:1981–1988
Kleinfeld RG, Creider MH, Frajula WJ (1956) Electron microscopy of intranuclear inclusions found in human and rat liver parenchymal cells. J Biophys Biochem Cytol 2 [Suppl]:435–438

Kleinfeld R, Koulish S (1957) Cytological aspects of mouse and rat liver containing nuclear inclusions Anat Rec 128:443–445
Klenk H-D, Rott R (1988) The molecular biology of influenza virus pathogenicity. Advances in Virus research, 34:247–281
Koenig S, Gendelman HE, Orenstein JM et al. (1986) Detection of AIDS virus in macrophages in brain tissue from AIDS patients with encephalopathy. Science 233:1089–1093
Kohler G, Milstein C (1975) Continuous cultures of fused cells secreting antibody of predetermined specificity. Nature 256:495–497
Kroh H, Cervos-Navarro J (1988) Back to the nuclear bridges. Acta Neuropathol 76:539–540
Kubai DF (1975) The evolution of the mitotic spindle. Int Rev Cytol 43:167
Kuhn CH (1967) Nuclear bodies and intranuclear globulin in Waldenströ's macroglobulinemia. Lab Invest 17:404–415
Lampert F, Lampert P (1975) Multiple sclerosis: morphologic evidence of intranuclear paramyxovirus or altered chromatin fibres? (Editorial) Arch Neurol 32:425–427
Lapis K, Bernhard W (1965) The effect of mitomycin C on the nucleolar fine structure of KB cells in cell culture. Cancer Res 25:628
Leduc EH, Wilson JW (1959) An electron microscope study of intranuclear inclusions in mouse liver and hepatoma. J Biophys Biochem Cytol 6:427–430
Leeson TS, Leeson CR (1968) The fine structure of Brunner's glands in man. J Anat 103:263
Lerner MR, Boyle JA, Mount SM, Wolin SL, Steitz JA (1980) Are snRNPs involved in splicing? Nature 283:220–224
Lodin Z, Blumajer J, Maires U (1978) Nuclear pore complexes in cells of the developing mouse cerebral cortex. Acta Histochem 63:74–79
Manuelidis L (1985) In-situ detection of DNA sequences using biotinylated probes. Focus 7:4–8
Martinez AJ, Ohya T, Jabbour JT, Duenas D (1974) Subacute sclerosing panencephalitis (SSPE): reappraisal of nuclear, cytoplasmic and axonal inclusions. Ultrastructural study of 8 cases. Acta Neuropathol 28:1–13
Maul GG (1977) The nuclear and the cytoplasmic pore complex: structure, dynamics, distribution and evolution. In: Bourne B, Danielli J, Jean K (eds) International review of cytology, Suppl 6 – studies in ultrastructure. Academic Press, New York, pp 74–184
Maul GG, Deave LL, Freed JJ (1980) Investigation of the determinants of nuclear pore number. Cytogenet Cell Genet 26:175–190
Mazlo M, Tariska I (1982) Are astrocytes infected in progressive multifocal leukoencephalopathy (PML)? Acta Neuropathol 56:45–51
McIntosh JR (1982) Microscopic methods for analysis of mitotic spindle structure. In: Wilson L (ed) Methods in cell biology, vol 25. Academic Press, New York, pp 33–56
Melnick JL (1984) Structure and classification of viruses. In Belshe RB (ed) Textbook of human virology, PSG Littleton, 1–28
Migliorisi G, Folkes E, Pawlowski N, Cramer EB (1987) In vitro studies of human monocyte migration across endothelium in response to leukotriene B and f-Met-Leu-Phe. Am J Pathol 127:157–167
Miller OJ, Tantravahi R, Miller DA, Yu L-C, Szabo P, Prensky W (1979) Marked increase in ribosomal RNA gene multiplicity in a rat hepatoma cell line. Chromosoma 71:183–195
Mills BG, Stabile E, Hosts PA, Graham C (1982) Antigens of two different viruses in Paget's disease of bone. J Dent Res 61:347
Mizusawa H, Hirano A, Llena JF, Kato T (1987) Nuclear bridges in multinucleated giant cells associated with primary lymphoma of the brain in acquired immune deficiency syndrome (AIDS). Acta Neuropathol 75:23–26
Monneron A, Bernhard W (1969) Fine structural organisation of the interphase nucleus in some mammalian cells. J Ultrastruct Res 27:266–288
Montgomery PO'B, Reynolds RC, Cook JE (1966) Nucleolar 'caps' induced by flying spot ultraviolet nuclear irradiation. Am J Pathol 49:555–567
Moroi Y, Peebles C, Fritzler MJ, Steigerwald J, Tan EM (1980) Autoantibody to centromere (kinetochore) in scleroderma sera. Proc Natl Acad Sci USA 77:1627–1631
Moyne G (1980) Methods in ultrastructural cytochemistry of the cell nucleus. Prog Histochem Cytochem 13:1–71

Murad TM, Scarpelli DG (1967) The ultrastructure of medullary and scirrhous mammary duct carcinoma. Am J Pathol 50:335–360
Murphy FA, Harrison AK, Gary GW, Whitfield SG Forrester FT (1968) St. Louis encephalitis virus infection of mice: electron microscopic studies of central nervous system. Lab Invest 19:652–662
Nishino H, Engel AG, Rima BG (1989) The mumps hypothesis of inclusion body myositis. Ann Neurol 25:260–264
Nistal M, Santamaria L, Paniglua RC (1985) Quantitative and ultrastructural study of Leydig cells in Klinefelter's syndrome. J Pathol 146:323–331
Norberg B (1969) Cytoplasmic microtubules and radial-segmented nuclei (Rieder cells). Scand H Haematol 6:312–318
Norberg B (1971) Contractile processes in human lymphocytes and monocytes from peripheral blood. Scand J Haematol 14 (Suppl 1)
Norberg B, Soderstrom N (1967) The effect of demecolcine on the oxalate-induced formation of radial-segmented nuclei (Rieder cells). Scand J Haematol 4:161–168
Nuova GJ, Richart RM (1989) Buffered formalin is the superior fixative for the detection of HPV DNA by in-situ hybridisation analysis. Am J Pathol 134:837–842
Payne CM, Nagle RB (1983) An ultrastructural study of intranuclear rodlets in a malignant extracranial neuroepithelial neoplasm. Ultrastruct Pathol 5:1–3
Pebusque MJ, Dupuy-Coin AM, Cataldo C, Seite R, Bouteille M, Moens PB (1981) Three-dimensional electron microscopy of the nucleolar organiser regions (NORs) in sympathetic neurons. Biol Cell 41:59–62
Penman S (1985) Virus metabolism and cellular architecture. In: Fields BN (ed) Virology. Raven Press, New York, pp 169–182
Ploton D, Bobichon H, Adnet JJ (1982) Ultrastructural localisation of NOR in nucleoli of human breast cancer tissues using a one-step Ag-NOR staining method. Biol Cell 43:229–232
Ploton D, Bendayan M, Adnet JJ (1983) Ultrastructural localisation of Ag-NOR proteins and nucleic acids in reticulated nucleoli. Biol Cell 49:29–34
Prineas JW (1972) Paramyxovirus-like particles associated with acute demyelination in chronic relapsing multiple sclerosis. Science 178:760–763
Propst A (1970) Über konzentrisch geschichtete Kerneinschlüsse in einem menschlichen Nebennieren-rindenadenom. Virchows Arch [cell Pathol]4:263–266
Raska I, Armbruster BL, Frey JR, Smetana K (1983) Analysis of ring-shaped nucleoli in serially sectioned human lymphocytes. Cell Tiss Res 234:707–711
Recher L, Sykes JA, Chan H (1976) Further studies on the mammalian cell nucleolus. J Ultrastruct Res 56:152–163
Reynolds RC, Montgomery PO (1967) Nucleolar pathology produced by acridine organe and proflavine. Am J Pathol 51:323–339
Rifkin BR, Heijl L (1979) The nuclear fibrous lamina of alveolar bone cell. J Periodontol Res 14:132–137
Riley D, Keller J (1978) Cell cycle-dependent changes in non-membranous nuclear ghosts from HeLa cells. J Cell Sci 29:129–146
Roitt I, Brostoff J, Male D (1985) Immunology. Churchill Livingstone, Edingburgh
Rueckert PR (1985) Picornaviruses and their replication. In: Fields BN (ed) Virology. Raven Press, New York, pp 705–738
Schatten G, Thoman M (1978) Nuclear surface complex as observed with the high resolution scanning electron microscope. Conformation of the membrane surfaces of the nuclear surfaces of the nuclear envelope and the pore complexes from *Xenopus aevis* oocytes. J Cell Biol 77:517–535
Scheer U, Rose KM (1984) Localisation of RNA polymerase I in interphase cells and mitotic chromosomes by light and electron microscopic immunocytochemistry. Proc Natl Acad Sci USA 81:1431–1435
Scheer U, Zentgraf H (1982) Morphology of nucleolar chromatin in electron microscopic spread preparations. In: Busch H, Rothblum L (eds) The cell nucleus, vol 11. Academic Press, New York, pp 144–176

Scheer U, Dabanvalle MC, Merkert H, Benevent R (1988) The nuclear envelope and the organisation of the pore complexes. Cell Biol Int Rep 12:669–689
Schel JH, Steenbergen LCA, Bekers AGM, et al. (1978) Change of the nuclear pore frequency during the nuclear cycle of *Physarum polycephalum*. J Cell Sci 34:225–232
Schmidt-Zachmann MS, Hugle B, Scheer U, Franke WW (1984) Identification and localisation of a novel nucleolar protein of high molecular weight by a monoclonal antibody. Exp Cell Res 153:327–346
Schoefl GI (1964) The effect of actinomycin D on the fine structure of the nucleolus. J Ultrastruct Res 10:224–243
Severs NJ, Jordan EG (1978) Nuclear pores. Can they expand and contract to regulate nucleocytoplasmic exchange? Experientia 34:1007–1011
Shah KV (1985) Papovaviruses. In: Fields BN (ed) Virology. Raven Press, New York, pp 371–391
Sharer LR, Cho E-S, Epstein LG (1985) Multinucleated giant cells and HTLV III in AIDS encephalopathy. Hum Pathol 16:760
Shelburne JD, Wisseman CL, Broda KR, Roggli VL, Ingram P (1983) Lung – non-neoplastic conditions. In: Trump BF, Jones RT (eds) Diagnostic electron microscopy, vol 4. J. Wileg & Sons, New York, pp 475–538
Simard R (1966) Specific nuclear and nucleolar ultrastructural lesions unduced by proflavin and similarly acting antimetabolites in tissue culture. Cancer Res 26:2316–2328
Simard R (1970) The nucleus: action of chemical and physical agents. Int Rev Cytol 28:169–211.
Simard R, Langelier Y, Mandeville R, Maestracci N, Royal A (1974) Inhibitors as tools in elucidating the structure and function of the nucleus. In: Busch H (ed) The cell nucleus. vol 3. Academic Press, New York, pp 447–487
Simard R, Bibor-Hardy V, Dagonais A, Bernard M, Pinard MF (1986) Role of the nuclear matrix during viral replication. In: Jasmin G, Simard R (eds) Methods and achievements in experimental pathology, vol 12, Karger, Basel, pp 172–199
Singer II (1975) Ultrastructural studies of H–1 parvovirus replication. II. Induced changes in the deoxyribonucleoprotein and ribonucleoprotein components of human NB cell nuclei. Exp Cell Res 95:205–217
Sirtori C, Bosisio M (1966) Oncolysis by herpes simplex. Lancet I:96
Smetana K (1977) Further cytochemical studies on the perichromatin region of the nucleus. Biol Cell 30:207–210
Smetana K, Busch H (1964) Studies on the ultrastructure of the nucleoli of the Walker tumour and rat liver. Cancer Res 24:537–557
Smetana K, Busch H (1974) The nucleolus and nucleolar DNA. In: Busch H (ed) The cell nucleus, vol 1. Academic Press, New York, pp 75–146
Smetana K, Gyorkey F, Gyorkey P, Busch H (1970) Comparative studies on the ultrastructure of nucleoli in human lymphosarcoma cells and leukaemic lymphocytes. Cancer Res 30:1149–1155
Smetana K, Gyorkey F, Gyorkey P, Busch H (1973) Ultrastructural studies on human myeloma plasmaycytes. Cancer Res 33:2300–2309
Sobel HJ, Schwartz R, Marquet E (1969) Non-viral nuclear inclusions. 1. Cytoplasmic invaginations. Arch Pathol 87:179–192
Somers KD, Murphey MM (1980) Cytochalasin B-induced multinucleation of human tumour and normal cell cultures. Cell Biol Int Rep 4:487–495
Spector DL, Ochs RL, Busch H (1984) Silver staining, immunofluorescence, and immunoelectron microscopic localisation of nucleolar phosphoproteins B23 and C23. Chromosoma 90:139–148
Sutton JS (1967) Ultrastructural aspects of in vitro development of monocytes into macrophages, epithelioid cells and multinucleated giant cells. Natl Cancer Monogr 26:71–141
Szekely JG, Copps TP, Morash BD (1980) Radiation-induced invagination of the nuclear envelope. Radiat Res 83:621–623
Tadano M, Makino Y, Fukunaga T, Okuno Y, Fukai K (1989) Detection of dengue 4 virus core protein in the nucleus. I. A monoclonal antibody to dengue 4 virus reacts with the antigen in the nucleus and cytoplasm. J Gen Virol 70:1409–1415

Tan EM (1979) Systemic lupus erythematosus. Immunologic aspects. In: McCarthy D.J. (ed). Arthritis and allied conditions, Lea & Febiger, Philadelphia

Terzakis J (1965) The nuclear channel system of human endometrium. J Cell Biol 27:293–304

Tome FMS, Fardeau M (1986) Nuclear changes in muscle disorders. In: Jasmin GT, Simard R (eds) Methods and Achievements in experimental pathology, vol 12. Karger, Basel, pp 261–296

Toner PG, Carr KE, Al Yassin TM (1980) Stomach. In: Johannessen JV (ed) Electron microscopy in human medicine, vol 7. Digestive system. McGraw-Hill, New York, pp 108–131

Torikata C, Ishiwata K (1977) Intranuclear tubular structures observed in the cells of an alveolar cell carcinoma of the lung. Cancer 40:1194–1201

Trump BF, Jesudason ML, Jones RT (1978) Ultrastructural features of diseased cells. In: Trump BF, Jones RT (eds) Diagnostic electron microscopy, vol 1. John Wiley, New York, pp 1–88

Trump BF, Laufer A, Jones RT (1983) Cellular pathobiology of human disease. Gustav Fischer, New York

Tsanev R, Tsanev I (1986) Molecular organisation of chromatin as revealed by electron microscopy. In: Jasmin G, Simard R (eds) Methods and achievements in experimental pathology, vol 12. Karger, Basel, pp 63–104

Vagner-Capodano AM, Mauchamp J, Stahl A, Lissitzky S (1980) Nucleolar budding and formation of nuclear bodies in cultured thyroid cells stimulated by thyrotropin, dibutyryl cyclic AMP, and prostaglandin $E_2$. J Ultrastruct Res 70:37–51

Valdivia MM, Tousson A, Brinkley BR (1986) Human antibodies and their use for the study of chromosome organisation. In: Jasmin G, Simard R (eds) Methods and achievements in experimental pathology. Karger, Basel, pp 200–223

Van der Velden HM, Wanka FC (1987) The nuclear matrix – its role in the spatial organisation and replication of eukaryotic DNA. Mol Biol Rep 12:69–77

Vazquez JJ, Ortuno G, Cervos-Navarro J (1970) An ultrastructural study of spheroidal nuclear bodies found in gliomas. Virchows Arch [Cell Pathol]5:288–293

Verheijen R, Van Venrooij W, Ramaekers F (1988) The nuclear matrix: structure and composition. J Cell Sci 90:11–36

Walker D, Padgett B (1983) Progressive multifocal leukoencephalopathy. In: Fraenkel-Conrat H, Wagner RR (eds) Comprehensive virology, vol 18. Plenum Press, New York, pp 161–193

Watson ML (1962) Observations on a granule associated with chromatin in the nuclei of cells of rat and mouse. J Cell Biol 13:162–167

Weiss L (1988) The cell. In: Weiss L (ed) Cell and tissue biology: a textbook of histology. Urban and Schwarzenberg, Baltimore, pp 1–65

Willison JHM, Johnston GC (1978) Altered nuclear pore diameters in G1-arrested cells of the yeast *Saccharomyces cerevisiae*. J Bacteriol 136:318–323

Wills EJ (1983) Ultrathin section electron microscopy in the diagnosis of viral infections. In: Sommers SC, Rosen PP (eds) Pathology annual 1983, part 1. Appleton-Century-Crofts Norwalk, pp 139–180

Wolff S (1960) Chromosome aberrations. In: Hollaender A (ed) Radiation protection and recovery. Pergamon Press, New York

Wolinsky JS (1979) Viral diseases. In: Johannesen JV (ed) Electron microscopy in human medicine, vol 6. Nervous system, sensory organs and respiratory tract. McGraw-Hill, New York, pp 54–84

Yagishita S, Itoh Y, Nakano T, Ono Y, Amano N (1979) Crystalloid inclusions reminiscent of Hirano bodies in autolyzed peripheral nerve of normal Wistar rats. Acta Neuropathol 47:231–236

Yamamoto T, Shahrabadi MS (1971) Enzyme cytochemistry and autoradiography of adenovirus-infected cells as determined with the electron microscope. Can J Microbiol 17:249–256

Yasuzumi G, Tsubo I, Sugihara R, Nakai Y (1964) Analysis of the development of Japanese B encephalitis (JBE) virus. I. Electron microscopic studies of microglia infected with JBE virus. Journal of Ultrastructure Research 11:213–229

Yeo JF (1986) Binucleated form of plasma cells in oral lesions – a review. Singapore Dent J 11:27–30

Yunis EJ, Hashida Y, Haas JE (1977) The role of electron microscopy in the identification of viruses in human disease. In: Sommers SC, Rosen PP (eds) Pathology Annual 1977, vol 12, part 1. Appleton-Century-Crofts, Norwalk, pp 311–330

Zbarsky J (1978) An enzyme profile of the nuclear envelope. Int Rev Cytol 54:295–360

Zimmerman A, Forer A (eds) (1981) Mitosis/cytokinesis. Academic Press, New York

Zucker-Franklin D, Melton JW, Quagliata F (1974) Ultrastructural, immunologic and functional studies on Sezary cells: a neoplastic variant of thymus-derived (T) lymphocytes. Proc Natl Acad Sci USA 71:1877–1881

Zu Rhein GM (1969) Association of papovavirions with a human demyelinating disease (progressive multifocal leukoencephalopathy). Prog Med Virol 11:185–247

Zu Rhein GM, Chou SM (1965) Particles resembling papovaviruses in human cerebral demyelinating disease. Science 148:1477–1479

# Nucleolar Organiser Regions

J. CROCKER

# 1 Introduction

Nucleolar organiser regions (NORs) are structures of central importance in the transcription of nucleic acid to protein. Thus, by means of hybridisation methodology, these regions have been shown to be loops of ribosomal DNA (rDNA) which transcribe to ribosomal RNA (rRNA) and thus ultimately to ribosomes and thence protein. In simple terms, the nucleolus may be regarded as a "ribosome factory". The NORs are transcribed to rRNA under the influence of RNA polymerase I. In view of the close relationship between NORs and cell activity, their size or number might reflect or predict cell proliferation, transformation or even overt malignancy. This exciting prospect has recently received extensive attention from pathologists and the techniques for NOR study are largely wholly novel to histopathologists. However, this reflects the sad truth that over-specialisation in science has led to lack of interdisciplinary communication (ANON 1987). In fact, NORs have been known to molecular biologists and cytogeneticists for many years. NORs reside on the short arms of the acrocentric human chromosomes 13, 14, 15, 21 and 22 (Fig. 1). Cytogeneticists have made use of the fact for over a decade for the investigation and identification of certain trisomies, notably that of chromosome 21, where NORs appear in inappropriate sites in metaphase spreads.

Although NORs were originally identified by in situ hybridisation with radiolabelled rRNA, they are most frequently studied by simple argyrophil methods, by virtue of the silver binding of the NOR-associated proteins. This enables ready recognition of NORs in chromosome spreads, whole cells and histological sections.

A persistent dilemma for the tumour histopathologists and, therefore, the clinician lies in the assessment of the degree of malignancy of neoplasms. This assessment is of quintessential importance in prognosis and therapy and therefore any techniques which may aid this often wholly subjective process are constantly being sought. Traditionally, the assessment of the grade of malignancy has relied upon tedious and subjective appraisal of features such as nuclear size and irregularity, loss of polarity,

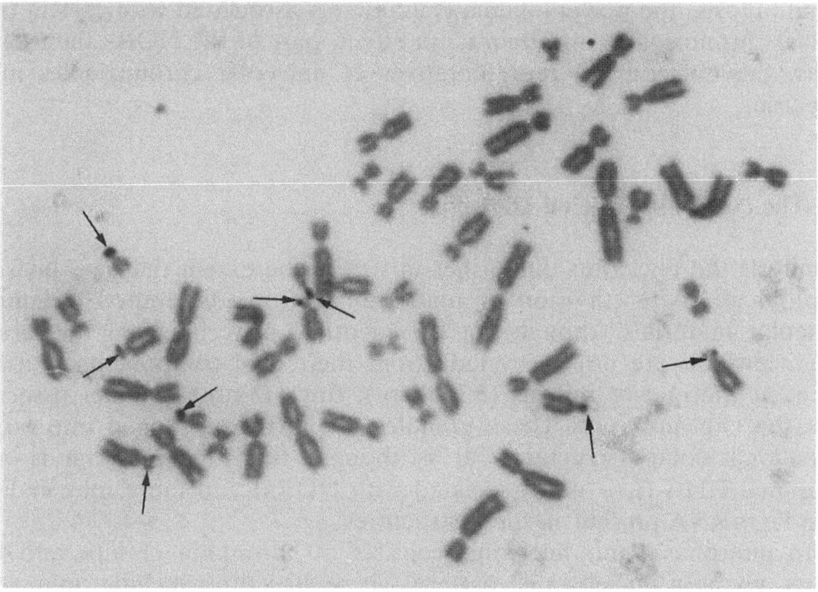

**Fig. 1.** A karyotype of a normal female human cell (46, XX), stained with the AgNOR technique, exhibiting NOR sites on the D and G group chromosomes. (Photograph kindly donated by Mr. Paul Leedham)

mitotic rate and, of course, invasion. Manifestly, more reliable and faster methods are required. A recent approach has been that of the application of antibodies to epitopes associated with cell proliferation, such as Ki 67 or transferrin receptors. A severe limitation to the usefulness of these methods lies in their requirement for fresh, frozen tissue. This chapter will describe the theoretical and practical basis for a novel approach to tumour histopathology, namely that of the study of NORs. In many instances, as will be seen, this technique overcomes some of these problems and is proving to be of considerable value to both the experimental and the diagnostic pathologist (ANON 1987).

## 2 Ultrastructural Aspects of the NORs

### 2.1 The Nucleolus

The numbers of nucleoli within cells may, of course, vary, as may their shape or form (thus, there may be oval, circular or strand-like nucleoli). The nucleolus is divided into two ultrastructurally distinct components: the rather filamentous and relatively electron-dense nucleolonema and the more extensive pars amorpha (THREADGOLD 1976). The extent of the latter depends, at least in part, upon the former.

In mitosis, the nucleolonema is intimately associated with certain (nucleolar) chromosomes and then is, in effect, part of the NORs themselves. Thus, the nucleoli are representatives of nucleolar chromosomes in interphase.

## 2.2 The Nucleolus in Cell Division

In mitosis the nucleolus diminishes in size to the extent that it is invisible in prophase. After division, it reassembles, from distributed premitotic nucleolar materials (though not the identical material from the latter). Thus, nucleoli are not "recycled" from their own components. Furthermore, at the end of mitosis (telophase), tiny (15-nm) granules associate with the chromosomes. These granules are further arranged into larger, "pseudonucleolar" structures. It is thought that this material is non-ribosomal RNA (it is not associated with NORs) and there may, indeed, even be mRNA present in these structures.

In interphase, the nucleolus consists of chromatin clumps and filaments, giving a nucleosomal pattern. These structures include unique nucleolar extended filaments (DERENZINI et al. 1982) which appear to be the same as the (metaphase) fibrillar component of the NOR (see below).

## 2.3 Ultrastructural Aspects of NORs in Relation to Nucleoli

### 2.3.1 *Nuleolus-Associated Chromatin*

Condensed chromatin is seen around nucleoli and is attached to the fibrillar centres (NORs). Thus, the fibrillar centres and intranucleolar dense chromatin merge and intermesh at the EM level (ASHRAF and GODWARD 1980; THIRY and GOESSENS 1986).

### 2.3.2 *Fibrillar Centres*

These relatively electron-lucent areas are surrounded by the dense fibrillar component and contain DNA adjoining the condensed chromatin described above. The cytochemistry of fibrillar centres, which may be multiple in some nucleoli, is described in Sect. 3. In interphase, fibrillar centres are equivalent to NORs, as has been confirmed biochemically. Fibrillar centres have been shown to vary in size, either in a resting state or in relation to alteration in nucleolar transcription (JORDAN and MCGOVERN 1981; LEPOINT and GOESSENS 1982; MIRRE and KNIBIEHLER 1982).

The numbers of fibrillar centres in a cell nucleus might be expected to relate to the numbers of chromosomes possessing NORs. However, three-dimensional reconstruction by MIRRE and KNIBIEHLER (1982) showed that

NOR ULTRASTRUCTURE

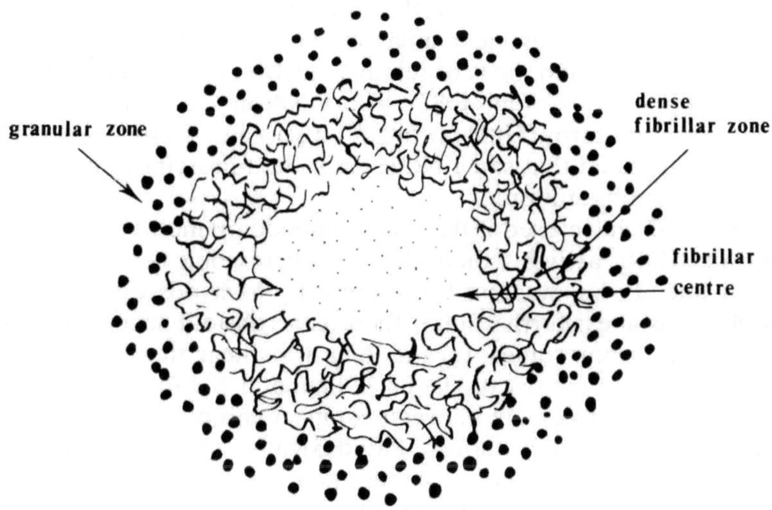


**Fig. 2.** The various layers of the fibrillar centre and its adnexae

this was not the case. CATALDO et al. (1985) further confirmed that there was no numerical relationship between rDNA sites and fibrillar centres, using stereo pairs of electron micrographs of human oocytes. It was also noted that large fibrillar centres were irregular rather than spherical, as had previously been supposed. An oocyte would be expected to contain five NORs, yet only two small and four large centres were detected. This observation was confirmed by VAGNER-CAPODANO and STAHL (1982), using silver staining of stimulated porcine thyroid cells. The lack of a definite relationship here is of great importance in the interpretation of NOR enumeration in tumour cells, in relation to ploidy. Furthermore, there may be more than one chromosomal NOR for each fibrillar centre and the numbers of fibrillar centres may increase with cell activation.

### *2.3.3 The Dense Fibrillar Component*

This RNA precursor-containing zone lies around (and partly within) the fibrillar centre (THIRY et al. 1985). DNA is also present here.

### *2.3.4 The Granular Component*

The granules of the granular component appear to represent pre-ribosomes, formed near to the fibrillar component. These are later transported to the cytoplasm as ribosomes proper (Fig. 2).

### 2.4 Ultrastructure of Chromosomes in Relation to NORs

At the electron microscope level, NORs appear as "secondary constrictions"; these zones appear as clear areas on Giemsa banding with light microscopic examination and areas of relative electron lucency at the ultrastructural level (GOESSENS et al. 1987). The DNA in secondary constrictions is in an extended form (HERNANDEZ-VERDUN and DERENZINI 1983) and it has been suggested that some NOR-associated proteins may help to maintain it in this configuration.

HSU et al. (1967) have used a flat-faced embedding method to enable visualisation of NORs at light, then electron microscopic levels. A Kangaroo rat cell line, Pt-$K_1$, was used for simplicity as it possesses a small number of chromosomes; furthermore, there is a constant secondary constriction on the X chromosomes, on the long arms near the kinetochore. The NOR was found to consist of 50–80 Å diameter filaments; the structure was considered intermediate between compact chromosome arms and loose, "lampbrush-like" kinetochores. Paradoxically, silver-stained NOR macromolecules appear *adjacent* to, rather than within, chromosomes at the electron microscope level (SCHWARZACHER and WACHTLER 1983). The argyrophilic NORs nestle in indentations in the chromosome, forming a pair on each chromatid (Fig. 3).

## 3 Cytochemistry and Biochemistry of NORs and Their Transcription

### 3.1 Detection and Visualisation of NORs

Much of the earlier work on NORs was, of course, performed on metaphase chromosome spreads and it is only more recently that these methods have been applied and adapted to the examination of whole cells or tissue

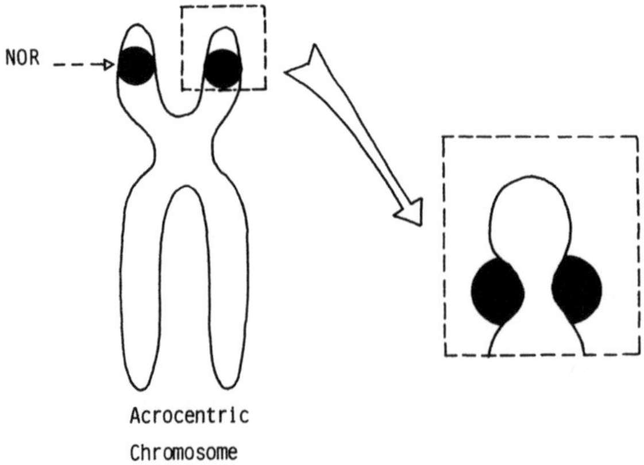


**Fig. 3.** The low-power and high-power structure of AgNOR sites in a human acrocentric chromosome. At high power the argyrophil sites can be seen to rest in grooves in the short arms of the chromosomes

**Table 1.** Techniques for the demonstration of NORs

| Reagent | Target |
| --- | --- |
| Radiolabelled rRNA | rDNA |
| Silver colloid (AgNOR) | NORAPs |
| Mercuridibromfluorescein | NORAPs |
| Bismuth ions | 100 K NORAP |
| Antibodies | Various NORAP epitopes |

sections by light and electron microscopy. The "classical" method was, perhaps, hybridisation of radiolabelled rRNA with the NOR rDNA. However, this is a relatively cumbersome, if highly specific, method. For general use, the argyrophil, AgNOR, method has found acceptance, either alone or with chemical modification of the tissue prior to staining. Other methods, using fluorescent or immune probes, have also been applied, as detailed below (Table 1).

### *3.1.1 Hybridisation of rRNA with NORs*

Early studies of rDNA utilised radiolabelled complementary rRNA; thus, NOR DNA was localised in *Drosophila melanogaster* by RITOSSA and SPIEGELMANN (1965). Similar localisation was shown by WALLACE and BIRNSTEIL (1966) and the method was also applied to mammalian genomes by HSU et al. (1975). In the latter study, sequential hybridisation and AgNOR staining were used, confirming the correspondence between the sites demonstrated with both techniques. This confirmed cytological evidence using $^3$H-rRNA hybridisation, where label was observed at "secondary constrictions" of chromosomes. The technique has even been applied at the level of individual genes of the rRNA type, disclosing that the genes for 18 S and 28 S RNA are localised on the short arms of the D and G group of acrocentric chromosomes (HENDERSON et al. 1972; HSU et al. 1975). Thus, concordance was shown between known genetic loci, morphologically recognisable feature ("secondary constrictions") and cytochemically demonstrable areas (AgNOR sites).

A problem may arise in rather small interphase nuclei (e.g. those of lymphocytes) since the grain sizes of silver in autoradiographic detection of hybridisation sites may hinder interpretation. In a novel approach to overcome or reduce this problem, WACHTLER et al. (1986) used a non-autoradiographic in situ hybridisation method. RNA coupled to mercury was the probe used, followed by sulphydryl-trinitrophenyl (mercury-binding) ligand then a fluorescein (FITC)-labelled antibody to trinitrophenol. In non-stimulated lymphocytes, the numbers of hybridisation sites outnumbered the AgNOR-stained loci. The converse was true in stimulation when several NORs commenced transcription, having previously been inactive.

This is not surprising since only transcriptionally active NOR areas are argyrophilic. The data concerning cell stimulation and NORs are further discussed in Sect. 5 below.

### *3.1.2 Argyrophil (AgNOR) Methods for the Demonstration of NORs*

The argyrophil, "AgNOR", technique, which demonstrates NOR-*associated proteins* (NORAPs), and thus indirectly NORs, is simple, reproducible and remarkably specific. The biochemistry of the NORAPs is discussed in detail in Sect. 3.2; however, in this section the mechanism of their argyrophilia is described.

The AgNOR reaction was first described by GOODPASTURE and BLOOM in 1975, and by HOWELL et al. in the same year. In the three-step AgNOR sequence, chromosomes were treated, sequentially, with: (a) 50 g/dl silver nitrate solution; (b) ammoniacal silver solution (40 g/dl silver nitrate in 50% aqueous ammonium hydroxide, pH 12–13); then (c) a developer composed of 3% formaldehyde adjusted to pH 7 with sodium acetate then adjusted to pH 5–6 with formic acid. GOODPASTURE and BLOOM (1975) applied this method to a series of chromosome spreads from animals ranging from *Carollia perspicillata* (Seba's fruit bat) to *Macaca mulatta* (the rhesus macaque). This range of specimens was selected as they corresponded to those used in previous hybridisation studies by PARDUE and HSU (1975) and HSU et al. (1975). Exact concordance was found between hybridisation sites and three-step AgNOR staining, suggesting that the latter could readily be regarded as an NOR-labelling method. This conclusion was also supported by STOCKER (1978). The method was subsequently abbreviated to a one-step sequence by HOWELL and BLACK (1980); this facilitated the procedure and rendered the results more reproducible. The method was usually run at a temperature of 60° C; however, PLOTON et al. (1986) recently showed further improvement in the results of the one-step method by lowering the temperature of the reaction to 20° C. The one-step method has other, ultrastructural advantages over the original three-step AgNOR sequence, as shown by PLOTON et al. (1982): first, the silver reaction product is of a very small mean particle size, giving good morphology; second "background" staining is minimal, and third, the reaction mixture diffuses quickly and deeply into the tissue sample. In the one-step method, silver nitrate and formic acid are mixed at appropriate dilution (see Sect. 6.10) and gelatin is used as a colloid stabiliser (Fig. 4).

An understanding of the *mechanism* of the AgNOR reaction is important. As long ago as 1961, DAS suggested that the process occurred essentially in two stages. First, submicroscopic nuclei of silver were supposed to deposit on "reactive sites" in nucleoli; this was followed by "nucleation" of further silver upon the initial moieties, in the presence of a developing agent (such as formaldehyde solution). This would then give a black colour

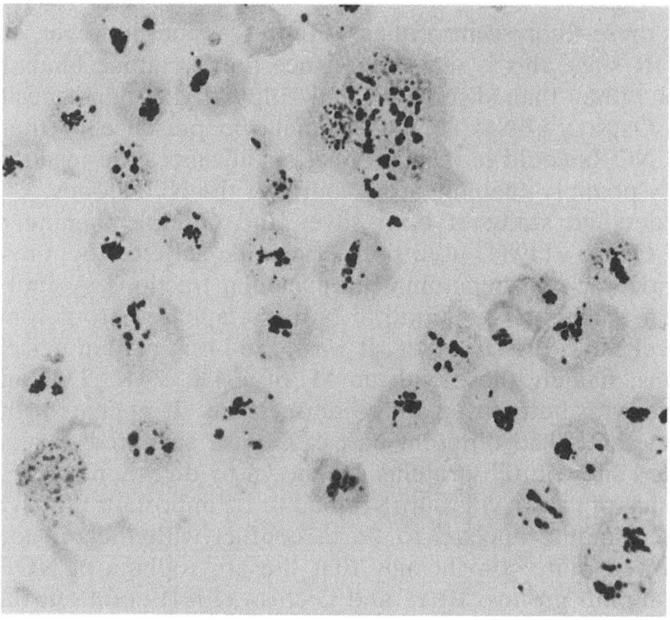

**Fig. 4.** Human promyelocytic leukaemia cells (HL 60) stained with the one-step method for AgNORs, in a cytocentrifuge preparation. The nuclei exhibit multiple black AgNOR "dots"

on light microscopy. Subsequently, these assumptions have been shown to be substantially correct.

In an exhaustive study of the chemistry of the AgNOR reaction, BUYS and OSINGA (1984) studied silver binding to metaphase spreads from human blood lymphocytes. For AgNOR staining, the three-step method was used, with and without labelling with fluorescein-tagged mercury probes and reagents such as dansylchloride or fluorescamine. In addition, specimens were treated by means of various hydrolytic procedures, namely, nucleic acid extraction by trichloroacetic acid, DNA removal by DNase I and RNA hydrolysis by perchloric acid or RNase. Proteolysis was performed with trypsin, pronase or papain. Sulphitolysis of -S-S and -S-H groups was also performed, using curpric sulphite. Silver- or fluorescence-tagged procedures were then assessed in the absence or presence of these pretreatments. It was found that histone extraction had no effect on the AgNOR sequence (as shown, previously, for example, by HOWELL et al. 1975; HOWELL 1977 and SCHWARACHER et al. 1978). After reduction of disulphide and sulphydryl groups by cupric sulphite or incubation with proteases, fluorescent tagging and silver staining were abolished, the concurrence of AgNOR and fluorescent-labelled -S-S and -S-H sites having previously been shown. Thus, it appears that AgNORs are coincident with disulphide and sulphydryl groups on NORAPs. Furthermore, it had been shown in this study that acrocentric chromosomes lacking AgNOR sites also failed to bind these fluorescent probes. (Fluorescamine, however, did not bind at AgNOR sites; this is reassuring since fluorescamine binds to protein in general rather than disulphide and sulphydryl groups specifically.) BUYS and OSINGA (1984) concluded that the presence of these sulphur groups at NORs could have functional significance, the resultant structural flexibility possibly enabling accessibility of the NOR loops.

In a further, detailed study of both silver and Giemsa staining of NORs, BUYS and OSINGA (1984) used polyacrylamide gel electrophoresis (PAGE) of rat Zajdela ascites hepatoma cell nucleolar fractions. A similar pattern of chemical reactivity was found in both metaphase chromosome spreads from the cells and PAGE bands. It was found that certain nucleolar phosphoproteins, namely those with an $M_r$ of 104 K, 78 K, 37 K and 29 K, were reactive with both silver and Giemsa stains. It appeared that the two staining methods had different bases: Giemsa appeared to react with phosphorylated sites on the proteins (as shown by digestion studies), whereas the presence of carboxyl groups seemed to be important for silver staining. This latter finding appears to be in conflict with the previous studies, which showed quite convincingly that the argyrophilia of NORs was the result of sulphur groups. BUYS and OSINGA (1984) point out that silver binding by proteins appears to result from high electron charge density, in turn related to phosphate or carboxyl groups. Since it is known that $C_{23}$ and $B_{23}$ proteins, the main NORAPs (see below), contain many carboxyl groups, it was concluded that these are responsible in the authors' system for the AgNOR binding sequence. The importance of the carboxyl group had previously been stressed by OLERT et al. (1979), who showed enhanced AgNOR staining after brief acidic or alkaline hydrolysis of metaphase spreads. It must be remembered, of course, that there may be disparities between results of silver binding to PAGE preparations, chromosomes and sections. Furthermore, the study of OLERT et al. (1979) was based on the abolition of the AgNOR reaction by diazomethane; it should be noted that this also alkylates sulphydryl groups (MEANS and FEENEY 1971). Perhaps it is safe to presume that carboxyl, sulphydryl and disulphide bonds are all involved in the argyrophilia of NORAPs. Certainly, on the basis of an extensive study of the influence of different fixation schedules on the AgNOR reaction in tissue sections, SMITH et al. (1988) suggested that silver binding could be *sequential*. Thus, these authors proposed initial binding of silver to carboxyl groups, followed by continuing nucleation around -S-S- and -S-H groups. The AgNOR sequence was found to be obliterated by fixation involving mercuric ions, which would "block" sulphur-containing groups from silver binding. Similar, dichromate-based fixatives abolished the AgNOR reaction, doubtless as a result of oxidation of sulphydryl groups, ultimately to cysteic acid molecules, Picric acid, known to bind to sulphydryl and disulphide groups, also greatly diminished argyrophilia (see Sect. 6.10.1).

In summary, it appears reasonable to presume, from current data, that both carboxyl and sulphur-containing groups are of essential importance in the AgNOR reaction. The functional significance of these groups is, as yet, uncertain, although the flexibility of NORAP tertiary structure afforded by the disulphide groups may be of importance in relation to rDNA configuration.

### *3.1.3 Ultrastructural Aspects of the AgNOR Reaction*

With increasing interest in and understanding of the basis of the AgNOR reaction and of the ultrastructure of the nucleolus, it was of considerable interest to investigate the localisation of AgNOR sites at the electron microscope level.

Initial attempts, such as those of HERNANDEZ-VERDUN et al. (1978) and BOURGEOIS et al. (1979), to localise the argyrophilia of NORs at the ultrastructural level, were hampered by relatively poor fixation. However, a pattern was found resembling that in interphase cells at the light microscope level. The poor morphology attained rendered precise interpretation difficult. However, certain observations were possible, as follows. First, the granular zone lacked silver deposition, and second, argyrophilia was observed in the fibrillar centre and dense fibrillar zone. Subsequently, HERNANDEZ-VERDUN et al. (1980) used modified fixation, namely a short exposure of the cells examined (TG human cell line) to glutaraldehyde followed by Carnoy's solution. Good morphological preservation was attained and argyrophilia was now seen clearly on the fibrillar centres in interphase (and on chromosomal NORs in mitosis). The findings confirmed that fibrillar centres corresponded to AgNORs but there was also some lighter staining of the dense fibrillar component. The authors also described nucleolar "blebs" which were argyrophilic and were represented by fibrillar structures lying throughout the nucleoplasm of certain cells; their significance was considered uncertain. Figure 5 shows fibrillar centres without silver staining (compare with Fig. 24).

To investigate the relationship between AgNOR proteins and DNA itself, HERNANDEZ-VERDUN et al. (1982) used simultaneous AgNOR staining and a Feulgen-like reaction for DNA at the ultrastructural level. The latter staining was based on a standard osmium-ammine complex reaction and the human TG cell line was studied. A constant finding was of overlap between fine intranucleolar chromatin fibres and the silver grains of the AgNOR reaction. Thus the AgNOR proteins were presumed to be in physical contact with thin chromatin fibres during interphase. Great clarity of staining was possible because of the osmium-ammine method, in which other nuclear/nucleolar structures and components are not visualised. The fact that the AgNOR-associated chromatin was observed to be dispersed is of significance because this form of chromatin appears to be that involved in active transcription. Thus, there was further evidence that the NORs are ribosomal transcription units.

In an attempt to investigate further the spatial relationships between NORAPs, RNA, DNA and other nucleoproteins, PLOTON et al. (1983)

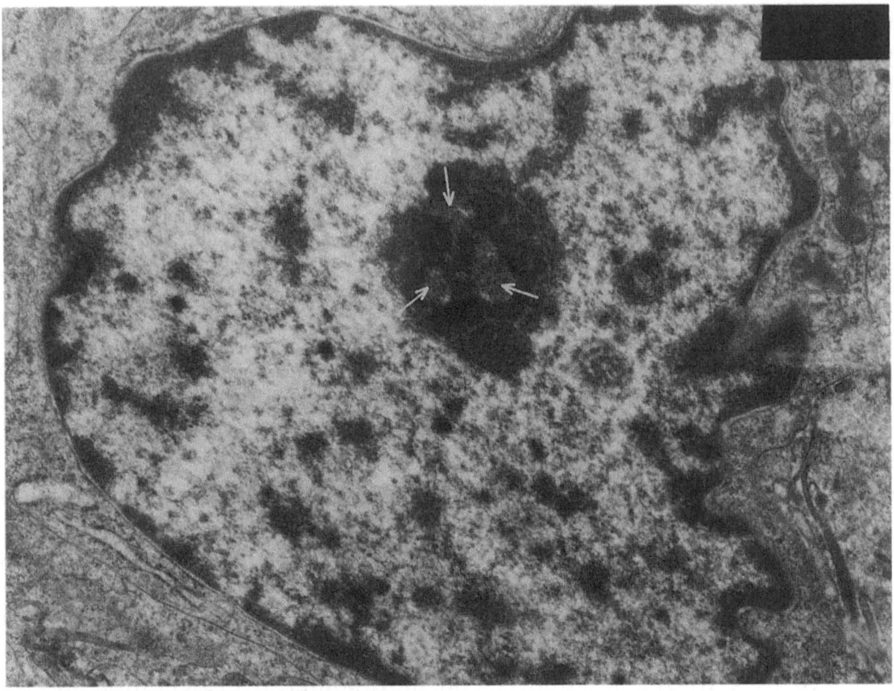

**Fig. 5.** An electron micrograph of fibrillar centres in a lymphoid cell from tonsil

applied a range of cytochemical methods to human breast carcinoma cells. Preferential (Bernhardt EDTA) staining was used for ribonucleoprotein (RNP), the osmium-ammine method for DNA and the RNase- and DNase-gold techniques for RNA and DNA. In addition the one-step AgNOR sequence was applied to the preparations. The cells examined were of interest as they possessed an unusual, reticular nucleolar structure, composed of thread-like cords with occasional rounded areas. The EDTA method for RNP stained all of the nucleolus and granules attached to it by RNP strands. Staining for DNA revealed clump-like activity peripherally in the nucleolar cords and RNA was localised in a pattern similar to RNP. The AgNOR reaction for NORAPs revealed activity at the edge of the nucleolar cords and rounded areas. It was concluded that nucleolar transcription occured at the edges of the cords, in view of the co-localisation of NORAPs, RNA and DNA. Further use of the osmium-ammine technique confirmed these general impressions when applied to resting human lymphocytes (review by HERNANDEZ-VERDUN 1983). In these cells, a single fibrillar centre is observed with peripheral RNP-containing fibrillar component. A central zone is demonstrable, containing extended, non-nucleosomal DNA. It appears that some of this DNA is transcriptionally *in*active, since in rat liver cells, stimulated by cortisol, these extended

DNA structures are visible even after actinomycin D-induced suppression of rRNA synthesis (DERENZINI et al. 1983).

PLOTON et al. (1984) further confirmed the identity of fibrillar centres and part of the dense fibrillar component with the AgNOR staining reaction. Again, there was no argyrophilia of perichromatin or interchromatin. As in previous studies, the EDTA method was used for localisation of RNP. Acetylation with acetic anhydride was performed to improve resolution (WASSEF et al. 1979). As before, AgNOR staining was always seen together with RNP. A further technical advance is attributed to MORENO et al. (1985), who applied the AgNOR method to semithin and ultrathin plastic ("Lowicryl"), and cryo-ultrathin, sections. Aldehyde-Carnoy fixation was used. This offers an advantage over some previous studies, where "block staining" was utilised. The protein nature of the AgNOR-bound structures was again confirmed, since it was abolished by pronase but was RNase-, DNase- and micrococcal nuclease-stable. Ultrastructural preservation was also excellent.

Coiled bodies have been shown in nuclei *outside* the nucleolus to be arygyrophilic, in rat sympathetic neurons (SEITE et al. 1982) where their pattern resembled fibrillar centres (PEBUSQUE and SEITE 1981; PEBUSQUE et al. 1981). This observation, if applicable to all cells, could have important implications, since the suggests that these extranucleolar coiled structures could be involved in preribosomal assembly.

It is clear, then, that the ultrastructural application of the AgNOR method has repeatedly demonstrated the identity of fibrillar centres (and, to a lesser extent, dense fibrillar component) as the site of NORAPs. It will be of future interest to examine the role played by "extranucleolar coiled bodies" in ribosomal processing. Another question to be solved is that of the role of transcriptionally inactive DNA in fibrillar centres.

### *3.1.4 Immunohistology of NORs*

Inevitably, despite the remarkably high specificity of the AgNOR reaction for NORAPs, several investigators have attempted to demonstrate NORAPs immunohistochemically, either on chromosomes or in sections. For routine or general experimental uses, such as those discussed later in this chapter, these methods probably have little advantage to offer over the AgNOR method. However, the use of immunohistology has served to demonstrate the precise localisation of certain protein components of NORs and confirmed the results of silver-binding studies performed in gels.

Probably the most extensively immunohistochemically investigated NORAP is *RNA polymerase I (RP I)*. SCHEER and ROSE (1984) developed rabbit antibodies to RP I which reacted with the nucleoli of various species in an immunofluorescence system. At the light microscope level, rat cells displayed multiple punctate areas of fluorescence which could be seen within nucleolar areas; this corresponded with the known ultrastructurally visible numerous fibrillar centres of these cells. When cells from *Xenopus* were similarly examined, one or two larger fluorescent spheres were observed, corresponding to the known large fibrillar centers in these cells (SCHEER and RASKA 1987). Both "pre-embedding" methodology (SCHEER and ROSE 1984) and immunolabelling of ultrathin frozen sections, using ultrastructural immunogold labelling for RP I, gave essentially identical results; gold particles were seen over fibrillar centres but were absent from the dense fibrillar component and other nucleolar zones. As well as rabbit antibodies to RP I, human autoantibodies (from scleroderma patients) have also been utilised (REINER et al. 1986 – cited by SCHEER and RASKA 1987) and show identical labelling results. Clearly, then, RP I is a component of the fibrillar centres. SCHEER and RASKA argue that this is a genuine phenomenon, since the absence of the enzyme from the dense fibrillar component is unlikely to be artefactual in their frozen sections. They also reject the possibility that their antibody exclusively recognises free rather then template-involved molecules, since it has previously been demonstrated that the antibody reacted with most subunits of RP I (ROSE et al. 1981).

Two phosphoproteins, designated *pp 135* ($M_r$, 135 K) and *pp 105* ($M_r$, 105 K), referred to in Sect. 3.3, were localised in various cell lines by PFEIFLE et al. (1986). Polyclonal rabbit anti-mouse pp 105 and pp 135 were used in light microscope (immunofluorescence) and electron microscopic (protein A-gold) examinations of these cells. The results were comopared with those of AgNOR staining in interphase and mitosis (Fig. 6 a). At the light microscope level, pp 135 was observed to be closely associated with AgNORs in mitosis, whereas pp 105 seemed to be associated with argyropohil activity areas only in telophase. Furthermore, pp 105 activity was found to cross-react with an antibody to rat $C_{23}$ protein. At the ultrastructural level, pp 105 was localised in the dense fibrillar component and granular zone, whereas pp 135 was found in the former only (Fig. 6 b). Neither protein was detected in the fibrillar centres. The staining for pp 135 was only possible with *cryo*-ultrathin sections; fixation and embedding in resin destroyed the labelling response (Fig. 6 c). The cross-reaction of pp 105 with $C_{23}$ suggested that the latter is the rat analogue of the former. The authors felt it to be unlikely that pp 135 was part of the RP I complex, since the known components of the latter possess quite different $M_r$ values.

The $M_r$ *100 K NORAP* has also recently been studied in interphase and mitosis by means of immunohistochemistry, by ESCANDE et al. (1985) and GAS et al. (1985). At the light microscope level, labelling was either with horseradish peroxidase or fluorescein and ultrastructurally with peroxidase or colloidal gold. The protein was localised in nucleoli at the optical level and by electron microscopy within the dense fibrillar component and granular zone but like pp 135 and pp 105 was absent from the fibrillar centres themselves. Permanent attachment of the 100 K protein to rDNA was shown by persistent chromosomal attachment through mitosis.

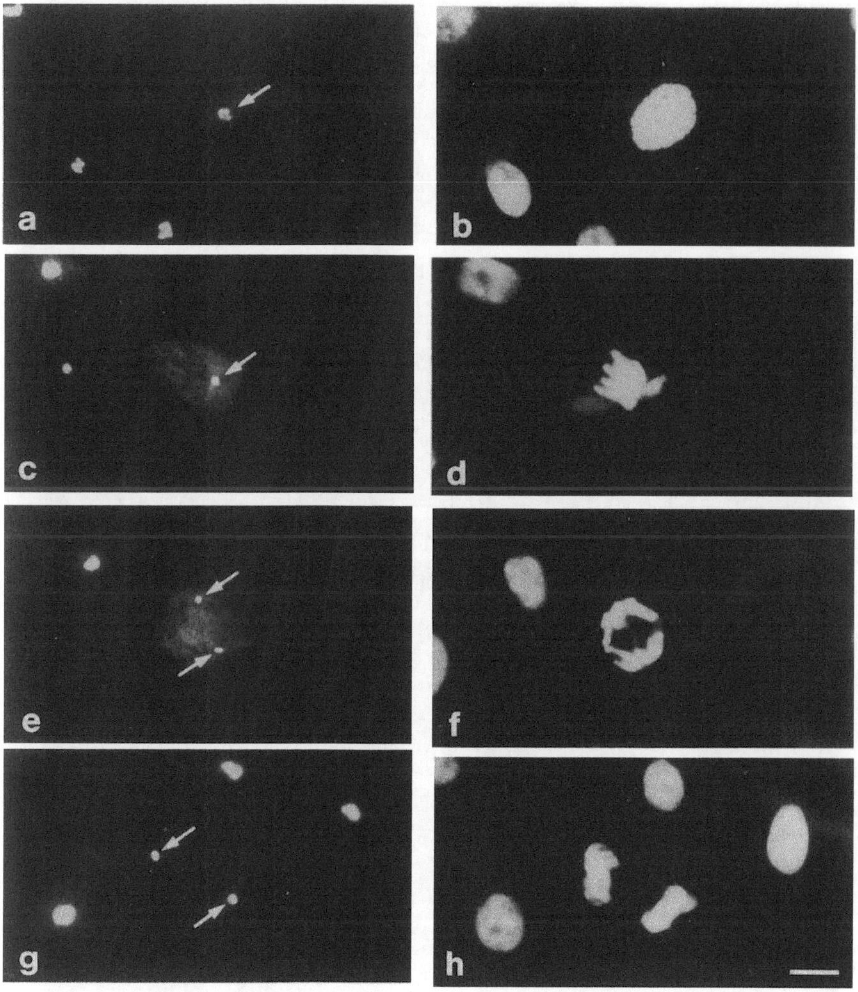


**Fig. 6a.** Localisation of pp 135 by means of immunofluorescence in a series of $PtK_2$ cells through their cycle (*a, c, e, g* show reaction for pp 135; *b, d, f, h* show DNA staining) (*a, b,* are prophase; *c, d* are prometaphase; *e, f* are anaphase and *g, h* are telophase).

The $M_r$ *80 K NORAP* was localised by means of an antibody produced by immunising a rabbit against a streptococcus (FAKAN and HERNANDEZ-VERDUN 1986). This reacted with the known NOR sites on the five human acrocentric chromosomes and at the ultrastructural level was observed in fibrillar centres (and was absent in the granular zone).

One of the major NORAPs, *$C_{23}$ protein* (nucleolin), as well as *$B_{23}$ protein*, was immunolocalised by SPECTOR et al. (1984), who showed co-localisation of argyrophilia and $C_{23}$ in the fibrillar centres and the dense fibrillar component. Co-localisation was also seen at the NORs in mitosis. This was an advance in localisation of the protein over the previous studies of

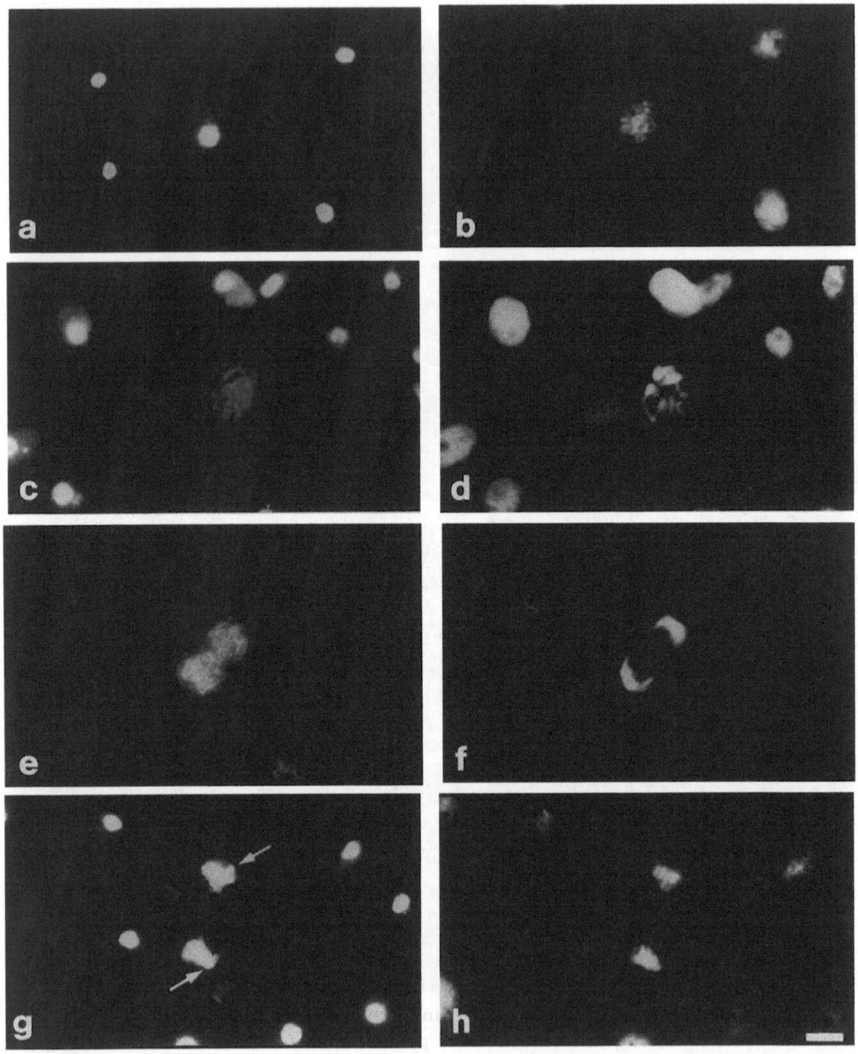


**Fig. 6b.** Localisation of pp 105 by means of immunofluorescence in a series of $PtK_2$ cells through their cycle (*a, c, e, g* show reaction for pp 105; *b, d, f, h* show DNA staining) (*a, b,* are prophase; *c, d* are prometaphase; *e, f* are anaphase and *g, h* are telophase).

LISCHWE et al. (1981), who found more general localisation in, for example, the granular regions but absence from the fibrillar centres. Double labelling for $C_{23}$ and $B_{23}$ at the light microscope level revealed diffuse nucleolar staining for $B_{23}$ but speckled nucleolar staining (akin to AgNOR staining) for $C_{23}$. It is of interest that OCHS and BUSCH (1984) had shown that reaction with a monoclonal antibody to $C_{23}$ blocked subsequent AgNOR staining on chromosomes and interphase nuclei; the same was not

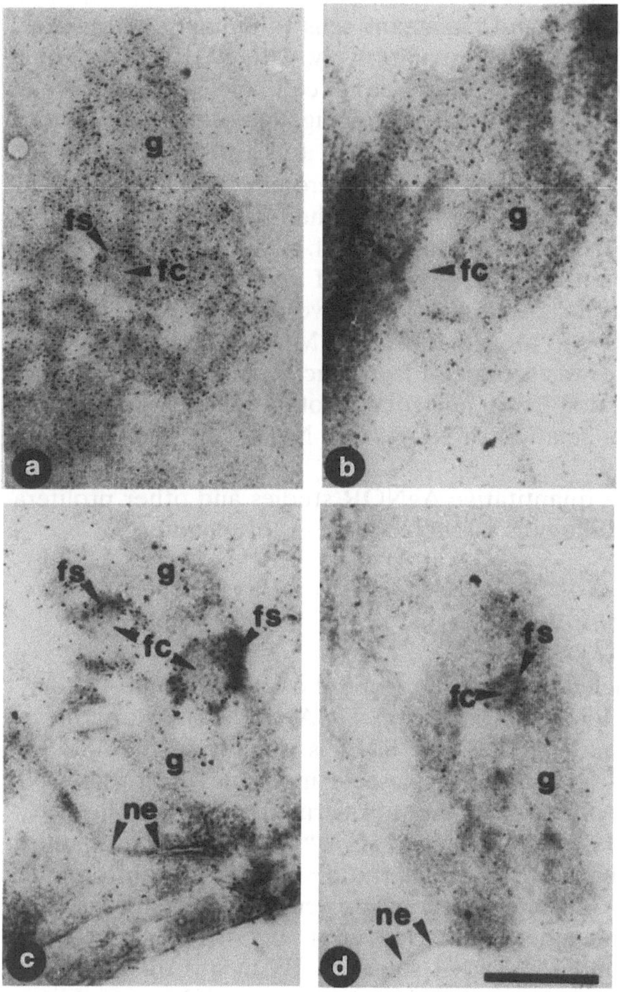


**Fig. 6c.** Immunoelectron microscopic localisation of pp 135 and pp 105 in F9 mouse teratocarcinoma cells (protein A-gold method). *a* is a section embedded in Lowicryl resin, labelled for pp 135, and *b–d* are cryo-ultrathin sections (*b* = antibody to pp 105; *c* = antibody to pp 135; *d* = control section). (Photographs **6a, b, c** reproduced by kind permission of Professor F.A. Anderer and Experimental Cell Research)

true with an antibody to $B_{23}$ protein. The relative importance of $C_{23}$ and $B_{23}$ proteins in argyrophilia was thus stressed.

### *3.1.5 Other Nucleolar Antigens*

It will, in future, prove of interest to consider the relationship between NORs, NORAPs and certain recently described nucleolar antigens. At present, the main difference between some of these latter is that, unlike NORAPs, they are not observed in resting cells. Furthermore, they may be used to distinguish benign and malignant cells (although, as described in Sect. 6, this may also be the case, quanitatively, with NORs). This group of molecules has been named "proliferating cell nuclear antigen" or "PCNA". PCNA may differ in normal and malignant cells, both qualitatively and quantitatively (*for example*, BUSCH and BUSCH 1977; DAVIS et al. 1978; SMETANA et al. 1983). A nucleolar protein, $M_r$ 58 K, has been shown in tumour cells but is absent from their normal counterparts (FREEMAN et al. 1985). In addition, BUSCH et al. (1979 a, b, 1981) demonstrated differences between nucleoli by means of immunohistochemistry.

More recently, CHATTERJEE et al. (1987) have described another antigen, P 40, associated with cell proliferation. This $M_r$ 40 K protein is of especial interest here, since it was separated from nucleolar $C_{23}$ protein, which is the immunologically dominant nucleolar protein. The biological and functional role of P 40 in relation to NORs and NORAPs is currently uncertain.

The relation between quantitative AgNOR studies and other proliferation markers in histopathology is discussed later in this chapter.

### 3.2 Biochemistry of rDNA Transcription

In simple terms, the nucleolus may be regarded as a "ribosome factory". The NORs, which are loops of ribosomal DNA (rDNA), transcribe to rRNA under the influence of RP I. The RNA is initially of the 45 S size variety and can be seen assembling as nascent molecules from the DNA core in chromosomal spreads, at the ultrastructural level. The RNA molecules grow outwards, being arranged in a "Christmas tree" pattern, the whole complex being surrounded at the 5′ end by protein granules (which contain about 70 different protein molecule types synthesised on previously assembled ribosomes) (ALBERTS et al. 1983). The 45 S RNA is then processed to 18 S and 28 S rRNA subunits. By means of radioactive pulse labelling, 18 S subunits appear in the cytoplasm in 1/2 hour, whereas the 28 S (+5.8 S and 5 S) rRNA subunits appear in an hour. The 45 S, 28 S, 18 S and 5.8 S rRNAs possess approximately 13 000, 5000, 2000 and 160 nucleotide bases, respectively. The sequential order of transcription is 18 S, 5.8 S, then 28 S rRNA subunits. About 50% of the assembled 45 S moiety ("transcribed spacer") is removed in processing. An extraordinary 6800 nucleotides are discarded in processing pre-rRNA in humans. The 5 S moiety is assembled *outside* the nucleolus and is then assembled with the other subunits within the nucleolus prior to transport out into the cytoplasm *via* the nuclear pores.

The five acrocentric chromosomes which, in human diploid cells bear NORs, contain together approximately 200 rRNA gene copies. These numbers vary in different species. On these chromosomes the rRNA genes are separated by spacer DNA, which can be seen as straight lines lying in tandem between the "Christmas tree" arrays of transcription (Fig. 7). An in-

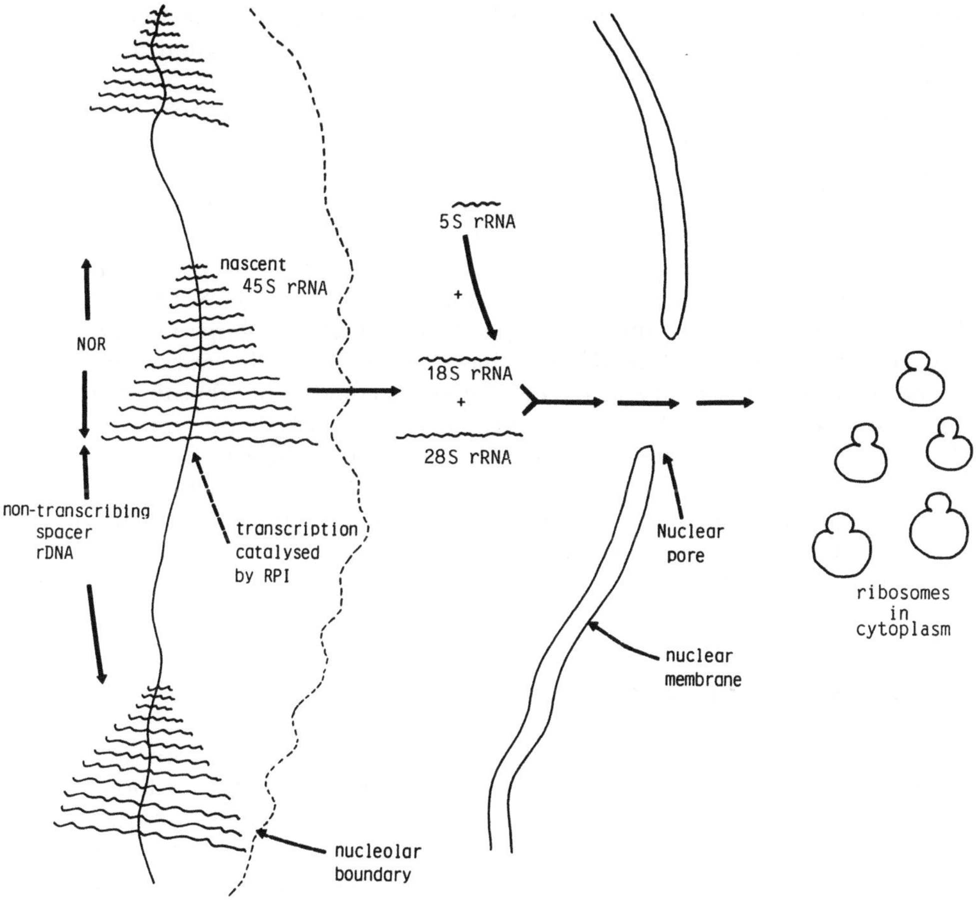


**Fig. 7.** A schematic diagram to illustrate the transcription of rDNA genes, ultimately to ribosomes

teresting technical approach to the demonstration of AgNOR proteins on nucleolar transcriptional units was described by ANGELIER et al. (1982). Use was made of "molecular spreads" from *Pleurodeles* (salamander) oocytes on EM grids, after chromosomes and nucleoli had been isolated by opening the oocyte nuclei, followed by microcentrifugation. A conventional AgNOR sequence was then applied and silver staining was observed over the axial regions of transcribing units but not over spacer rDNA. The spacer rDNA repeat is *non-transcribing*; the lengths of these spacers varies but there is homology between shorter and longer spacers, implying internal repetition.

In some animals, some rRNA genes are extrachromosomal. For example, in *Xenopus laevis*, during oogenesis, additional rRNA genes are made by means of gene amplification and these extrachromosomal units are usually present as circular molecules of DNA. This phenomenon had first been observed in this situation when multiple nucleoli were seen in oocytes. Amplified rDNA differs from chromosomal rDNA in that it possesses a different pattern of non-transcribing spacers.

The role of *nucleolin* ($C_{23}$ protein) in transcription is increasingly understood in relation to its binding to non-transcribing rDNA and is discussed in Sect. 3.3.1.

Finally, the recent sequencing studies of the 28 S and 18 S rRNA genes have afforded interesting information. There is enormous variation in the size of 28 S RNA. Thus, for example, the human molecule (5025 base pairs) is about 80% larger then that of *E. coli* (review by SCHMICKEL 1987). Indeed, across all animal and plant phyla there are large non-variant length sequences of 28 S rRNA. It has been proposed that variation in evolutionary rate of change in rRNA sequences could imply that the variable sequences are situated at the periphery of the ribosome and could thus interact with extraneous factors such as viruses (SCHMICKEL 1987). The size of the 18S rRNA is much less variable from species to species, and shows a remarkable degree of homology in many mammals in relation to the human (1870 base pairs) molecule. The reasons for this "molecular conservation" are obscure.

## 3.3 The NOR Associated Proteins (NORAPs)

It will be appreciated from the foregoing sections that the means by which NORs are visualised and assessed is that of labelling NORAPs. Although the majority of those NORAPs whose molecular size is known are of uncertain function; the principal protein, $C_{23}$ (recently ascribed the name "nucleolin"), has an increasingly understood functional role in rDNA transcription. The other NORAPs are described below but, sadly, little is yet known of their function (Table 2).

### *3.3.1 Nucleolin ($C_{23}$ Protein)*

Nucleolin is the most extensively investigated and characterised of the NORAPs. PRESTAYKO et al. (1974) described this 110 K $M_r$ (pI 5.3) "$C_{23}$" protein from Novikoff hepatoma ascites cells, by means of two-dimensional PAGE analysis. It was also found in the 60 S ribosomal subunit. In fact, the $M_r$ of the protein originally named $C_{23}$ has been ascribed various values by different authors. LISCHWE et al. (1979) assigned it an $M_r$ of 100 K, whilst JONES et al. (1981) attributed it with a size of 103 K and LISCHWE et al. (1981) gave a value of 110 K. These minor discrepancies may presumably by explained by differing gel techniques and purities of preparations. Subsequently, the argyrophilia of nucleolin was reported in fibrillar centres (DASKAL et al. 1980) and the protein was also seen to be silver binding in PAGE preparations (LISCHWE et al. 1979). The im-

**Table 2.** The known NORAPs

| |
|---|
| RNA polymerase I (RP I) |
| Nucleolin ($C_{23}$ protein) |
| $B_{23}$ protein |
| 100 K protein |
| 80 K protein |
| Phosphoproteins pp 135 and pp 105 (the latter probably = nucleolin) |

munocytochemical localisation of nucleolin has been described earlier in the chapter. (The protein is localised in the dense fibrillar component in interphase and on NORs on mitotic chromosomes.)

The role of nucleolin is increasingly well understood. It has recently been shown to bind to the non-transcribing spacer region of rDNA. OLSON et al. (1983) demonstrated binding of this protein to a segment of (non-transcribing) 3500-nucleotide DNA fragment which lies next to the site of transcription initiation for pre-rRNA. The significance of this finding has recently been stressed and expanded by JORDAN (1987, in a review of the 10th Nucleolar Workshop, held in Stevensbeck. The Netherlands). The structural organisation of nucleolin leads to various certainties or speculations regarding its role in transcription. Firstly, nucleolin has been sequenced and it has been shown that the 713-amino acid protein possesses three regions, as follows: (a) the N-terminal end, which could be analogous to molecules known as HMG which may be essential in removing H1 (histone) from condensed chromatin to form nucleosomes in pretranscription; (b) the central region, which binds to RNA (it has previously been shown that the protein can be seen accompanying early RNA in nucleoli); and (c) the C-terminal end region, which is probably in extended form. JORDAN (1987) has proposed that nucleolin might regulate RP I. In ingenious experiments, rRNA genes have been transfected into cells of other species and amplified, and they may then be transcribed by RP II rather than RP I. When this is done, no argyrophilia can be seen on the chromosomes of the transfected donor. Since nucleolin is the principal argyrophilic NORAP, this suggests that it is involved in the selection and control of RP I.

It appears, then, that nucleolin may be of quintessential importance in the transcription of rRNA genes. The precise relationship of the protein to other NORAPs, including RP I, is, as yet, undetermined but this will surely be a field of rapid and exciting development in the near future.

### *3.3.2 $B_{23}$ Protein*

$B_{23}$ is, like nucleolin, a phosphoprotein but with a more widespread immunolocalisation throughout the nucleolus. Its homology, if any, with nucleolin is uncertain, as is its role. It does, however, appear to be associated with RNA. The $M_r$ of $B_{23}$ is 37 K at pI 5.1.

### *3.3.3 The $M_r$ 100 K Protein*

On performing PAGE analysis of the non-histone proteins of nucleoli, approximately 10% of their mass is a moiety of $M_r$ 100 K. The protein is associated with preribosomal structures, recently transcribed RNA and rDNA. As is often the case with these NORAPs, the precise nature of this "p 100" is not wholly clear; however, it may be related to nucleolin and, like the $M_r$ 110 K, 125 K and 130 K proteins, is highly phosphorylated. It is readily demonstrated in nucleoli by binding of $Bi^{3+}$ (HERNANDEZ-VERDUN 1983). As with nucleolin, the situation of "p 100" in the dense fibrillar component and granular component and its absence from the fibrillar centre suggests a role in transcription and in vitro evidence exists to imply that maturation of 45 S rRNA precursor formation is also assisted.

### *3.3.4 The $M_r$ 80 K Protein*

This molecule has been demonstrated by means of immunoblotting proteins from nucleolar extracts (COURVALIN et al. 1983). Its role is uncertain but is is localised in the fibrillar centres, being absent from the granular component.

### *3.3.5 Phosphoproteins pp 135 and pp 105*

These two polyphosphorylated proteins have been isolated from nuclei and their immunolocalisation has been discussed above (Sect. 3.1.4). When these proteins are exposed to $^{32}$P-ATP they bind the substrate strongly, and large amounts of pp 135 and pp 105 may be seen in exponentially growing cells; pp 105 is especially abundant (cited by PFEIFLE et al. 1986). However, the two molecules are apparently not structurally related. It is possible that nucleolin ($C_{23}$) and pp 105 may be similar if not identical (PFEIFLE et al. 1986). The function of these two proteins is unknown.

### *3.3.6 NORAPs in Relation to RNA Synthesis Blockage Experiments*

A method which has proved to be of great value in the study of NORAPs and their localisation has been that of blockage of rRNA synthesis in vitro. The agent most widely used for this purpose has been actinomycin D (AMD), although adriamycin and D-galactosamine exert the same effect (SPECTOR et al. 1984). Cells subjected to treatment by these agents display altered nuclear morphology, with "segregated" nucleoli, where the fibrillar centres and dense fibrillar component are not distinct from the granular regions. After this treatment, the nucleoli have been reported to comprise only a fibrillar and a granular zone. The association of NORAPs with these structures or with RNP after treatment of cells with AMD can thus afford information regarding the relationship between the NORAP examined and rDNA transcription. After such treatment, argyrophilia has been observed in the fibrillar zone (DASKAL et al. 1980; DIMOVA et al. 1982). The fact that AMD did *not* decrease NOR argyrophilia was in keeping with the association of NORAPs with rDNA rather than rRNA (RAMAN and SPERLING 1981).

RP 1 has also been shown to redistribute after AMD treatment (of *Xenopus* cells), the movement being from an intranucleolar to peripheral area (SCHEER and RASKA 1987). The $M_r$ 100 K NORAP also changes its site of immunolocalisation after AMD suppression, the labelling vanishing from the dense fibrillar component, supporting a role for the protein in rDNA transcription. Interestingly, in this study, it was noted that part of the $M_r$ 100 K protein was retained in the segregated nucleoli and this has been taken to imply a possible structural as well as transcriptional role (FAKAN and HERNANDEZ-VERDUN 1986). If cells are treated with the protease inhibitor leupeptin, rRNA does not fully mature, and staining for the $M_r$ 100 K protein and the size of the fibrillar zone are both correspondingly increased; thus, again, the role of this NORAP in rDNA transcription is confirmed.

Thus, the treatment of cells by AMD or other rRNA blocker provides us with a powerful means for the examination of the role of NORAPs in rDNA transcription. It is of interest to note that whereas treatment with puromycin (which blocks protein synthesis) leads to a decrease in the volume of fibrillar centres in actively growing cells, AMD does not alter the size of these structures. Thus, there appears to be a protein turnover in these centres which is governed by rDNA transcription.

## 4 NORs and Chromosomal Abnormalities

It is certainly not, of course, the remit of this chapter to present a detailed and comprehensive review of the cytogenetic aspects of NORs. However, some pertinent information regarding the chromosomal sites of NORs and their value in the investigation of certain chromosomal disorders is outlined below, together with a consideration of NOR positions on abnormal chromosomes of malignant cells.

### 4.1 Chromosomal Trisomies, NORs and Gene Amplification

The position of "secondary constrictions", later shown to be NORs, on chromosomes 13, 14, 15, 21, and 22 in the human is fortuitous, since it enables the investigation of certain chromosomal alterations, such as

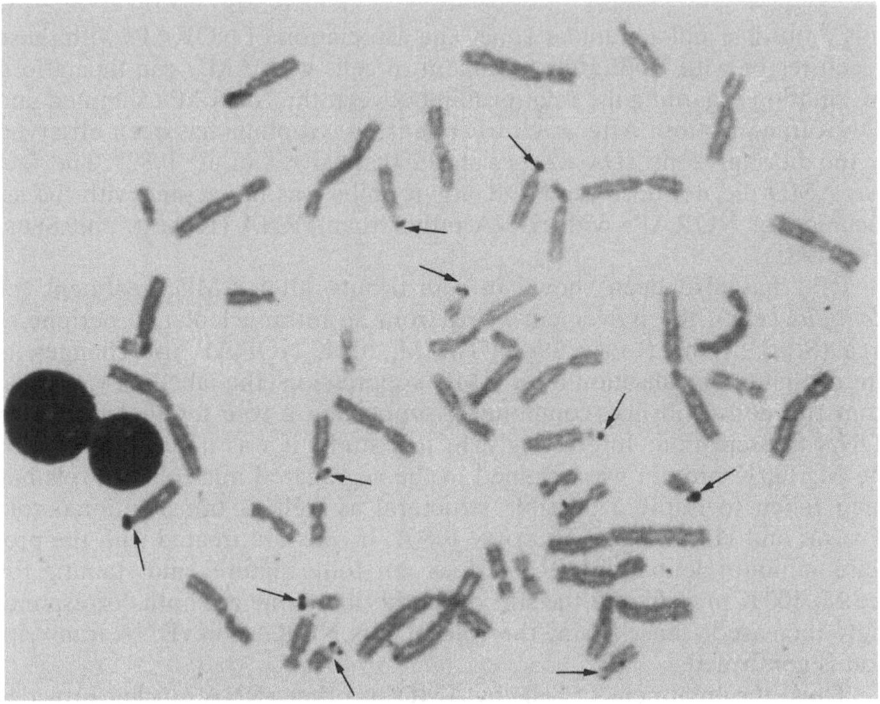

**Fig. 8.** A karyotype preparation, stained for NORs by means of the AgNOR method, showing, in a case of Down's syndrome, (47, XY,+ 21), atypical placement of NOR sites on D and G group chromosomes. (Photograph kindly donated by Mr. Paul Leedham)

trisomies, in metaphase spreads (Fig. 8). There is good evidence that the degree of argyrophilia and total number of chromosomal AgNORs varies from individual to individual, but is constant for each subject. Indeed, these features appear to be inheritable (review by BABU and VERMA 1985).

Meiotic non-dysjunction, causing trisomies, such as Down's and Patau's syndromes, may result from chromosomal associations resulting at NOR sites and, indeed, these conditions may be detected by AgNOR staining of metaphases. Some 40% of lethal trisomies occur in acrocentric NOR-containing chromosomes. The non-dysjunction here results from a close association of rDNA in the pachytene fibrillar centres (i.e. a sort of "overlap" between pachytene genetic material in meiosis). In trisomy 21, where NORs are gained (per chromosomal complement), increased rDNA has been detected biochemically, as would be expected. Another sort of disorder which can occur, between homologous or heterologous acrocentric chromosomes, results from nucleolar or NOR associations. In these cases, one of two changes occurs: first, the whole chromosomes fuse, including the (short arm) NOR zones; second, the centromeres may fuse and reject the NORs. In these cases, the amount of biochemical measured rDNA is diminished, as anticipated.

Thus, studies of NORs in meiotic or metaphase chromosomes may help to elucidate the origin and type of certain trisomies. The picture is, however, complicated by the phenomenon of *gene amplification*. This is where excess tandem copies of rDNA are formed, probably as a result of unequal chromosomal crossover or duplication. This could result from hormonal and other stimuli or be a direct genetic affect. A specific chromosome, 14p+ has been noted in a human family, which has 6–8 times more rDNA than the normal 14 chromosome (MILLER et al. 1978). Interestingly, not all of this rDNA appears to be transcribing (as shown by AgNOR staining). These amplified rRNA genes are in a G + C-rich, homogeneously staining region (HSR). It is thought that this HSR may contain genes other than rDNA. Thus, much of this HSR is transcriptionally inactive and, typically, contains much 5-methylcytosine (5 MeC) (MILLER et al. 1981). It is known that inactive genes, whether mammalian, host or, for example, viral, are highly methylated if non-transcribing. Indeed, MILLER et al. (1981), using immunoperoxidase-tagged antibodies, showed that the DNA of the 14p+ chromosome is rich in 5 MeC. In certain transformed or malignant cells, there may be much more pronounced rRNA gene amplification than in 14p+ chromosomes and clearly this phenomenon is worthy of extensive further investigation. The relevance of these observations to the quantitative tumour-histopathological findings described in Sect. 6 will clearly be complex, especially in view of possible "feedback" mechanisms regulating gene amplification phenomena in malignancy.

## 4.2 NORs and Chromosomal Changes in Malignancy

In view of the above comments, it is perhaps not surprising that there have been sporadic reports and descriptions of abnormal NORs in malignant cells. Thus, for example, CROSSEN and GODWIN (1985) described rearrangement and amplification of rRNA genes in the chronic myeloid leukaemia cell line, K 562. The authors refer to the fact that other genes, such as c-*abl* and *c*λ, had previously been shown to be amplified in K 562 cells. In a study of human testicular tumours, DELOZIER-BLANCHET et al. (1986) showed ectopic, possibly insertional NORs in four out of seven cases. There appeared to be no direct evidence of gene amplification per se in these tumours, merely of NOR ectopia. However, in analysis of a series of nine cell lines established from squamous cell carcinomas of human head and neck, four of the lines possessed NORs 10–30 times larger than normal (HAUSER-URFER and STAUFFER 1985). Since these NORs were detected by argyrophilia and were thus transcriptionally active, it is interesting to speculate that this may have occured in these proliferating, malignant cells at the expense of 5 MeC-rich inactive areas. Thus, the examination of NORs in chromosome spreads may be of more than passing interest in malignant cells. This is certainly borne out by *interphase* NOR studies in such cells, as described below.

## 5 Growth Regulation, the Cell Cycle, Tumour Cells and NORs

### 5.1 NORs in Mitosis and the Cell Cycle

The morphological arrangement of NORs in mitosis has been reviewed by GOESSENS et al. (1987). Briefly, in *prophase*, the dense fibrillar and the granular components disappear, and the nucleolus itself decreases in size. In *metaphase and anaphase*, argyrophilia demonstrates the fibrillar centres on some chromosomes, in cavities on the edges of the latter, and the nucleolus per se is no longer visible. Then, in *telophase* the fibrillar component again becomes apparent, related to the fibrillar centres. This is followed by the reappearance of the granular component, leading to re-emergence of a recognisable nucleolus. These observations were expanded by FIELD et al. (1984), who examined nucleolar conformation in phytohaemagglutinin (PHA)-stimulated human lymphocytes. Sequential silver staining, Feulgen reaction, DNA microdensitometry and tritiated-thymidine autoradiography were applied and it was found that the cell phase was related to the size and position of the NORs, as was the number of divisional generations. Resting ($G_0$) lymphocytes had a single argyrophilic nucleolar area; this enlarged in $G_1$ phase, sometimes with smaller associated granules. In the first generation in culture this pattern persisted in S and $G_2$ phases. However, in the second and third generation divisions, multiple smaller granules appeared. The fusion of NORs in the first generation but their subsequent failure to do so is not understood, but it could be that the attenuated $G_1$ phase in such cell cultures does not allow time for fusion from the second generation onwards. It has also been proposed that multiple micronucleoli could be helpful, teleologically, in rapidly dividing cells, for efficient transcription. Certainly, the amount of RNA synthesis depends on the number of nucleoli in a cell (KURATA et al. 1978). Figures 9 and 10 show the light microscopic appearance of NORs in phases of the cell cycle.

Interestingly, it has been shown that the AgNOR stain intensity reflects rRNA gene activity, as measured by [$^3$H]uridine incorporation into 18 S and 28 S moieties, by actively growing cultured human fibroblasts (MORTON et al. 1983). It was concluded that the "NOR score" could reflect the actual rRNA synthesis level and from one individual to another there was up to a threefold difference in uridine uptake.

### 5.2 Modulation of rRNA Activity and NORs

It is, of course, essential that there should be regulatory mechanisms for rRNA transcription. In the most dramatic, positive sense this is evident as gene amplification. However, the more subtle details and controlling factors are increasingly well understood.

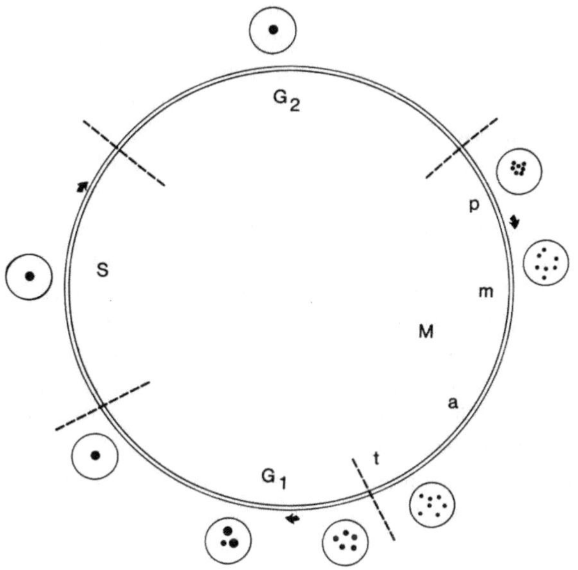


**Fig. 9.** Segregation and aggregation of NORs through the cell cycle

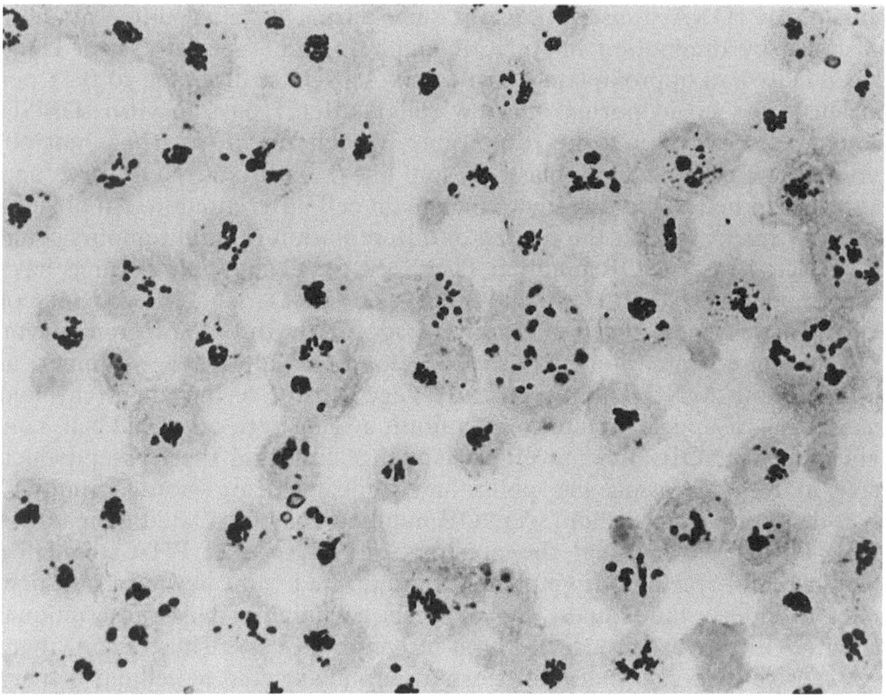

**Fig. 10.** Human J 111 histiocytic cells labelled by the AgNOR technique. Nuclei with NOR segregation and aggregation are seen. (Photograph from a preparation, courtesy of Mr. Stephen Edwards)

Predictably in this context, the effects of certain hormones have been investigated (DE CAPOA et al. 1985 a). These authors selected growth hormone (GH) and dexamethasone as suitable hormones, since both are known to affect rRNA synthesis; thus, GH is known to increase synthesis of new rRNA and dexamethasone has been shown to stimulate genes transcribing to metallothioneins and certain "acute phase" proteins. After incubations with these two agents, AgNORs were studied in metaphase spreads from the cells used (human fibroblasts) and were significantly increased. It can therefore be concluded that both agents activate r-gene clusters. In a similar study, DE CAPOA et al. (1985 b) showed that rRNA synthesis was increased by increased total serum concentration. Fluorescence was used to identify the rRNA gene clusters and it was observed that different r-gene clusters were activated at differing frequencies and to varying extents of transcription. Also, the degree of r-gene activation was related to the amount of NOR argyrophilia.

In relation to these findings, it is interesting that cellular *differentiation* has been related to NOR activity. With that much used model of cellular differentiation, the HL60 human promyelocytic leukaemic cell line, REEVES et al. (1984) counted AgNORs per cell after treatment with dimethylsulphoxide (DMSO). This agent is known to induce (myeloid) differentiation in HL60 cells (review by CROCKER and BURNETT 1986). Suppression of rDNA transcription was seen during terminal differentiation, as shown by diminution in AgNOR numbers and, especially, size. Uninduced cells had approximately five or six AgNORs, dropping to 0–1 per nucleus (as seen in normal marrow cells), after 5 days in vitro. DMSO alone was shown not to be responsible for this effect, as it was not observed in control lymphoblastoid cell lines, and thus NOR size and number can be related directly to malignant cell differentation. An obvious corollary might be that the degree of differentation in solid tumours could be reflected by AgNOR numbers (see Sect. 6). Compatible findings have been reported by SMETANA and LIKOVSKY (1984), who showed loss of AgNOR sites in advanced maturation stages of erythroid and granulocytic cells compared to earlier phases of development. Interestingly, numerical differences in AgNORs between early stage cells of erythroid and myeloid maturation disappear in late cell maturation. Thus, pro-erythroid cells contain more AgNORs than myeloblasts, suggesting that the former (which have more cytoplasmic basophilia and ribosomes) are more "primitive" than the latter. Furthermore, AgNOR numbers may be related to *age*. DAS et al. (1986) showed that the numbers of AgNORs in PHA-stimulated human peripheral blood lymphocytes diminished from newborn and neonates to elderly individuals. This suggests declining rDNA transcriptional activity with increasing age, at least in lymphoid cells. Thus, again, there is extensive evidence that NORs are essentially related to cellular activity. This may readily be observed in proliferating plant cells, grown in cultures, where more NORs are observed than in the normal cell counterparts (Fig. 11).

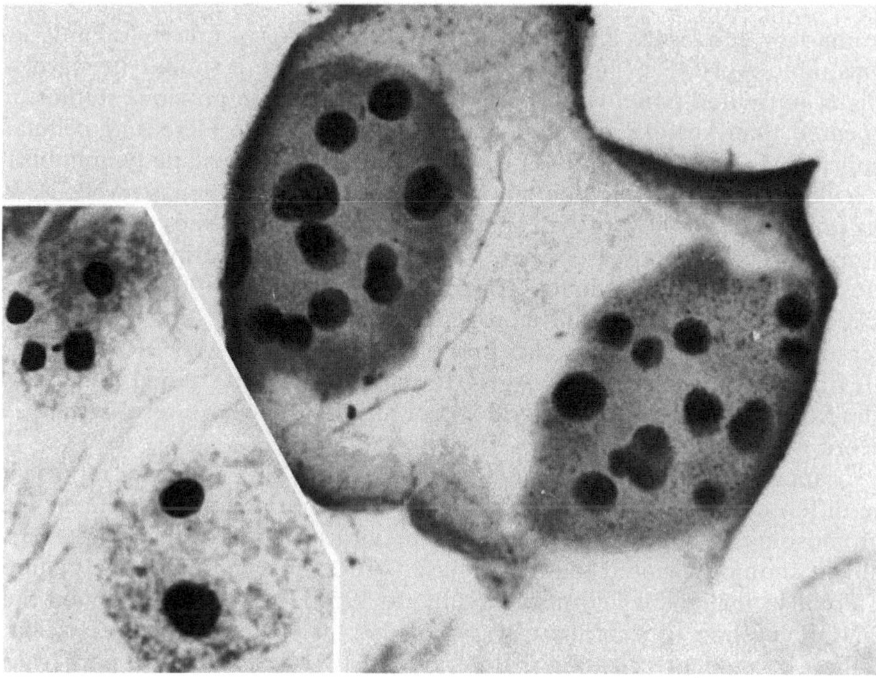

**Fig. 11.** NOR sites stained by the AgNOR method in cells of *Allium sativum* (garlic) growing on a callus maintenance medium. Some cells possess large numbers of AgNORs. These cells are proliferating and presumably there is lack of selection for normal chromosome number. *Inset:* cells of *Allium sativum* growing on normal medium. The usual number of AgNORs (i.e. up to four) are present. (Photographs by courtesy of Mrs. Susan Armstrong)

## 5.3 NOR Acitivity in Tumour Cells

Several studies of AgNOR activity have been undertaken in patients suffering from certain malignancies. The results of some of these studies have been disappointing, perhaps because in some the peripheral blood lymphocytes have been examined, rather than the malignant cells themselves. The reason for this experimental decision is obscure. SCHULZE et al. (1984) showed no difference between the numbers of AgNORs in peripheral blood lymphocytes in lymphoid malignancies. When bone marrow cell NOR activity was studied by ARDEN et al. (1985), lower *metaphase* AgNOR counts were observed in normal cells than in acute lymphoblastic leukaemic cells. In a sense, this corresponds to the studies of differentiation described above. MAMAEV et al. (1985) studied AgNORs in metaphase and interphase preparations in patients with chronic myeloid leukaemia (CML), using both PHA-stimulated peripheral blood lymphocytes and bone marrow cells. The AgNOR numbers in peripheral lymphocytes were not significantly different from those in normal individuals and the marrow cells were heterogeneous, with up to 67% not containing demonstrable AgNORs. The numbers of the latter were similar in marrow cells of untreated patients to those in blast crisis. As in previous studies, it appeared that AgNOR expression was suppressed with increasing cellular maturity. MAMAEV et al. (1987) also studied AgNORs in acute lymphoblastic leukaemia (ALL) and acute non-lymphoblastic leukaemias (ANLL). It was found that in 90% of ALL patients, fewer AgNOR sites were present in marrow cell metaphases than in PHA-stimulated peripheral lymphocytes from the same patients. A variable number (0%–90%) of mitoses lacked AgNORs, with great heterogeneity, perhaps suggesting varying proliferation pool sizes. It was also proposed that lack of AgNORs could result from protease activation. AgNOR activity was also studied in marrow cells from Ph'-positive CML by SATO et al. (1986), whose results were in accord with the above data.

It appears, then, that the numbers of AgNOR sites in nuclei represent the differentation state of individual cells and/or tumours. This reflects the chromosomal changes described in the previous section of this chapter. Finally, it should be noted that interphase NORs may be qualitatively quite different in malignant and normal cells. In 1979, BUSCH et al. applied the AgNOR method to imprints of Novikoff heptatoma, KB and HeLa cells. In these malignant cells, AgNOR arrays in two to three rows each with three to five granules were observed, although such complexes were not seen in normal and regenerating liver cells. It would be of considerable interest to examine this phenomenon at the ultrastructural level.

## 6 Histopathological Aspects of NORs

It should be apparent, from the experimental results described previously regarding NORs and tumour differentation and cell development in vitro, that interesting information could be available from quantitive studies of neoplastic cells. It was highly significant, therefore, when PLOTON et al. (1986) applied the AgNOR method successfully to paraffin sections of human tissue, as well as cell smears, chromosomes and plastic semithin sections. These authors showed that improved staining was obtained by performing the argyrophil reaction at 20° C (rather than the conventional 60° C) and examined the AgNORs stained by means of various optical methods. Paraffin sections of an unspecified number of human prostates revealed large (5 μm) nucleoli containing "numerous" black dots, whereas hyperplastic gland cells contained only two dots and intervening lymphocytes contained one dot. It was concluded that the AgNOR method might find a use in tumour pathology and grading.

The initial promise of this preliminary study has been extensively fulfilled, with studies of numerous types of human neoplasm revealing interesting and potentially useful quantitative data. These are discussed individually below and summarised in Table 3.

**Table 3.** Histopathological value of AgNOR enumeration in the distinction between benign and malignant neoplasms, or their grading

| Type of neoplasms | NOR value in grading |
|---|---|
| Non-Hodgkin's lymphoma | +++ |
| Hodgkin's disease | N/A |
| Skin melanocytic tumours | +++ |
| Skin adnexal tumours | ++ |
| Oat cells vs lymphocytes | +++ |
| Mesothelial (pleural) cells | +++ |
| Breast | ++ |
| Thyroid | – |
| Prostate | – |
| Paediatric tumours | ++ to +++ |
| Salivary glands | +++ |
| Inverted nasal papillomas | +++ |
| Stomach | + |
| Cervix | + |
| Gliomas | +++ |
| Liver | +++ |

– = of no diagnostic value
\+ = may be useful, but much "overlap" of values
++ = may be useful, but much "overlap" of values
+++ = highly recommended as a reliable diagnostic method

## 6.1 Non-Hodgkin's Lymphomas

The assessment and grading of non-Hodgkin's lymphomas (NHLs) is a complex yet clincally most important procedure; prognosis and therapy are in part dependent upon this and for some years there have been attempts to facilitate and improve the grading. Methods such as electron microscopy, enzyme histochemistry, immunohistochemistry, morphometry and DNA flow cytometry have all played a role in this rapidly developing field and it now appears that NORs may be of considerable value in the grading of NHLs.

The importance of nuclear and/or nucleolar measurement in the morphometric assessment of NHLs has been repeatedly stressed (for example, CROCKER et al. 1983; VAN DER VALK et al. 1983; CROCKER 1984) and, indeed, subjective review of nuclear and nucleolar pattern is an essential part of the Kiel classification (LENNERT 1978). These observations led CROCKER and NAR (1987) to investigate the distribution of argyrophilic NOR sites in NHLs. Routinely formol-saline fixed, paraffin-embedded sections were stained with the standard silver formate preparation at 20° C and the mean numbers of AgNOR sites per lymphoma cell nucleus were enumerated. Seventy-five specimens of NHL were included, including representatives of both high- and low-grade Kiel types. In addition, five palatine tonsils and five "reactive" lymph nodes were studied. The AgNOR sites were clearly visualised as black dots in the nuclei and were counted by means of an interactive image analyzer. Striking results were obtained; in interfollicular and mantle zone areas and all other parts of the tonsils and reactive nodes, apart from germinal follicle centre cells and tonsillar epithelium, there was a mean count of approximately one AgNOR per nucleus. In contrast, follicular centroblasts possessed 4.6–4.8 AgNORs and, interestingly, the basal cells of the tonsillar epithelium possessed twice as many (mean 2.2) AgNORs as the upper epithelial cells (mean 1.0). Both centroblasts and basal epithelial cells have, of course, a high proliferation level compared to the other cells of tonsils and lymph nodes. Thus, the AgNOR content seemed to relate to division rate, either causally or indirectly. When the specimens of NHL were studied, a total separation was observed between high- and low-grade lesions; the former had mean AgNOR counts ranging from 4.4 to 6.8 (Fig. 12) and the latter from 1.0 to 1.5 (Fig. 13). There was no overlap between the two groups and the

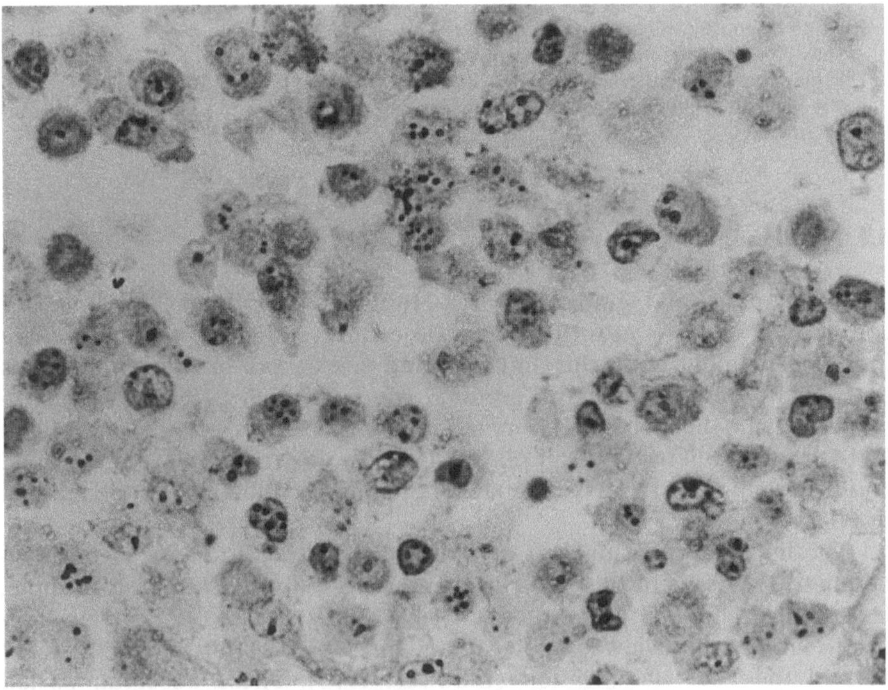

**Fig. 12.** A section of high-grade NHL stained for AgNORs. Each nucleus contains numerous black dots

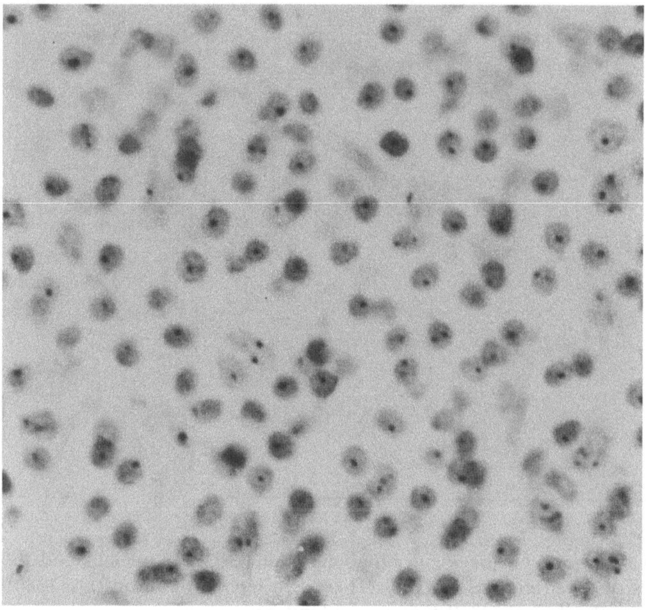

**Fig. 13.** A section of low-grade NHL stained for AgNORs. Each nucleus contains one to two NOR dots

method was considered to be potentially diagnostically useful. In this context, the author has found the AgNOR reaction particularly useful in the distinction between B-lymphoblastic NHLs and low-grade lymphomas; this may otherwise be a histological problem since the nuclear size and shape of lymphoblastic lymphoma nuclei are often similar to those of low-grade lesions. It is important to note that AgNOR staining sites do *not* always correspond to nucleoli themselves and the reaction should not be regarded as being simply a "nucleolar stain". The observations of CROCKER and NAR (1987) were broadly supported by those of POGORELOV et al. (1987), who examined AgNORs in peripheral blood lymphoid cells in NHL, chronic lymphocytic leukaemia and hairy cell leukaemia. It was shown that the amount of AgNOR area related to grade of malignancy.

A recurrent question which arises in histopathological studies of AgNORs, whatever their potential diagnostic usefulness, is that of the fundamental significance of the findings. It might be expected that the numbers of AgNORs per nucleus would reflect the degree of ploidy, since a normal diploid cell should contain only ten NORs (two sets of five; i.e. chromosomes 13, 14, 15, 21 and 22). However, it must be recalled that AgNOR counts on 3-μm sections will not give *absolute* counts. Furthermore, acrocentric associations may lower interphase NOR numbers. To test for a relationship between ploidy and AgNOR numbers, CROCKER et al. (1988c) performed DNA flow cytometry on a series of ten low-grade and ten high-grade NHLs, using paraffin wax-embedded material, and compared the

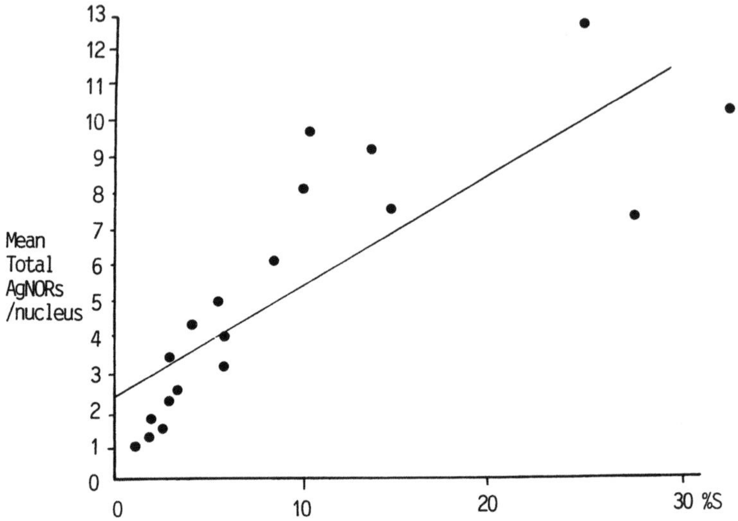


**Fig. 14.** Linear correlation plot between the mean number of AgNORs per nucleus and the % S phase cells for a series of 20 specimens of NHL. There is a high correlation ($r = 0.86$). (Photograph reproduced from the *Journal of Pathology*, by kind permission)

DNA index and the %S phase cells (dividing cell percentage) with the mean AgNOR count for each specimen. Perhaps suprisingly, there was no correlation between the AgNOR numbers and DNA aneuploidy ($r = 0.14$). In contrast, there was a high correlation between the AgNOR count and %S phase cells ($r = 0.86$) (Fig. 14). The immediate conclusion to be drawn from these observations is that AgNOR numbers in some way reflect or even direct the cells' proliferation. Certainly, from the foregoing chapter section there is evidence to suggest this. However, their numbers are still likely to reflect chromosomal abnormalities, even if the numbers of hyperdiploid or aneuploid cells do not suffice to give an abnormal DNA index. Indeed, the modal value of DNA index for DNA aneuploid NHL is only approximately 1.1 and it is unlikely that cytometry would easily distinguish such a value from, say, the 1.0 of diploid cells. Clearly, there is a need for a study comparing AgNOR counts in tissue sections with those performed on tumour metaphase spreads.

To investigate further the relationship between cell proliferation counts and AgNOR numbers in NHL, HALL et al. (1988) and CROCKER et al. (1988b) performed a comparative study of Ki 67 labelling and AgNOR staining in a series of 80 cases of lymphoma. The monoclonal antibody Ki 67 recognises an antigen in the nuclei of cells in all phases of the cell cycle except $G_0$ (GERDES et al. 1984). HALL et al. (1988) labelled the 80 specimens of NHL in frozen sections with Ki 67 and in paraffin sections for AgNORs. The percentage of cells reacting with Ki 67 and the mean numbers of AgNORs per nucleus for each case were compared and corre-

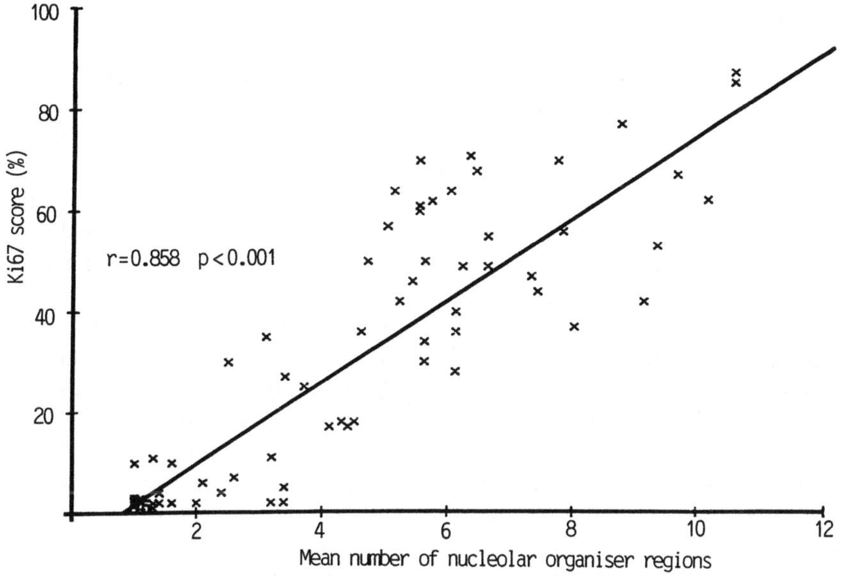


**Fig. 15.** Linear regression plot between the mean numbers of AgNORs per nuclear profile and the numbers of Ki 67$^+$ cells ("Ki 67 score") for a series of 80 cases of NHL. There is high correlation ($r = 0.86$). (Photograph reproduced from *Histopathology*, by kind permission)

lated remarkably well ($r = 0.86$; $P < 0.001$) (Fig. 15). It thus appears that NOR labelling can be substituted for Ki 67 binding, as a result of causal or non-causal relationships between the two. The advantage of the AgNOR method here is considerable, since, unlike Ki 67, it does not require frozen sections. Similarly, the method is cheaper, quicker and less cumbersome than DNA flow cytometry (CROCKER et al. 1988 b). CROCKER et al. (1989 b) have now shown, in normal and neoplastic lymphoid tissues, that Ki 67-positive cells (for example, centroblasts) also have high AgNOR counts; conversely, those which are Ki 67-negative (such as lymphocytes) have only one to two nucleoli. Thus, the findings of HALL et al. (1988) have been confirmed at a cellular level. (This study was facilitated by the use of sequential immunostaining and AgNOR reaction on the same section.) It will now be of interest to attempt a direct correlation between clinical survival and AgNOR counts in NHL.

Another technique which discriminates clearly between high- and low-grade NHLs is that of the measurement of AgNOR *sizes*. EGAN and CROCKER (1988), using image analysis and light microscopy, showed clear separation of both groups of NHL; high-grade lymphomas have much smaller AgNORs than those of low-grade histology. This method, both at light and electron microscope levels (as in DERENZINI et al. 1986), may well provide diagnostic discrimination in tissues where enumeration alone will not do so.

## 6.2 Hodgkin's Disease

Early studies of parameters of cellular proliferation in Hodgkin's disease indicated that although there were cell populations with different kinetics, in general little cell turnover was occuring (PECKHAM and COOPER 1969). Preliminary studies of the number of NORs in Hodgkin's disease have been performed (HALL and CROCKER 1988) and three populations of cells can be identified. First, Sternberg-Reed cells and similar multinucleate cells which contain large numbers of AgNORs; these are characteristically distributed around the periphery of the nuclei and also in the nucleoli. A second population of cells also contain large numbers of AgNORs (5–15). These cells do not have the morphology of Sternberg-Reed cells, but resemble large and medium-size blast cells (presumably Hodgkin's cells). A third population of cells with the morphology of small lymphocytes have only one, or occasionally two, AgNORs per cell. The proportion of cells in each of these three groups varies in each histological subgroup. These results are consistent with the recent study using Ki 67 immunostaining as a marker of cellular proliferation and CD 30 immunoreactivity as a marker of Sternberg-Reed and Hodgkin's cells. This suggested that the majority of CD 30 immunoreactive cells possess nuclear Ki 67 staining (GERDES et al. 1987). It is possible to immunostain AgNOR preparations (HALL 1987, unpublished observations) by taking AgNOR-stained sections, washing in buffer and performing conventional immunostaining procedures. In Hodgkin's disease (and possibly other forms of lymphoma) where there is complex cellular heterogeneity the expression of results of NOR staining as mean AgNOR counts may be very misleading. It may be more appropriate to represent data as histograms of number of AgNORs (abscissa) against number of cells (ordinate). It is conceivable that the shape of the resulting histogram may be characteristic of the type of lymphoma and provide more information than a single mean of a *non-parametric* distribution. Further studies of Hodgkin's disease will require this approach and careful analysis of the various cell populations, perhaps aided by double staining.

## 6.3 Skin Neoplasms

Another problem with obvious clinical importance lies in making the distinction between benign naevocellular naevi and melanocarcinomas. Accordingly, CROCKER and SKILBECK (1987) studied a series of proven examples of these lesions. A very well-defined difference in AgNOR numbers was found between the two groups; thus, the naevus group (including intradermal, juvenile, compound, junctional and cellular blue naevi) possessed approximately one AgNOR per nucleus. Conversely, lentigo maligna, superficial spreading melanoma and melanocarcinomas had pooled AgNOR counts ranging from 7.4 to 8.23, there being no overlap between the two groups of lesions (Fig. 16). In a novel ultrastructural ap-

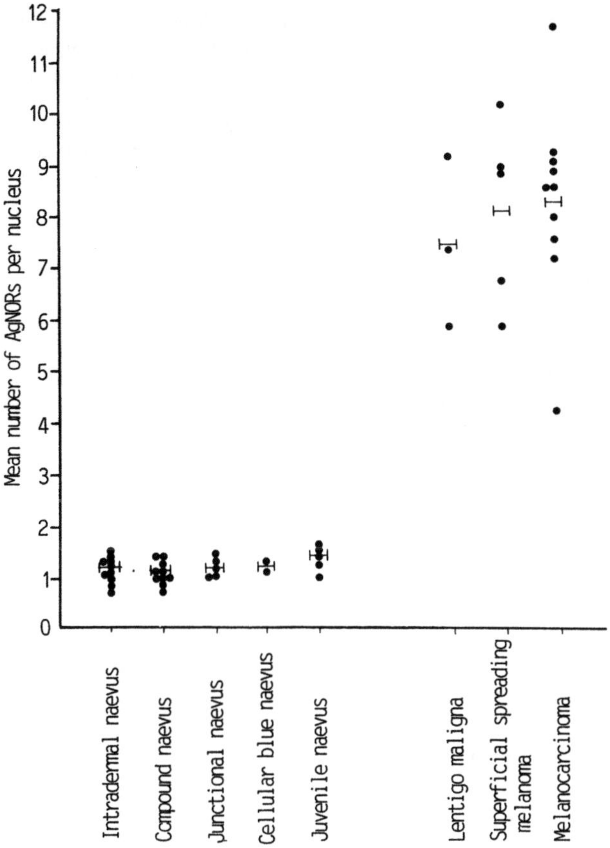


**Fig. 16.** Values of mean AgNOR count per nucleus for melanocarcinomas and naevocellular naevi of skin. There is clearly a highly significant difference between the values for both groups. (Reproduced from *J. Clin. Pathol.*, by kind permission)

proach, DERENZINI et al. (1986) examined the nucleoli of benign naevocellular naevi and melanocarcinomas. The former contained one or two large fibrillar centres (mean area of 0.482 $\mu m^2$ in profile) whereas the latter possessed multiple, smaller (0.221 $\mu m^2$ mean area) fibrillar centres. It was suggested that this method could be of diagnostic value.

A larger range of cutaneous and adnexal tumours was subsequently studied by EGAN and CROCKER (1988). Sixty-eight specimens were studied including eccrine and apocrine tumours, hair follicle tumours, mixed tumours, clear cell acanthomas and squamous and basal cell carcinomas. Dermatofibromas were also included. The numbers of AgNORs per nucleus appeared to be related to the degree of differentation in squamous cell carcinoma and the cellularity in dermatofibromas. However, basal cell carcinoma possessed may more AgNORs than did all of the other speci-

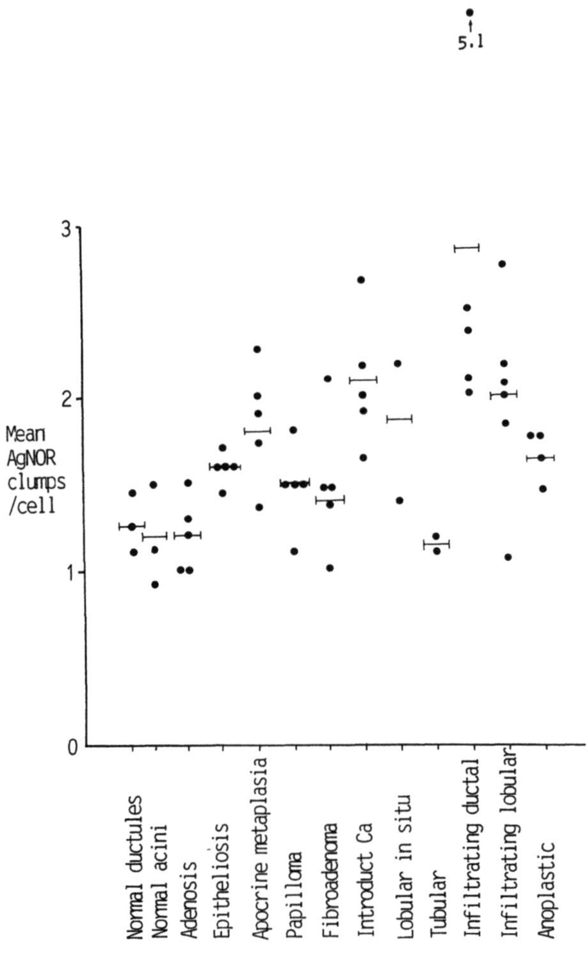


**Fig. 17.** Values of mean AgNOR counts per nuclear profile for a range of breast lesions. In this series, the numbers of NOR "clumps" (i.e. nucleoli) were counted and are clearly poor discriminators

mens and it was considered that this could be of diagnostic value. This is of interest since it may reflect the fact that basal cells of the normal epidermis possess more AgNORs than the more superficial cells (CROCKER and SKILBECK 1987).

## 6.4 Breast Tumours

SMITH and CROCKER (1988) studied AgNOR counts in a series of breast lesions, both benign and malignant, and in normal breast tissue. The AgNOR sites were enumerated in three different ways: (a) the total numbers of AgNORs were counted for each nucleus; (b) "clumps" of AgNORs were examined (these probably corresponded to clusters of NORs in nucleoli); (c) the numbers of AgNORs per "clump" were counted. In addi-

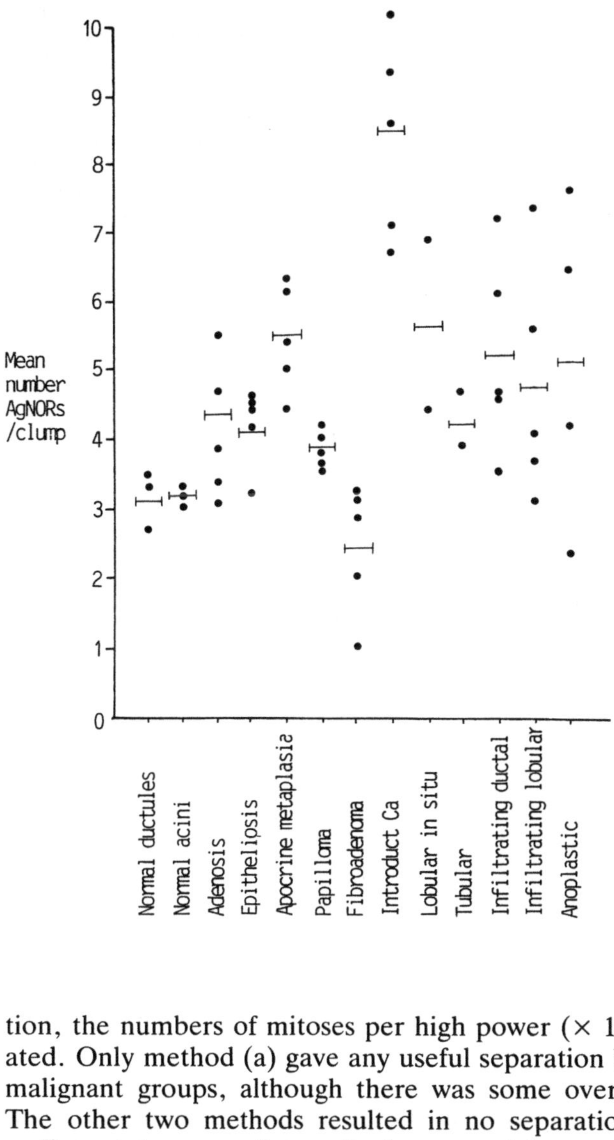


**Fig. 18.** The same series as in Fig. 17, the numbers of AgNORs per "clumps" having been recorded. Again, there is poor discrimination between benign and malignant groups

tion, the numbers of mitoses per high power (× 1000) field were enumerated. Only method (a) gave any useful separation between the benign and malignant groups, although there was some overlap in individual cases. The other two methods resulted in no separation between benign and malignant tumours. Interestingly, normal breast possessed similar mean AgNOR counts to hyperplastic mammary tissue; this is perhaps explicable since premenopausal specimens were used, which were presumably under hormonal stimulus (Figs. 17–19). Similar "overlaps" between groups are seen in other endocrine tissues (see below).

A surprising but most interesting observation was made by EGAN and SMITH (1987), who showed that myoepithelial cells in malignant breast lesions have mean cellular AgNOR count of 2.6 as opposed to 1.0 in benign lesions. This may suggest an increased turnover or proliferation of myoepithelial cells in breast malignancy.

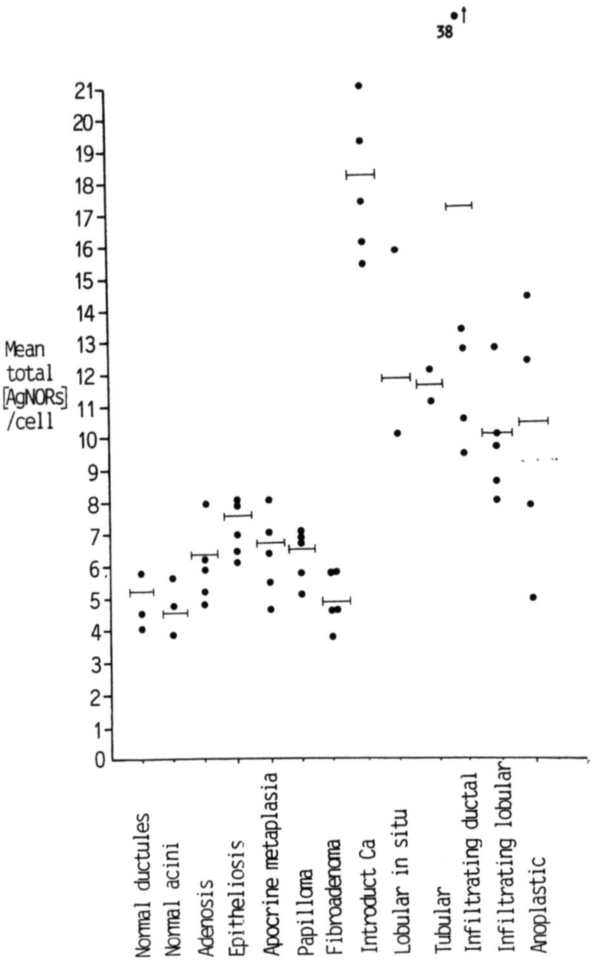


**Fig. 19.** The same series again, on this occasion the *total* AgNORs per nuclear profile being enumerated. Now, there is relatively good separation between benign and malignant specimens. (Figures 17–19 reproduced from *Histopathology*, by kind permission)

## 6.5 Thoracic Neoplasms

The enumeration of AgNOR sites has afforded diagnostically helpful data in two contexts in chest tumours. First, CROCKER et al. (1987) were able to distinguish between lymphocytic infiltrates and oat cell carcinoma infiltrates in bronchial material. To avoid tautology, the nature of the specimens was confirmed by positive staining for leukocyte common antigen (lymphocytes) and neurone-specific enolase (oat cells). The lymphoid specimens contained 0.9–1.7 AgNORs per nuclear profile, as opposed to 4.2–7.3 for the oat cell carcinomas. Thus, a frequent problem with biopsy interpretation can be avoided by rapid, subjective assessment of AgNOR numbers, since the ranges are so widely separated. A caveat exists, however, since, of course, low-grade (small cell) NHLs could not be distin-

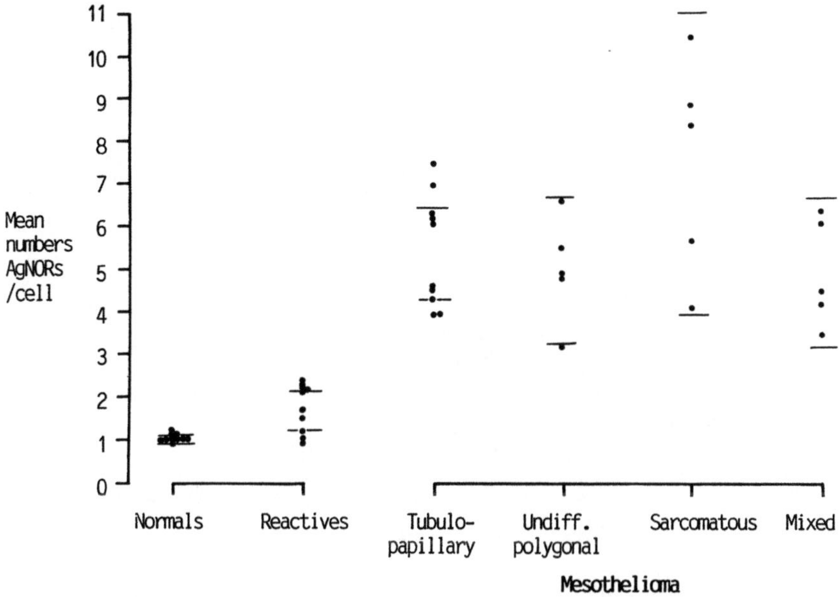


**Fig. 20.** The mean numbers of AgNORs per nuclear profile for a series of benign and malignant pleural mesothelial specimens. There is good separation (confidence intervals shown by *bars*). (Photograph reproduced from *Thorax*, by kind permission)

guished from benign lymphoid infiltrates. Nonetheless, the usefulness of the method in detecting oat cell carcinoma is self-apparent.

The second major diagnostic application in the field of thoracic tumours lies in the differentiation between benign and malignant pleural mesothelial cells. It was shown that whereas normal mesothelial cells contained a mean of 1.04 AgNORs per nucleus, malignant mesotheliomas of various histological subtypes contained a mean of 4.94–7.52 AgNORs (AYRES et al. 1988; CROCKER et al. 1988a) (Fig. 20). Again, formal counting would not be necessary to make this important distinction, the groups being so widely numerically separate. Interestingly a statistically highly significant difference was detected between AgNOR numbers in normal and "reactive" pleural mesothelium. In the author's experience, diagnostically difficult cases can be prospectively separated by this method.

Finally, in a preliminary study of the relation between AgNOR numbers and prognosis in squamous cell carcinoma of the bronchus, BOLDY et al. (1988) showed a general trend to increasing AgNOR counts with decreasing differentiation of the neoplasms. There was, however, considerable group "overlap". It will be interesting to study these cases in relation to survival/prognosis.

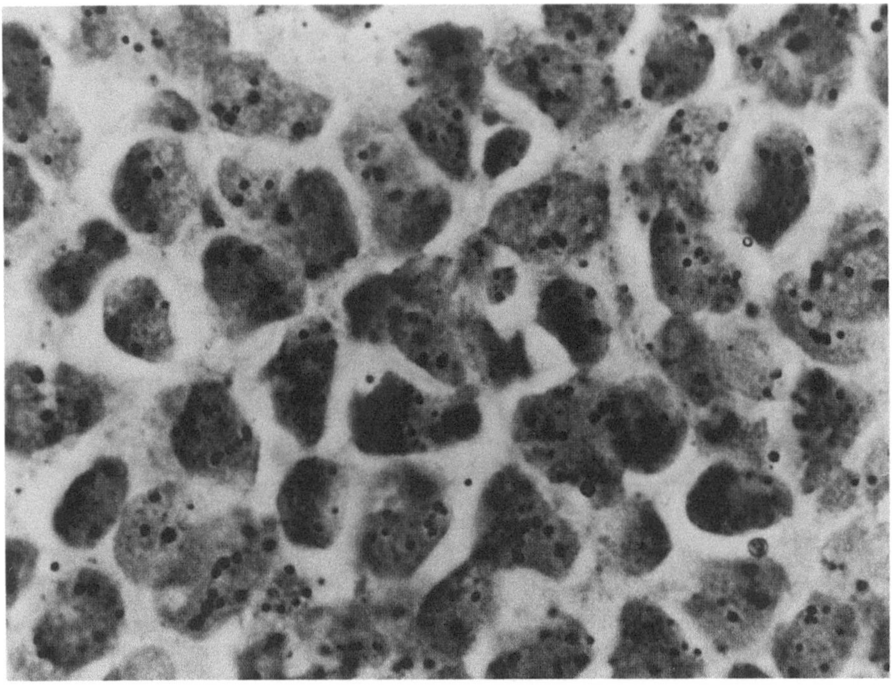

**Fig. 21.** AgNOR preparation of childhood neuroblastoma. Each nucleus contains numerous "dots". (Photograph reproduced from *the Journal of Pathology*, by kind permission)

## 6.6 Paediatric Neoplasms

Numerous paediatric tumours have been studied, including some investigations of prognosis. EGAN et al. (1987) studied AgNOR numbers in a series of small cell tumours of childhood, and observed that the ranges of counts were different for neuroblastoma, rhabdomyosarcoma and Ewing's tumour, the differences being statistically highly significant. Thus, the mean number of AgNORs per nuclear profile in Ewing's tumour was 9.7 (range 7.0–10.7), in neuroblastoma 12.96 (range 12.0–18.5) and in rhabdomyosarcoma 4.7 (range 3.0–7.1) (Figs. 21–23).

Fibrous proliferations of childhood and infantile fibrosarcomas were also studied by EGAN et al. (1988c). The former, benign conditions included infantile digital fibromatosis, desmoid fibromatosis and myofibromatosis and possessed a pooled mean AgNOR count of 3.7 per nuclear profile. In contrast, the fibrosarcomas had a mean count of 11.5.

In terms of *prognosis*, EGAN et al. (1988e) found no relation between AgNOR numbers and survival in Ewing's sarcoma. However, the same authors (EGAN et al. 1988a, f) showed a definite relationship between AgNOR counts and prognosis in childhood neuroblastomas. The NOR

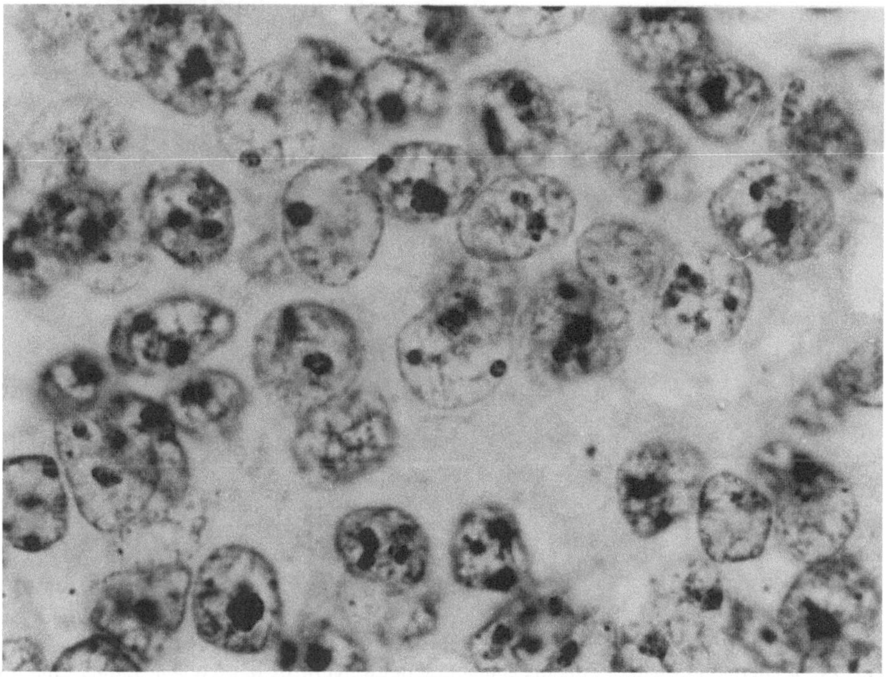

**Fig. 22.** AgNOR stain performed on a childhood rhabdomyosarcoma. Several "dots" are seen in each nucleus, but fewer than in neuroblastoma (Fig. 21). (Photograph reproduced from *the Journal of Pathology*, by kind permission)

numbers also correlated with the degree of differentiation and the mitosis-karyorrhexis index in the stroma-poor variety of these neoplasms. However, the method was again shown to be of no prognostic significance in embryonal rhabdomyosarcoma (EGAN et al. 1988 d). The reasons for these relationships, or lack of them, are obscure but may relate to cell cycling in individual tumours, or to local effects such as gene amplification.

## 6.7 Tumours of Ear, Nose and Throat

Salivary gland neoplasms were studied by MORGAN et al. (1988), who enumerated AgNORs in normal glands as well as pleomorphic adenoma, adenoid cystic carcinoma, mucoepidermoid tumour, acinic cell tumour, adenocarcinoma and adenolymphoma. The malignant group of tumours had AgNOR counts much greater than the benign neoplasms, with a mean range of 3.92 (adenoid cystic carcinoma) to 9.23 (adenocarcinoma) as opposed to 1.41 (pleomorphic adenoma) to 1.75 (acinic cell tumour). There was no overlap between the benign and malignant tumour counts. The differentiation between pleomorphic adenoma and adenoid cystic carcinoma,

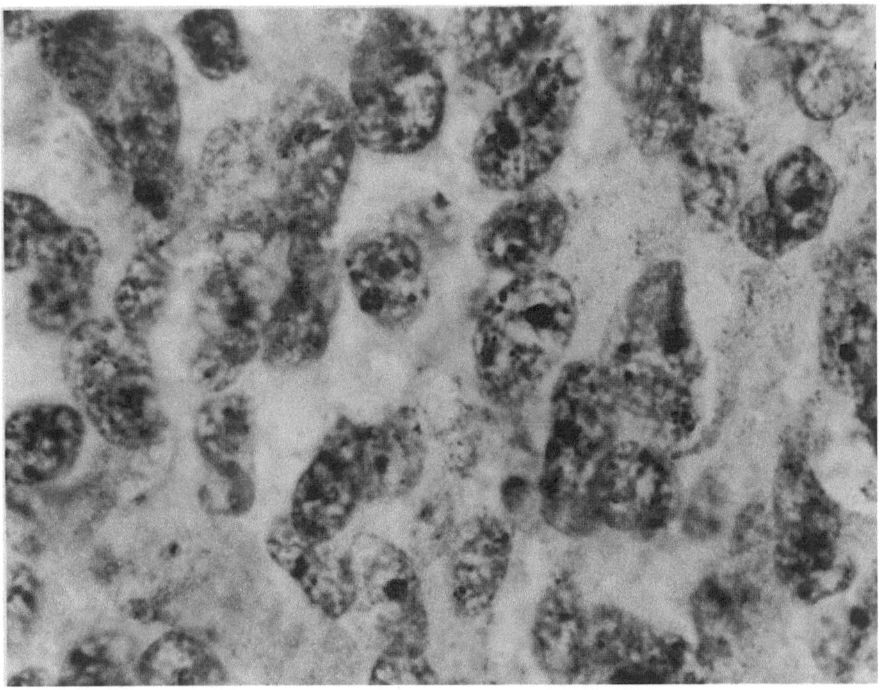

**Fig. 23.** A childhood Ewing's sarcoma stained by the AgNOR method. When counted, the Ag-NORs are intermediate in number between those in neuroblastoma and rhabdomyosarcoma. (Photograph reproduced from *the Journal of Pathology,* by kind permission)

which is sometimes most difficult (especially in small biopsies), is thus facilitated. The series also included nine "problem cases" where AgNOR counts led to benign or malignant diagnoses which were confirmed on major excision, thus showing the prospective value of NOR counts in these tumours.

"Inverted transitional cell papillomas" of nose have been investigated by EGAN et al. (1988 g), who showed a much higher pooled mean AgNOR count (6.22 per nuclear profile) in frankly malignant invasive cases than in benign lesions (2.2 per nucleus). Most significantly, those cases shown to be undergoing malignant transformation on the basis of standard cytological criteria possessed an intermediate number of AgNORs (3.25 per nuclear profile). The diagnostic value of this finding is self-evident.

Lesions of the larynx have also been studied by means of the AgNOR method. ASHWORTH and HELLIWELL (1988) examined normal, dysplastic and malignant laryngeal epithelium, both in the basal and suprabasal areas. However, they were unable to distinguish dysplastic from normal or malignant epithelium on the basis of NOR counts. (This reflects the "continuum" of numbers seen, for example, in squamous lesions of the bronchus, as described in Sect. 6.5.)

Recently, Copper and Michaels (1989) have shown that AgNOR enumeration is a dramatically useful and reliable means of detecting cholesteatoma in aural biopsies.

## 6.8 Gastrointestinal Tumours

Suarez et al. (1988) demonstrated a generally increasing trend in the numbers of AgNORs in gastric lesions with increasing malignancy. However, there was considerable "overlap" between the different groups, although malignant lesions were statistically significantly different from the rest.

Derenzini et al. (1988) examined colonic hyperplastic and adenomatous polyps together with adenocarcinomas, using electron and light microscopy. The malignant tumours possessed numerous small scattered fibrillar centres, whereas in the polyps the centres were larger and clustered. Interestingly, in two polyps, cells were observed with an intermediate distribution.

## 6.9 Miscellaneous Tumours

### *6.9.1 Thyroid Tumours*

As in other endocrine tumours, the diagnostic value of AgNOR enumeration in thyroid tumours is limited. Nairn et al. (1988) showed a trend to increased AgNOR numbers with thyroid malignancy but there was great overlap between the counts in thyroid hyperplasia and follicular carcinoma. Only anaplastic carcinoma had clearly greater AgNOR counts.

### *6.9.2 Prostatic Tumours*

Colloby and Crocker (1988) showed a considerable overlap in mean AgNOR numbers between well-differentiated prostatic adenocarcinoma and hyperplastic prostate, although poorly differentiated carcinomas had many more AgNORs per nuclear profile.

### *6.9.3 Tumours of the Uterine Cervix*

Rowlands (1988) examined a series of cervical biopsies, including herpetic changes and various grades of intraepithelial hyperplasia. Again, although there was overlap between the ranges, there were some statistically significant differences between the groups. The method was considered to be of limited diagnostic value. These results were largely confirmed by Egan et al. (1988 b), although they were able to distinguish CIN 3 from

HPV infected basal cells. It is of interest to note that no relationship was shown between AgNOR counts and DNA flow cytometry (ROLLASON 1988, personal communication).

### 6.9.4 *Gliomas of the CNS*

CAREY and CROCKER (1988) showed striking differences between "reactive" glial cell AgNOR counts and AgNOR numbers in grade I–IV astrocytomas. There were differences in the counts between the different grades of astrocytoma and the technique appears to be of diagnostic value.

### 6.9.5 *Pituitary Adenomas*

MCNICOL et al. (1988) showed a significantly higher number of AgNORs per nuclear profile in macroadenomas than microadenomas. It was also claimed that some tumour types, for example prolactinomas and corticotroph adenomas, could be distinguished by the technique, although the numbers quoted appear close.

### 6.9.6 *NORs and the Liver*

CROCKER and MCGOVERN (1988) have shown a well-defined difference between numbers of AgNORs per nucleus in livers exhibiting chronic active hepatitis, cirrhosis and hepatocellular carcinoma. A "test series" of equivocal biopsies supported these data and the method appears to be diagnostically useful.

## 6.10 NOR Techniques in Histopathological Practice

Of the various methods for the demonstration of NORAPs, in the author's experience by far the least cumbersome and most satisfactory is the one-stage argyrophil method, performed at 20° C (or room temperature). This should be performed as follows, using routinely processed tissue, embedding in paraffin wax:

1. Cut sections at 3 μm thickness, dewax and take to distilled, deionised water.
2. Prepare $Ag^+$ mixture thus: dissolve gelatine at 2 g/dl in 1 g/dl aqueous formic acid. Mix (1:2 volumes) with 50 g/dl aqueous silver nitrate solution.
3. Pour this mixture over sections and leave, under safe-light (Kodak No. 1 A) conditions, for 30–60 min, at room temperature.
4. Wash *thoroughly* in distilled, deionised water.
5. Counterstain, if required, in Mayer's haemalum.
6. Dehydrate, clean and mount in synthetic medium.

Several specific points must be stressed:

1. The section thickness suggested is a compromise between thinner slices (which would give finer detail of AgNORs) and those of greater thickness (where more of the total nuclear AgNOR component would be seen but overlap of AgNORs would be a problem).
2. Deionised water may be insufficiently pure for the procedure (including reagent dilution etc.) since here may be residual $Cl^-$ ions which will precipitate out $Ag^+$ by the common ion effect. The use of *distilled*, deionised water is a useful precaution.
3. It may be necessary to vary the time of incubation depending upon various factors, including the method of tissue fixation and the tissue type, and a "titration" is worthwhile before any large-scale procedure.
4. The use of a counterstain is a matter of choice. In the author's hands it is generally unnecessary, since in the absence of counterstaining the cells and their nuclei take on a golden yellow hue, enabling morphological evaluation. Once mounted, the preparations appear, in effect, to be permanent.

For the *ultrastructural* demonstration of NORs by means of their argyrophilia, the following method, (after PLOTON et al. 1982) is recommended (Fig. 24):

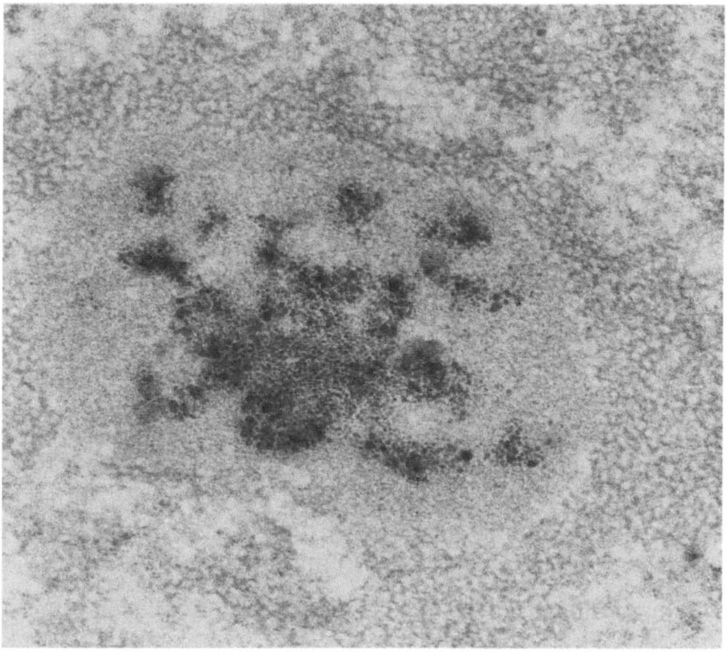

**Fig. 24.** A fibrillar centre at EM level, showing numerous silver grains

1. Finely mince fresh tissue.
2. Fix in 1.2%–1.8% glutaraldehyde in 0.1 M phosphate buffer (10 min at 4° C).
3. Rinse in buffer.
4. Fix in Carnoy's solution (5 min at 4° C)
5. Rinse in 70% alcohol.
6. Wash in pure water twice.
7. Stain with usual AgNOR solution for 10 min at 70° C in dark.
8. Rinse several times in pure distilled $H_2O$.
9. Dehydrate through graded alcohols to pure alcohol.
10. Two changes of pure ethanol (30 min).
11. Two changes of "Inhibisol" (30 min).
12. 50:50 "Inhibisol": TAAB EM resin (overnight).
13. 25:75 "Inhibisol": TAAB EM resin (6 h).
14. Pure TAAB EM resin overnight.
15. Embed tissue in resin and polymerise at 70° C.
16. Cut sections onto 300 mesh copper/rhodium grids.
17. Stain for 10 min in uranyl/acetate/acetone solution.
18. Wash in acetone.
19. Stain for 10 min in Reynold's lead citrate stain.
20. Wash.
21. Dry and view.

The author has also found the binding of $Bi^{3+}$ ions to NORs, by virtue of the $M_r$ 100 K NORAP, to be a useful light microscope method. The following method, after that of LOCKE and HUIE (1977) is satisfactory:

1. Fix tissue in formalin, process to wax and cut at 3 μm. Heat sections onto glass slides.
2. Dewax sections in xylene and hydrate to *pure* distilled water.
3. Wash sections in 0.2 M Tris buffer, pH 7.6.
4. Stain in bismuth staining solution for 2 1/2 h.
5. Wash in 0.1 M Tris buffer. Three changes, 10 min each.
6. Wash in buffer with several drops of saturated amonium sulphide added – 10 min.
7. Dehydrate, clear and mount.

*Bismuth staining solution*

Sol. A 400 mg sodium tartrate added to 10 ml of 1 N NaOH – solution added to 200 mg bismuth oxynitrate.

Sol. B Tris buffer pH 7.6. Add sol. B to sol. A in a 3.5:1 ratio. Adjust to pH 7 (from 14) by adding 10% HCl, dropwise. At pH 7 a cloudy precipitate forms. Place sections in humidity chamber.

Thus, the $Bi^{3+}$ band is localised by means of formation of bismuth sulphide (Fig. 25).

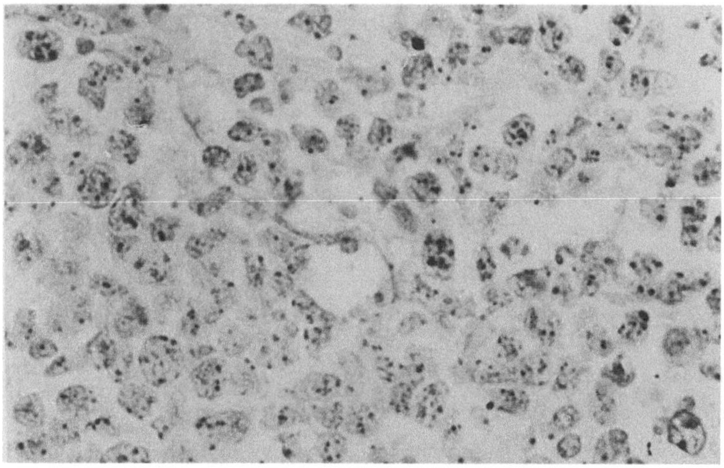

**Fig. 25.** An NHL stained for NORs by means of the bismuth technique. In this high-grade specimen, numerous NORs are seen, as in an argyrophil specimen

### 6.10.1 *Tissue Fixation and the AgNOR Reaction*

In view of the increasing use of the AgNOR technique in histopathology, SMITH et al. (1988) considered it timely to investigate a series of fixatives in relation to their suitability for the subsequent reaction. Striking and consistent findings were obtained. In summary, alcohol-based fixatives gave optimal results; fixatives containing mercury or dichromate were very unsatisfactory. "Routine" 10% formol saline was satisfactory but 10% neutral buffered formalin gave results approaching those of alcohol-based fixatives. Formaldehyde solution forms methylene bridges between -SH and $>C=0$ groups and could thus inhibit the AgNOR reaction; however, the wash in 70% ethanol in processing would break down these bridges and thus enable the silver binding to occur. Dichromate ions oxidise sulphydryl groups to cysteic acid and prevent silver uptake, thus rendering dichromate-containing fixatives useless for the AgNOR sequence. Glutaraldehyde is a poor fixative in this context, since the cross-links it forms are not broken by ethanol (in contrast to formaldehyde). The reasons for these findings are in general readily followed. Thus, mercury will compete with the silver ions in the AgNOR reaction substrate and block attachment of the latter to NORAP -SH, -S-S- and $>C=0$ groups (Figs. 26, 27).

### 6.10.2 *Enumeration of AgNOR Sites in Histological Sections*

It must, of course, be remembered that the counts of NORs achieved in, say, 3-μm sections, can only be *representative* of the true AgNOR content of whole nuclei. Nonetheless, as a standardised comparative method this

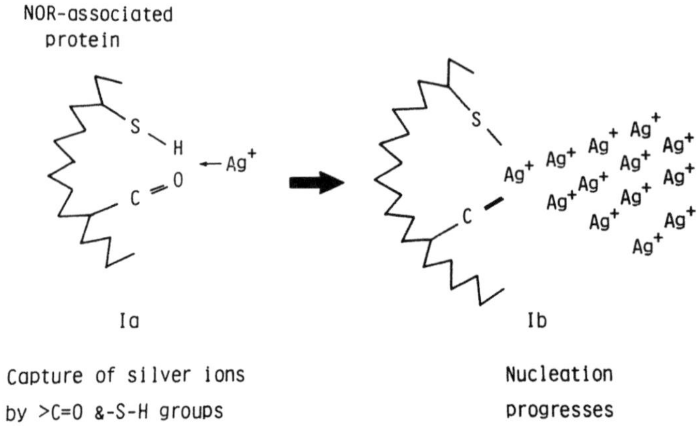


**Fig. 26.** The proposed binding of silver ions to carboxyl and sulphydhryl groups of a NORAP. (Photograph reproduced from *the Journal of Pathology*, by kind permission)

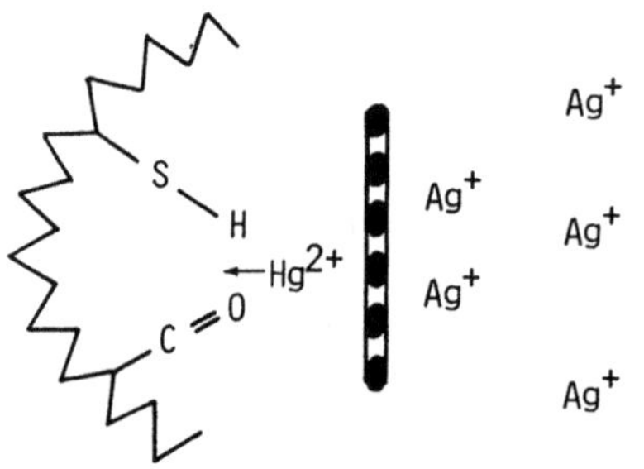


**Fig. 27.** Blockage to the process seen in Fig. 26 by mercury ions, preventing the binding of silver ions to -SH groups

is quite acceptable and the variations in numbers between different groups of cells could not readily be accounted for by nuclear volumetric differences. Nonetheless, it would be of considerable interest to attempt *absolute* counts, using whole cell preparations. In tissue sectioning it is necessary to achieve a compromise between optimal AgNOR "dot" resolution (at, say 1-μm thickness), where counts would be unacceptably low compared to the true number of NORs, and between grossly thick (for example, 10-μm) sections where NOR overlap would render enumeration impossible. In the author's experience, 3-μm sections are a tenable compromise. Unfortunately, the size of AgNOR dots cannot be reliably measured with the light microscope since their dimensions are at the limit of optical resolution; clearly, the measurement of NOR (fibrillar centre) size can only be performed at the ultrastructural level.

Ideally, of course, for the AgNOR method to be applied in *routine* histophathology, formal counting should be unnecessary. This is the case, for example, in the field of naevoceullular naevi and melanocarcinomas or benign and malignant pleural mesothelial cells, where there is no overlap between groups. Thus, a simple visual estimate of AgNOR numbers (for example, one as opposed fo four to six) would suffice to make the diagnosis. However, in some areas (for example, small cell paediatric tumours) formal counting is necessary. When this is the case, enumeration can be performed in two ways. Either automatic or interactive image analysis may be employed or simple visual counting can be performed, ideally with an eyepiece graticule to prevent recounting of AgNORs. In the author's experience, either method is satisfactory.

The next decision confronting the investigator is that of which nuclear dots to count. This should be straightforward, but SMITH and CROCKER (1988), in a study of NORs in breast lesions, approached the problem in three ways. AgNOR sites were counted in three ways: (a) total AgNOR dots; (b) "clumps" of AgNORs and (c) "dots" within clumps were summated. It was found that only (a) gave discriminating results. "Clumps" of AgNORs presumably represented nucleoli reaggregating after division. In the case of epithelial tumours, especially in situ carcinomas, a new approach is perhaps needed, with AgNOR counts being performed at known suprabasal levels in the epithelium. This might "amplify" the counts seen in abnormally high levels of proliferation in cells above the basal level. The method could be applied readily by means of image analysis, or manually using "masked" photographs of the epithelium under study.

CROCKER et al. (1989a) have recently outlined proposals for a standardised method of AgNOR enumeration in tissue sections. Counting is especially easy when performed with cell imprints, as demonstrated by BOLDY et al. (1989).

### 6.10.3 Conclusions and Prospects for the Future

Clearly, the first priority is that there should be more quantitative studies of NORs in histological sections in relation to "problem cases" and to clinical prognosis. The former has been shown to be of well-defined value in the case of salivary gland tumours by MORGAN et al. (1988) and the latter *indirectly* in NHLs by HALL et al. (1988). Furthermore, NOR *size* is a feature to be studied in detail in relation to tumour type or grade. In addition, NOR activity must be increasingly related to other tumour markers, including those of proliferation, in a large range of lesions. Indeed, the whole field of chronobiology is to be examined in detail in relation to NORs (FAKAN and HERNANDEZ-VERDUN 1986). Certainly, in rat sympathetic ganglion cells there is a 24 h circadian rhythm where nucleolar volume alters. It is not impossible that, for example, fibrillar centre number, size and activity may follow circadian (daily) or perhaps menstrual cycles. This could lead to other "variables" in tumour pathology, although such rhythms observed in mitotic rates in normal tissues may be absent in malignant tissues.

Time will tell which is or are the best methods for NOR visualisation. The AgNOR method is accurate, quick and time-tested but empirical

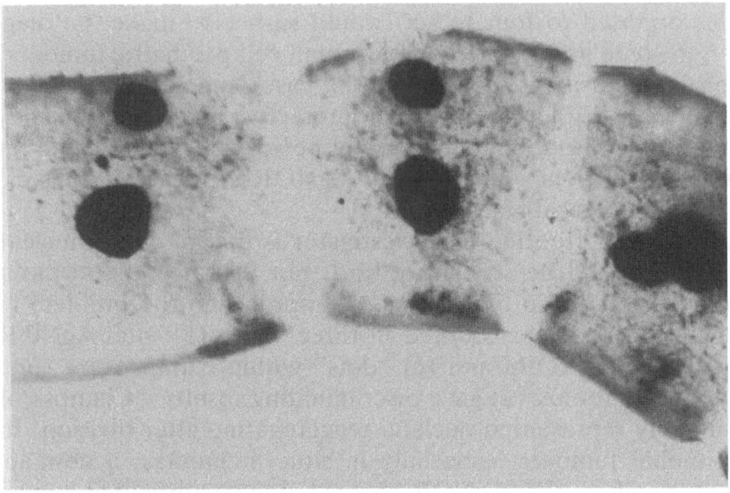

**Fig. 28.** Argyrophil preparation of interphase cells of an $F_1$ hybrid of *Allium cepa* (common onion) x *Allium fistulosum* (Japanese bunching onion). The latter possesses more rDNA gene copies than the former, and forms the larger of the NORs displayed. (Photograph by courtesy of Mrs. Susan Armstrong)

studies of other probes, such as mercurifluorescein and bismuth will be of interest. In addition, the fields of exfoliative and aspiration cytology are ideal for application of NOR methodology.

However, by far the most important questions still lie unanswered. First, are nuclear AgNOR counts related to ploidy? A recent study by CROCKER et al. (1989c) has shown clearly that, in NHLs, when chromosome methaphases are analysed and stained for AgNORs, there is no relationship between their numbers and the AgNOR content of interphase cells from the same specimens. This is true of both high- and low-grade tumours and appears to have shown with certainty that ploidy is *not* directly related to interphase NOR numbers. Perhaps plant cells, with known stable polyploidies may provide at least part of an answer here. Certainly, plant "chimaeras" or $F_1$ hybrids retain the original NOR sizes of their progenitors (Fig. 28). Another main problem to be tackled is that of nuclear AgNOR numbers in relation to the cell cycle. It is known that these two features are related and variations in the latter in any given tumour must affect the former. It would be interesting to relate the two phenomena in individual cells of a given tumour, perhaps with flow cytometric techniques using cell suspensions. Finally, the use of antibody or other probes in relation to known NORAPs, such as nucleolin, may tell us much about transcriptional activities of individual tumour cells. It is to be hoped that some answers to these questions and further diagnostic applications may emerge in the next half-decade.

*Acknowledgements*. The author is especially grateful to Mrs. Ruth Fry for her patient and painstaking wordprocessing. Mrs. Pamela Jackson supplied skilful graphics assistance and photographic work was kindly performed by Mr. Michael J. Chard, Mr. Christopher Davies and Miss Tracey Bissell. Helpful advice and discussion were provided by the staff of the West Midlands Regional Cytogenetics Department, East Birmingham Hospital (Director: Dr. M. Hulten), especially Mrs. Susan Armstrong and Mr. Paul Leedham. The author is also indebted to his wife, Dr. Catherine Crocker, for reading the manuscript carefully. Finally, I am deeply thankful to the technical staff of the Department of Histopathology, East Birmingham Hospital, for their constant skilful histological support, without which little or none of this would have been possible.

## References

Alberts B, Bray D, Lewis J, Raff M, Roberts K, Watson JD (1983) Molecular biology of the cell. Garland, New York

Anon (1987) NORs – a new method for the pathologist. Lancet I:1413–1414

Angelier N, Hernanedez-Verdun D, Bouteille M (1982) Visualisation of Ag-NOR proteins on nucleolar transcriptional units in molecular spreads. Chromosoma 86:661–672

Arden KC, Pathak S, Frankel LS, Zander A (1985) Ag-NOR staining in human chromosomes: differential staining in normal and leukemic bone-marrow samples. Int J Cancer 36:647–649

Ashraf M, Godward MBE (1980) The nucleolus in telophase, interphase and prophase. J Cell Sci 41:321:329

Ashworth MT, Helliwell TR (1988) Nucleolar organizer regions in benign, dysplastic and malignant laryngeal squamous epithelium. J Pathol 154:64 A

Ayres JG, Crocker J, Skilbeck NQ (1988) Differentiation of malignant from normal and reactive mesothelial cells using the argyrophil technique for nucleolar organizer region associated proteins. Thorax 43:366–370

Babu KA, Verma RS (1985) Structural and functional aspects of nucleolar organizer regions (NORs) of human chromosomes. Int Rev Cytol 94:151–176

Boldy D, Ayres JG, Crocker J (1988) Nucleolar organizer regions in squamous lung tumours. (in preparation)

Boldy DAR, Crocker J, Ayres JG (1989) Application of the AgNOR method to cell imprints of lymphoid tissue. J Pathol 157:75–80

Bourgeois CA, Hernandez-Verdun D, Hubert H, Bouteille M (1979) Silver staining of NORs in electron microscopy. Exp Cell Res 123:449–452

Busch RK, Busch H (1977) Antigenic proteins of nucleolar chromatin of Novikoff hepatoma ascites. Tumori 63:347–357

Busch H, Daskal Y, Gyorkey F, Smetana K (1979 a) Silver staining of nucleolar granules in tumor cells. Cancer Res 39:857–863

Busch H, Gyorkey F, Busch RK, Davis FM, Gyorkey P, Smetana K (1979 b) A nucleolar antigen found in a broad range of human malignant tumor specimens. Cancer Res 39:3024–3030

Busch H, Busch RK, Chan PK, Isenberg W, Weigand R, Russo J, Furmanki P (1981) Results of a blind study on the presence of the human tumor nucleolar antigen in breast carcinomas, benign breast tumors and normal breast tissues. Clin Immunol Immunopathol 18:155–167

Buys CHCM, Osinga J (1982) A relation between G-, C- and N-band patterns as revealed revealed by progressive oxidation of chromosomes and a note on the nature of N-bands. Genetics 58:3–9

Buys CHCM, Osinga J (1984) Selective staining of the same set of nucleolar phosphoproteins by silver and Giemsa. A combined biochemical and cytochemical study on staining of NORs. Chromosoma 89:387–396

Carey MP, Crocker J (1988) Differentiation of gliomas and gliosis by the enumeration of NORs. (in preparation)

Cataldo C, Souchier C, Vasserot M, Calisti A, Vagner-Capodano A–M, Stahl A (1985) Three-dimensional analysis of fibrillar centres and associated chromatin in the nucleolus of human oocytes in primordial follicles. Biol Cell 54:91–194

Chatterjee A, Freeman JW, Busch H (1987) Identification and partial characterisation of a $M_r$ 40,000 nucleolar antigen associated with cell proliferation. Cancer Res 47:1123–1129

Colloby PS, Crocker J (1988) Nucleolar organizer regions in prostatic malignancy. (in preparation)

Cooper J, Michaels L (1989) Argyrophilic nucleolar organizer region-associated proteins (Ag-NORs) in stratified squamous epithelia of congenital and acquired cholesteatomas and the external ear. J Pathol (in press)

Courvalin JC, Maunoury R, Hernandez-Verdun D, Maro B, Bornens M (1983) Une proteine de 80 kD est associee a l'organisateur nucleolaire (NOR) des cellules humaines. Biol Cell 49:10 a

Crocker J, (1984) Morphometric and related quantitative techniques in the study of lymphoid neoplasms. A review. J Pathol 143:69–80

Crocker J, Burnett D (1986) Granulocyte markers in histophatology. A review. J Pathol 150:77–88

Crocker J, Egan MJ (1988) Correlation between NOR sizes and numbers in non-Hodgkin's lymphomas. J Pathol 156:233–239

Crocker J, McGovern J (1988) Nucleolar organizer regions in normal, cirrhotic and carcinomatous livers. J Clin Pathol 41:1044–1048

Crocker J, Nar P (1987) Nucleolar organizer regions in lymphomas. J Pathol 151:111–118

Crocker J, Skilbeck N (1987) Nucleolar organizer region associated proteins in cutaneous melanotic lesions: a quantitiative study. J Clin Pathol 40:885–889

Crocker J, Jones EL, Curran RC (1983) A comparative study of nuclear form factor, area and diameter in non-Hodgkin's lymphomas and reactive lymph nodes. J Clin Pathol 36:298–302

Crocker J, Ayres J, McGovern J (1987) Nucleolar organiser regions in small cell carcinoma of the bronchus. Thorax 42:972–975

Crocker J, Ayres JG, Skilbeck NQ (1988 a) Evaluation of the AgNOR technique in the diagnosis of malignant mesotheliomas. J Pathol 154:43 A

Crocker J, Hall PA, Macartney JC, Stansfeld AG (1988 b) A comparative study of nucleolar organizer regions (AgNORs), Ki 67 staining and DNA flow cytometry in non-Hodgkin's lymphomas. J Pathol 154:37 A

Crocker J, Macartney JC, Smith PJ (1988 c) Correlation between DNA flow cytometric and nucleolar organizer region data in non-Hodgkin's lymphomas. J Pathol: 154:151–156

Crocker J, Boldy DAR, Egan MJ (1989 a) How should we count AgNORs? Proposals for a standardised approached. J Pathol (in press)

Crocker J, Murray PG, Boldy DAR (1989 b) Sequential labelling with monoclonal antibodies (including Ki 67) and demonstration of AgNORs in frozen sections. J Pathol (in press)

Crocker J, Janmohamed R, Armstrong S, Leyland MJ, Hulten M (1989 c) The relationship between numbers of interphase NORs and NOR-bearing chromosomes in non-Hodgkin's lymphomas. J Pathol 157:166 A

Crossen PE, Godwin JM (1985) Rearrangement and possible amplification of the ribosomal RBA gene sites in the human chronic myelogenous leukemia cell line K 562. Cancer Genet Gytogenet 18:27–30

Das NK (1961) Demonstration of a non-RNA nucleolar fraction by silver staining. Exp Cell Res 26:428–431

Das BC, Rani R, Mitra AB, Luthra UK (1986) The number of silver-staining NORs (rDNA) in lymphocytes of newborns and its relationship to human development. Mech Ageing Dev 36:117–123

Daskal Y, Smetana K, Busch H (1980) Evidence from studies on segregated nucleoli that nucleolar silver-staining proteins $C_{23}$ and $B_{23}$ are in the fibrillar component. Exp Cell Res 127:285–291

Davis FM, Busch RK, Yeoman LC, Busch H (1978) Differences of nucleolar antigens of rat liver and Novikoff hepatoma ascites cells. Cancer Res 38:1906–1915

de Capoa A, Baldini A, Marlekaj P et al. (1985a) Hormone-modulated rRNA gene activity is visualized by selective staining of the NOs. Cell Biol Int Rep 9:791–796

de Capoa A, Marleka P, Baldini A (1985b) The transcriptional activity of individual ribosomal DNA gene clusters is modulated by serum concentration. J Cell Sci 74:21–35

DeLozier-Blanchet CD, Watt H, Engel E (1986) Ectopic nucleolus organizer regions (NORs) in human testicular tumours. Cytogenet Cell Genet 41:107–113

Derenzini M, Hernandez-Verdun D, Bouteille M (1982) Visulisation in situ of extended DNA filaments in nucleolar chromatin of rat hepatocytes. Exp Cell Res 141:463–469

Derenzini M, Pession A, Betts-Eusebi CM, Novello F (1983) Relationship between the extended, non-nucleosomal intranucleolar chromatin in situ and ribosomal RNA synthesis. Exp Cell Res 1145:127–143

Derenzini M, Betts CM, Ceccarelli C, Eusebi V (1986) Ultrastructural organization of nucleoli in benign naevi and malignant melanomas. Virchows Archiv [Cell Pathol] 52:343–352

Derenzini M, Romagnoli T, Mingazzini P, Marinozzi V (1988) Interphasic NOR distribution as a diagnostic parameter to differentiate benign from malignant epithelial tumors of human intestine. Virchows Arch [Cell Pathol] 54:334–340

Dimova RN, Markov DV, Gajdardjieva KJ, Dabeva MD, Hadjiolov AA (1982) Electron microscopic localisation of silver staining NOR-proteins in rat liver nucleoli upon D-galactosamine block of transcription. Eur J Cell Biol 28:272–277

Egan MJ, Crocker J (1988) Nucleolar organizer regions in cutaneous tumours. J Pathol 154:247–253

Egan MJ, Smith KE (1987) S100 protein and myoepithelial cells of breast. J Clin Pathol 40:1485

Egan MJ, Raafat F, Crocker J, Smith K (1987) Nucleolar organizer regions in small cell tumours of childhood. J Pathol 153:275–280

Egan MJ, Crocker J, Raafat F, Williams D (1988a) The prognostic significance of nucleolar organizer regions in neuroblastoma and the relationship to established prognostic indices. J Pathol 154:109 A

Egan MJ, Freeth MG, Crocker J (1988b) CIN, HPV infection and nucleolar organiser regions in cervical epithelium. J Pathol 154:80 A

Egan MJ, Raafat F, Crocker J, Smith K (1988c) Nucleolar organiser regions in fibrous proliferations of childhood and infantile fibrosarcoma. J Clin Pathol 41:31–33

Egan MJ, Raafat F, Crocker J, Williams D (1988d) Prognostic significance of nucleolar organiser regions in embryonal rhabdomyosarcoma. J Clin Pathol 41:477

Egan MJ, Raafat F, Crocker J, Williams D (1988e) Prognostic importance of nucleolar organiser regions in Ewing's sarcoma of childhood. J Clin Pathol 41:232

Egan MJ, Raafat F, Crocker J, Williams D (1988f) A comparative study of the degree of differentiation of neuroblastoma and the mean numbers of nucleolar organiser regions. J Clin Pathol 41:527

Egan MJ, Ramsden K, Crocker J (1988g) Nucleolar organizer regions in inverted transitional cell papillomas of nose. Histopathology 13:579

Escande ML, Gas N, Stevens B (1985) Immunolocalization of the 100K nucleolar protein in CHO cells. Biol Cell 53:99–110

Fakan S, Hernandez-Verdun D (1986) The nucleolus and the nucleolar organizer regions. Biol Cell 56:189–206

Fazekas de St. Groth F, Webster RG, Datyner A (1963) Two new staining procedures for quantitative estimation of proteins on electrophoretic strips. Biochem Biophys Acta 71:377–379

Field DH, Fitzgerald PH, Sin FYT (1984) Nucleolar silver-staining patterns related to cell cycle phase and cell generation of PHA-stimulated human lymphocytes. Cytobios 41:23–33

Freeman JW, Busch RK, Ross B, Busch H (1985) Masking of nontumorous antigens for development of human tumor nucleolar antibodies with improved specificity. Cancer Res 45:5637–5642

Funaki K, Matsui S, Sasaki M (1975) Localisation of nucleolar organizers in animal and plant chromosomes by means of an improved N-banding technique. Chromosoma 49:357–370

Gas N, Escande ML, Stevens B (1985) Immunolocalisation of the 100 kDa nucleolar protein during the mitotic cycle in CHO cells. Biol Cell 53:209–218

Gerdes J, Schwab U, Lemke H, Stein H (1983) Production of a monoclonal antibody reactive with a human nuclear antigen associated with cell proliferation. Int J Cancer 31:13–20

Gerdes J, Lemke H, Baisch H, Wacker HH, Schwab U, Stein H (1984) Cell cycle analysis of a cell proliferation associated human nuclear antigen defined by the monoclonal antibody Ki 67. J Immunol 133:1710–1716

Gerdes J, van Baarlen J, Pileri S, Schwarting R, van Unnik JAM, Stein H (1987) Tumor cell growth fraction in Hodgkin's disease. Am J Pathol 129:390–393

Goessens G, Thiry M, Lepoint A (1987) Relations beween nucleoli and nucleolus-organizing regions during the cell cycle. In: Stahl A, Lucini JM, Vagner-Capodano AM (eds) Chromosomes today, vol 9. Allen and Unwin, London, pp 261–271

Goodpasture C, Bloom SE (1975) Visualisation of nucleolar organizer regions in mammalian chromosomes using silver staining. Chromosoma 53:37–50

Hall PA, Crocker J (1988) Nucleolar organizer regions and Hodgkin's disease. (in preparation)

Hall PA, Crocker J, Watts A, Stansfeld AG (1988) A comparison of nucleolar organizer region staining and Ki 67 immunostaining in non-Hodgkin's lymphoma. Histopathology 12:373–381

Hauser-Urfer IH, Stauffer J (1985) Comparative chromosome analysis of nine squamous cell carcinoma lines from tumors of the head and neck. Cytogenet Cell Genet 39:35–39

Henderson AA, Warburton D, Atwood KC (1972) Location of ribosomal DNA in the human chromosome complement. Proc Natl Acad Sci USA 69:3394–3398

Hernandez-Verdun D (1983) The nucleolar organizer regions. Biol Cell 49:191–202

Hernandez-Verdun D, Derenzini M (1983) Non-nucleosomal configuration of chromatin in nucleolar organizer regions of metaphase chromosomes in situ. Eur J Cell Biol 31:360–365

Hernandez-Verdun D, Hubert J, Bourgeois CA, Bouteille M (1978) Identification ultrastructurale de l'organisateur nucleolaire par le technique a l'argent. CR Acad Sci (Paris) 287:1421–1423

Hernandez-Verdun D, Hubert J, Bourgeois CA, Bouteille M (1980) Ultrastructural localistion of Ag-NOR stained proteins in the nucleolus during the cell cycle and in other nucleolar structures. Chromosoma 79:349–362

Hernandez-Verdun D, Derenzini M, Bouteille M (1982) The morphological relationship in electron microscopy between NOR-silver proteins and intranucleolar chromatin. Chromosoma 85:461–473

Howell WM (1977) Visualisation of ribosomal gene activity: silver stains proteins associated with rRNA transcribed from oocyte chromosomes. Chromosoma 62:361–367

Howell WM, Black DA (1980) Controlled silver staining of nucleolus organizer regions with a protective colloidal developer: a one-step method. Experientia 36:1014

Howell WM, Denton TE, Diamond JR (1975) Differential staining of the satellite regions of human acrocentric chromosomes. Experientia 31:260–262

Hsu TC, Brinkley BR, Arrhighi FE (1967) The structure and behavior of the nucleolus organizers in mammalian cells. Chromosoma 23:137–153

Hsu TC, Spirito SE, Pardue ML (1975) Distribution of 18 + 28 S ribosomal genes in mammalian genomes. Chromosoma 53:25–36

Jones CE, Busch H, Olson MOJ (1981) Sequence of a phosphorylation site in nucleolar protein $B_{23}$. Biochim Biophys Acta 667:209–212

Jordan G (1987) At the heart of the nucleolus. Nature 329:489–490

Jordan EG, McGovern JH (1981) The quantitative relationship of the fibrillar centres and other nucleolar components to changes in growth conditions, serum deprivation and low doses of actinomycin D in cultured human fibroblasts (strain MRC-5). J Cell Sci 52:373–389

Kurata A, Misumi Y, Sakaguchi, Shiokawa K, Yamana K (1978) Does the rate of ribosomal RNA synthesis vary depending on the number of nucleoli in the nucleus? Exp Cell Res 115:415–419

Lau Y-F, Wertelecki W, Pfeiffer RA, Arrighi FE (1979) Cytological analyses of a 14p+ variant by means of N-banding and combinations of silver staining and chromosome bandings. Hum Genet 46:75–82

Lennert K (1978) Malignant lymphomas other than Hodgkin's disease. Springer, New York Heidelberg

Lepoint A, Goessens G (1982) Quantitative analysis of Ehrlich tumour cell nucleoli during interphase. Exp Cell Res 37:456–459

Lewin B (1983) Genes. John Wiley, New York, 369–373

Lischwe MA, Smetana K, Olson MOJ, Busch H (1979) Proteins $C_{23}$ and $B_{23}$ and the major nucleolar silver staining proteins. Life Sci 25:701–708

Lischwe MA, Richards RL, Busch R, Busch H (1981) Localisation of phosphoprotein $C_{23}$ to nucleolar structures and to the nucleolus organiser regions. Exp Cell Res 136:101–109

Locke M, Huie P (1977) Bismuth staining for light and electron microscopy. Tissue Cell 9:347–371

Mamaev N, Mamaeva S, Liburkina I, Kozlova T, Medvedeva N, Makarkina G (1985) The activity of nucleolar organizer regions of human bone marrow cells studied with silver staining. I. Chronic myelocytic leukemia. Cancer Genet Cytogenet 16:311–320

Mamaev NN, Mamaeva SE, Grabovskaya IL, Makarkina GN, Kozlova TV, Medvedeva NV, Marynets OV (1987) The activity of nucleolar organizer regions of human bone marow cells studied with silver staining. II. Acute leukaemia. Cancer Genet Cytogenet 25:65–72

Matsui S, Sasaki M (1973) Differential staining of nucleolus organizers in mammalian chromosomes. Nature 246:148–150

McNicol AM, Colgan J, McMeakin W, Teasdale GM (1988) Nucleolar organizer regions in pituitary adenomas. J Pathol 154:106 A

Means GE, Feeney RE (1971) Chemical modification of proteins. Holden-Day, San Francisco, p 254

Miller DA, Breg WR, Warburton D, Dev VG, Miller OJ (1978) Regulation of rRNA gene expression in a human familial 14p+ marker chromosome. Hum Genet 43:289–297

Miller OJ, Tantravahi U, Katz R, Erlanger BF, Guntaka RV (1981) Amplification of mammalian ribosomal RNA genes and their regulation by methylation. In: Arrighi FE, Rao PN, Stubblefield E (eds) Genes, chromosomes and neoplasia. Raven, New York, pp 253–270

Mirre C, Knibiehler B (1982) A reevaluation of the relationships between the fibrillar centres and the nucleolus-organizing regions in reticulated nucleoli: ultrastructural organization, number and distribution of the fibrillar centres in the nucleolus of the mouse Sertoli cell. J Cell Sci 55:261–276

Moreno FJ, Hernandez-Verdun D, Masson C, Bouteille M (1985) Silver staining of the nucleolar organizer regions (NORs) on Lowicryl and cryo-ultrathin sections. J Histochem Cytochem 33:389–399

Morgan DW, Crocker J, Watts A, Shenoi PM (1988) Salivary gland tumours studied by means of the AgNOR technique. Histopathology 13:553

Morton CC, Brown JA, Holmes WM, Nance WE, Wolf B (1983) Stain intensity of human nucleolus organiser region reflects incorporation of uridine into mature ribosomal RNA. Exp Cell Res 145:405–413

Nairn R, Crocker J, McGovern J (1988) Limited value of AgNOR enumeration in the assessment of thyroid neoplasms. J Clin Pathol 41:1136

Ochs RL, Busch H (1984) Further evidence that phosphoprotein $C_{23}$ (110 kD/pI 5.1) is the nucleolar silver staining protein. Exp Cell Res 152:260–265

Olert J, Sawatzki G, Kling H, Gebauer J (1979) Cytological and histochemical studies on the mechanism of the selective silver staining of nucleolus organizer regions (NORs). Histochemistry 60:91–99

Olson MOJ, Rivers ZM, Thompson BA, Kao W, Case St (1983) Interaction of nucleolar phosphoprotein $C_{23}$ with cloned segments of rat ribosomal deoxyribonucleic acid. Biochemistry 22:3345–3351

Pardue ML, Hsu TC (1975) Locations of 18s and 28s ribosomal genes on the chromossomes of the Indian muntjac. J Cell Biol 64:251–254

Pebusque MJ, Seite R (1981) Electron microscopic studies of silver-stained proteins in nucleolar organizer regions: location in nucleoli of rat sympathetic neurons during light and dark periods. J Cell Sci 51:85–94

Pebusque MJ, Vio-Cigna M, Seite R (1981) Ultrastructural location of Ag-NOR stained proteins in nucleoli of rat sympathetic neurons during the dark period. Biol Cell 40:151–154

Peckham MJ, Cooper EH (1969) Proliferation characteristics of the various classes of cells in Hodgkin's disease. Cancer 24:135–146

Pfeifle J, Boller K, Anderer FA (1986) Phosphoprotein pp 135 is an essential component of the nucleolus organizer regions (NOR). Exp Cell Res 162:11–22

Ploton D, Bobichon H, Adnet J-J (1982) Ultrastructural localization of NOR in nucleoli of human breast cancer tissues using a one-step Ag-NOR staining method. Biol Cell 43:229–232

Ploton D, Bendayan M, Adnet J-J (1983) Ultrastructural localization of Ag-NOR proteins and nucleic acids in reticulated nucleoli. Biol Cell 49:29–34

Ploton D, Menager M, Adnet J-J (1984) Simultaneous high-resolution localisation of Ag-NOR proteins and nucleoproteins in interphasic and mitotic nuclei. Histochem J 16:897–906

Ploton D, Menager M, Adnet J-J (1985) Simultaneous ultrastructural localisation of Ag-NOR (nucleolar organizer region) proteins and ribonucleoproteins during mitosis, in human breast cancerous tissues. J Cell Sci 74:239–256

Ploton D, Menager M, Jeannesson P, Himber G, Pigeon F, Adnet J-J (1986) Improvement in the staining and in the visualization of the argyrophilic proteins of the nucleolar organizer region at the optical level. Histochem J 18:5–14

Pogorelov VM, Darovsky BM, Kotelnikov VM et al. (1987) Morphometric characteristic of argentaffine structures of lymphoid cells nucleoli in peripheral blood of patients with non-Hodgkin's lymphomas and chronic lymphatic leukaemia. Exp Oncol (USSR) 9:43–48

Prestayko AW, Klomp GR, Schmoll DJ, Busch H (1974) Comparison of proteins of ribosomal subunits and nucleolar preribosomal particles from Novikoff hepatoma ascites cells by two-dimensional polyacrylamide gel electrophoresis. Biochemistry 13:1945–1951

Raman R, Sperling K (1981) Patterns of silver staining on NORs of prematurely condensed muntjac chromosomes following RNA inhibition. Exp Cell Res 135:373–378

Reeves BR, Casey G, Honeycombe JR, Smith S (1984) Correlation of differentiation state and silver staining of nucleolar organizers in the promyelocytic leukaemia cell line HL-60. Cancer Genet Cytogenet 13:159–166

Ritossa FM, Spiegelman S (1965) Localization of DNA complementary to ribosomal RNA in the nucleolus organized region of *Drosophila melanogaster.* Proc Natl Acad Sci USA 53:737–745

Rose KM, Stetler DA, Jacob ST (1981) Protein kinase activity of RNA polymerase I purified from a rat hepatoma: probable function of M 42000 and 24600 polypeptides. Proc Natl Acad Sci USA 78:2833–2837

Rowlands D (1988) Nucleolar organising regions in cervical intraepithelial neoplasia. J Clin Pathol 41:1200–1202

Sato Y, Abe S, Kubota K, Sasaki M, Miura Y (1986) Silver-stained nucleolar organizer regions in bone marrow cells and peripheral blood lymphocytes of Philadelphia chromosome-positive chronic myelocytic leukaemia patients. Cancer Genet Cytogenet 23:37–45

Scheer U, Raska I (1987) Immunocytochemical localization of RNA polymerase I in the fibrillar centres of nucleoli. In: Stahl A, Luciani JM, Vagner-Capodano AM (eds) Chromosomes today, vol 9. Allen and Unwin, London, pp 284–300

Scheer U, Rose KM (1984) Localization of RNA polymerase I in interphase cells and mitotic chromosomes by light and electron microscopic immunocytochemistry. Proc Natl Acad Sci USA 81:1431–1435

Schmickel RD (1987) The molecular organization of the human ribosomal gene. In: Stahl A, Luciani JM, Vagner-Capodano AM (eds) Chromosomes today, vol 9. Allen & Unwin, London

Schulze B, Golinski C, Fonatsch C (1984) Heterochromatin and nucleolus organizer regions in cells of patients with malignant and premalignant lymphatic diseases. Hum genet 67:391–395

Schwarzacher HG, Wachtler F (1983) Nucleolus organizer regions and nucleoli. Hum Genet 63:89–99

Schwarzacher HG, Mikelsaar A-V, Schnedl W (1978) The nature of the Ag-staining of nucleolus

organizer regions. Electron- and light microscopic studies on human cells in interphase, mitosis and meiosis. Cytogenet Cell Genet 20:24–39
Seite R, Pebusque MJ, Vio-Cigna M (1982) Argyrophilic proteins on coiled bodies in sympathetic neurons identified by Ag-NOR procedure. Biol Cell 46:97–100
Smetana K, Likovsky Z (1984) Nucleolar silver-stained granules in maturing erythroid and granulocytic cells. Cell Tissue Res 237:367–370
Smetana K, Gyorkey F, Chan PK, Tan E, Busch H (1983) Proliferating cell nuclear antigen (PCNA) and human malignant tumor nucleolar antigen (HMTNA) in nucleoli of human hematological malignancies. Blut 46:133–141
Smith R, Crocker J (1988) Evaluation of nucleolar organizer region-associated proteins in breast malignancy. Histopathology 12:113–125
Smith PJ, Skilbeck NQ, Harrison A, Crocker J (1988) The effect of a series of fixatives in the AgNOR technique. J Pathol 155:109–112
Spector DL, Ochs RL, Busch H (1984) Silver staining, imunofluorescence and immunoelectron microscopic localization of nucleolar phosphoproteins $B_{23}$ and $C_{23}$. Chromosoma 90:139–148
Stocker AJ (1978) Correspondence of silver banding with rRNA hybridisation sites in polytene chromosomes of *Rhynchosciaria hollaenderi*. Exp Cell Res 114:429–434
Suarez V, Newman J, Hiley C, Crocker J, Collins M (1988) The value of NOR numbers in neoplastic and nonneoplastic epithelium of the stomach. Histopathology (in press)
Taylor EF, Martin-DeLeon PA (1980) Comparison of N-banding and silver staining of human NORs. Hum Genet 54:217–219
Thiry M, Goessens G (1986) Ultrastructural study of the relationships between the various nucleolar components in Ehrlich tumour and Hep-2 cell nucleoli after acetylation. Exp Cell Res 164:232–242
Thiry M, Lepoint A, Goessens G (1985) Re-evaluation of the site of transcription in Ehrlich tumour cell nucleoli. Biol Cell 54:57–64
Threadgold LT (1976) The ultrastructure of the animal cell, 2nd edn. Pergamon, Oxford, pp 42–43
Vagner-Capodano AM, Stahl A (1982) Numerical relationship between nucleolar organizer regions and fibrillar centres in porcine thyroid cells cultivated in vitro. Biol cell 45:111
Van der Valk P, Mosch A, Kurver PJ, Meijer CJLM (1983) Morphometric characterisation of 52 B cell non-Hodgkin's lymphomas. J Clin Pathol 36:289–297
Wachtler F, Hopman AHN, Wiegant J, Schwarzacher HG (1986) On the position of nucleolus organizer regions (NORs) in interphase nuclei. Studies with a new, non-autoradiographic in situ hybridization method. Exp Cell Res 167:227–240
Wallace H, Birnsteil ML (1966) Ribosomal cistrons and the nucleolar organizer. Biochim Biophys Acta 114:296–310
Wang HC, Juurlink BHJ (1979) Nucleolar organizer regions (NORs) in Chinese hamster chromosomes as visualised by Coomassie brilliant blue. Chromosoma 75:327–332
Warburton D, Henderson AS (1979) Sequential silver staining and hybridisation in situ on nucleolus organizing regions in human cells. Cytogenet Cell Genet 24:168–175
Wassef M, Burglen J, Bernhard W (1979) A new method for visualisation of preribosomal granules in the nucleolus after acetylation. Biol Cell 34:153–158

# Quantitation of the Nucleus

T. J. STEPHENSON

# 1 Introduction

Known to be the location of the control data which dictate benign or malignant behaviour of the cell (LASKEY 1987), the nucleus has been at the centre of the quest for the pathognomic features of malignancy for over a century (HALL and FU 1985). In practical histopathology, interpretation of the features of nuclear size, shape, density and chromatin distribution is usually made with routinely stained sections examined by light microscopy. Diagnostic histo- or cytopathologists have to base their impressions on total accumulated experience. This expertise is based in turn on peer review, expert sources and clinicopathological correlation.

This complex pattern recognition occurs principally at the subconscious level and thus does not lend itself readily to verbal description. Terms commonly used to describe features typical of the nuclei in malignant rather than benign neoplasms include: possession of a high mitotic "rate", aberrant mitoses, pleomorphism, hyperchromaticism, nucleolar abnormalities and an atypical chromatin pattern (UNDERWOOD 1987). All of these features are *subjective* observations (BECK and ANDERSON 1987).

Despite this limitation, in many practical instances diagnoses based on such criteria are sufficiently accurate to dictate clinical decisions. The same learning skills which enabled the histopathologist to embark on diagnostic practice constantly allow adaptation to new information. For example, the rapid advances in lymphoma classification (STANSFELD 1985) made possible by techniques including monoclonal antibody immunohistochemistry (GOUDIE 1987) are absorbed into the pathologist's experience. Suddenly, individual lymphocytes become recognisable as "probably a T cell", where previously the morphological features of such a cell were unknown (BARANSKA et al. 1983). The pathologist has automatically learned the correlation between nuclear morphology and immunophenotype and has used this correlation for subsequent reference (CROCKER 1984).

For this reason it is unlikely that quantitative microscopy of nuclei, including even sophisticated computer-assisted methods, will render human interpretation of nuclear features redundant. Just as data from previously described techniques have refined the pathologist's diagnostic criteria, so the information from new molecular biological techniques is likely to be used to similar advantage.

Human diagnostic assessment of histological and cytological preparations has well-documented limitations arising out of subjectivity. These can be improved upon by quantitative microscopy. Serious intra- and interobserver differences have been documented in reporting premalignant cells from urine (OOMS et al. 1983a, b; SHERMAN et al. 1984), cervical smears (SEYBOLT and JOHNSON 1971; DERMAN 1981; BACUS et al. 1984), tissue samples (SIEGLER 1956; COCKER et al. 1968), and sputum samples (EVANS and SHELLEY 1982).

To understand why such variability should exist within and between "expert" diagnosticians, the strengths and weaknesses of the human visual system must be considered (HALL and FU 1985). Edges and boundaries are easily detected due to the arrangement of ganglion cells at the retinal level and the presence of occipital cortical neurons sensitive to frequency distribution. Higher analysis of these frequency distributions enables complex pattern recognition which is both rapid, due to parallel data processing, and accurate (GONZALES and THOMASON 1978). Limited by serial data processing, machines cannot presently match this speed and accuracy. Despite this, there are conspicuous deficiencies in human visual perception. Absolute spatial distribution is poorly judged; the visual system concentrates on relative features (MARR 1982). Optical density cannot be assessed accurately due to adaptation mechanisms to light intensity in the iris and retina. Higher-order statistical relationships in an image, such as those within the texture of nuclear chromatin, are not well recognised.

### 1.1 The Potential Advantages of Quantitative Techniques

The potential advantages of the various methods of quantitative microscopy of the nucleus will now be discussed against the background of the strengths and weaknesses of the human visual system. While the eye easily detects all-or-none phenomena, quantitative techniques are better for assessment of continuous parameters such as size and density. The potential advantages of *objective* nuclear measurement have long been recognised. HALL and FU (1985) record that, in 1851, Lebert measured cells and nuclei while describing the changes occurring in malignancy, and that Virchow measured cancer cells and described pleomorphism in their nuclei. Nuclear quantitation offers objectivity in measurement of these nuclear variables and the numerical data produced are readily communicated and are suitable for statistical analysis. For this reason it is an area of intense interest (BAAK et al. 1982; BAAK and OORT 1983; BAAK 1987a, b), with long-established stereological principles (UNDERWOOD 1970; AHERNE and DUNNILL 1982) being incorporated into novel technology for nuclear quantitation.

## 2 Instrumentation

The equipment needed in quantitative studies of the nucleus ranges from a simple ruler to sophisticated computerised image analysis systems; the appropriate choice depends on the question to be answered. A ruler held against a scaled photomicrograph can give valid measurements of diameter, but more sophisticated systems may offer advantages of automation and data handling speed, allowing large sample sizes to be measured.

### 2.1 Basic Planimetric Techniques

The best-established and most basic form of quantitative measurement in diagnostic pathology is point counting. One form of point counting is the manual counting of "all-or-none" discrete events in a section, for example, the counting of mitotic figures in a section of a uterine smooth muscle tumour to assess malignancy (CHRISTOPHERSON et al. 1972). Even with such an apparently simple technique, consideration has to be given to observer disagreement in what constitutes a mitotic figure, the effects of temperature and fixation on the completion of mitoses, section thickness (BROWN et al. 1986) and tissue shrinkage (ABERCROMBIE 1946).

The most familiar type of point counting is overlaying a grid pattern on a photomicrograph. This basic technique remains a standard against which modern digital image processing systems are tested. Although point counting techniques have most commonly been applied to relatively large structures within tissue sections, for example gland lumina and epithelial volume fractions in endometrial tissue (BAAK et al. 1981b, 1987, 1988a), they remain applicable to suitably magnified images of nuclei.

Well-recognised mathematical relationships have been developed from the work of the geologist Delesse in 1847 describing correlation between appearances in transection planes and underlying three-dimensional geometry. These deductions from a two-dimensional section to a three-dimensional tissue are termed stereology. The mathematical considerations underlying stereology have been well reviewed (WEIBEL 1963, 1979, 1980; AHERNE and DUNNILL 1982). Generally, the stereological assessment of volume fraction of constituents within a cell is easier than the estimation of surface areas and shapes.

In the agreed standard notation for stereological measurements (WEIBEL 1979), the parameter being studied is denoted by an upper case letter (V = volume) while the reference parameter is shown by an upper case suffix. For example, the volume of a cell occupied by a particular component such as the nucleus would be shown as $V_V$.

### *2.1.1 Volume Proportions*

The volume proportion of a tissue component, such as the nucleolus within the nucleus determined from an electron micrograph, can be determined from its volume in a section, applying the fundamental relationship noted by Delesse:

$$A_A = V_V.$$

There are three main ways in which this can be carried out: (a) an eyepiece graticule can be used during direct microscopy; (b) a grid can be superimposed on a photographic print, a projected section, or a projected transparency; or (c) a digital image analysis system can superimpose a set of points over a video image which may then be counted manually or automatically.

The work involved is made more easy if only one point falls on any one feature (UNDERWOOD 1970). It is therefore important to combine an appropriate tissue magnification with the point density of the graticule. The suitabilities of different formats of graticules in practical histomorphometry have been discussed (WEIBEL 1963). Although high density lattices, perhaps with 100 points, thrown over a relatively low magnification image offer the superficial attraction of more rapid area coverage, many observers prefer to work with fewer points, perhaps only 25, over a higher magnification image (BECK and ANDERSON 1987). Where several features are present with widely disparate $V_V$ values, double lattice graticules may be used for simultaneous counting of the features.

As with all stereological and morphometric techniques, extreme caution must be taken to avoid the introduction of bias through the selction of "favourite" fields for point counting.

#### 2.1.1.1 Avoiding Bias

Random section field selection, perhaps using random number tables or dice to choose the microscopic stage micrometer coordinates of individual fields, is a simple method for avoiding observer bias. Even with this method, there is the theoretical risk of multiple counting of individual features through the inclusion of overlapping fields. A good compromise between convenience and precision is termed "stratified random" sampling (WEIBEL 1979, 1980; WEIBEL and GOMEZ 1962). In this, a starting point is chosen using random numbers, then successive fields are selected using a regular lattice pattern.

#### 2.1.1.2 Anisotropic Structures

Only spherical or randomly orientated non-spherical structures are free from the problems of anisotropism. The nucleus may be elongated and sometimes has a clearly polar orientation, for example in various simple epithelia (RIGAUT et al. 1985). Problems of anisotropism may therefore occur. Point-counting techniques have the advantage over other planimetric methods that their validity is not impaired by anisotropism (WEIBEL 1979, 1980).

### *2.1.2 Surface Area Measurements*

Surface area measurement by graticule superimposition is more laborious than volume fraction measurement (computer-assisted image analysis may offer a less time-consuming alternative). The graticule used contains a series of lines, the lengths of which are calibrated using a stage micrometer or a scale bar in an electron micrograph. The observer counts the number of times these lines are intersected by the surface being investigated. The surface density (surface area per unit volume) is given by:

$$S_V = \frac{2I}{L}$$

where I is the number of intercepts and L is the represented length of the test lines multiplied by the number of fields measured. Provided that a curvilinear graticule line such as that described by AHERNE and DUNNILL (1982) is used, surface areas can still be assessed validly in anisotropic structures. The total surface area of the structure being studied can be obtained by multiplying $S_V$ by the total volume of the tissue.

Estimates of surface area obtained from studies carried out at different magnifications are not directly comparable due to the Mandelbrot effect (WEIBEL 1979). Progressively higher magnifications of the same structure reveal increasing numbers of surface indentations and details leading to apparently increased surface area. Studies of surface area may only therefore be compared directly when performed under identical conditions (BROWN et al. 1986). A potential field where the Mandelbrot effect may operate in nuclear analysis is in the examination of complex shaped nuclei such as those in T-cell lymphomas (BARANSKA et al. 1983; SIMON 1987), the apparent degree of nuclear convolution being dependent on magnification.

### 2.1.3 Number of Objects in an Organ

Counting the number of objects in an organ, for example the number of nuclei in a tissue or the number of nucleoli in a nucleus, is more complex than estimation of $V_V$ or $S_V$. Most biological particles are variable in size or shape. The extent of either of these variations may not be apparent at the start of the experiment. The description of shape is itself a complex and specialised area of morphometry (TSUBAKI and JIMBO 1979; GSCHWIND et al. 1986; STEVENS et al. 1987). The formal mathematical elimination of distributional error requires preliminary analysis of the structures under investigation (ADAMS 1968). Simplifying assumptions, such as those described by DE HOFF and RHINES (1961), can be used to allow estimates to be made from relatively simple measurements, but all have limitations. For example the method of DE HOFF and RHINES (1961) is only strictly valid where objects are relatively large, spherical and vary only in diameter. Different procedures apply for particles much larger than the section thickness, for those similar in size to the section thickness and for those substantially smaller than the section thickness.

Where objects are not necessarily spherical but are thought to have a regular shape the method of WEIBEL and GOMEZ (1962) may be used:

$$N_V = \frac{K N_A^{3/2}}{b\sqrt{V_V}}$$

In this formula K is a size distribution coefficient, b is a "shape factor" and $N_A^{3/2}$ is the cube of the square root of the number of particles per unit area in the section. The values of b for particles of commonly encountered biological shapes have been described by WEIBEL (1979). Detailed application of this and other formulae has been described by BROWN et al. (1986).

As can be seen, the formal methods for estimating the true number of objects per unit area are complicated. In practice, given repetitive tissue processing by identical techniques leading to identical shrinkage, counts of particles per unit area of sections carefully prepared at similar thickness may be adequate for reference purposes. In this way, for example, inflammatory cell nuclei per unit area have been counted to compare the immune response to different antigen preparations (BECK et al. 1986).

### 2.1.4 Mitotic Indices

Proliferation indices are estimates of the fraction of the cell population occupying any part of the cell cycle. The two most commonly measured are the mitotic index ($I_M$) and the flash labelling index ($I_S$), which is usually measured by high resolution autoradiography after tritiated thymidine (WRIGHT and ALISON 1984) or bromodeoxyuridine exposure. The advent of flow cytometry is rendering the measurement of such indices from tissue sections, with their inherent problems, less common.

$I_M$, the proportion of the cell population in mitosis at any time, can be estimated by simple observation. The methods for converting simple mitotic counts into the $I_M$ have been reviewed (WRIGHT and ALISON 1984). Several problems need to be considered in designing this type of study.

The target cell population must be carefully defined. Mitoses are not distributed evenly throughout tissues; in the intestine they are concentrated in the lower parts of the crypts (TUTTON 1973) and in tumours there may be profound variations in relationship to tumour periphery (HERMENS and BARENDSEN 1969; AHERNE et al. 1977) and to tumour blood vessels (TANNOCK 1968, 1970).

A second problem is that of the relationship between the mitotic count and the interphase nuclear population; there may be non-normal distribution. Methods for testing for normal distribution have been reviewed (WRIGHT and ALISON 1984).

Geometrical artefacts must be considered. For example, nuclei in the different phases of mitosis are unlikely to be the same size as those in interphase. ABERCROMBIE (1946) understated the problem when he wondered "whether the size of mitotic figures is not sometimes relevant in the very numerous investigations of mitotic rate". Clearly the effect of inclusion of nuclear fragments in tissue sections depends on the relationship between nuclear diameter and section thickness. Several methods have been published for correction for this effect (see Sect. 3.2), of which that devised by SIMNETT (1967) is readily applicable.

Although $I_M$ is generally a reliable indicator of proliferative rate, like other state parameters it is dependent not only on the rate of transit of cells into mitosis, but also upon how long they remain in that phase. There is evidence that circadian variations occur in the time cells spend in mitosis (EVENSEN 1965; MOLLER and KEIDING 1982). In some circumstances, stathmokinetic techniques such as the metaphase arrest method can be used to compensate for this effect (TANNOCK 1967).

Because $I_M$ is so time dependent, it can be influenced by what mitotic phases are included in the count. Prophase and telophase may be difficult to recognise in tissue sections. In tissues with high proliferative rates some investigators have included only metaphase and anaphase counts.

There is evidence that a delay in fixation may lead to a decreased proportion of mitoses, presumably by allowing existing mitoses to go to completion (BULLOUGH 1950; RAJEWSKY 1965; DENEKAMP and KALLMAN 1973; GRAEM 1979). A possible method for avoiding this artefact is to "explode" mitoses in ice-cold hypotonic saline, which has the useful effect, besides inhibiting mitoses, of facilitating recognition of mitotic figures.

Provided that time taken for fixation and methods of fixation remain the same, there is no reason why crude mitotic counts should not be used comparatively. They should be quoted for a given area of section at a constant thickness. In practice, mitotic counts taken in this way have proved clinically useful in predicting tumour behaviour, notably in smooth muscle tumours of the uterus (TAYLOR and NORRIS 1966).

## 2.2 Morphometry and Its Relationship to Stereology

Most simply defined, morphometry is the quantitative description of shape (BAAK 1987a, b). WEIBEL (1979, 1980) used a wider definition of morphometry as "the quantitative description of a structure". This definition could theoretically also include both static and flow DNA cytometry, but in practice the word morphometry indicates a specific area of quantitative pathology requiring relatively simple equipment for interactive analysis. *Morphometry* is restricted to the estimation of distance, area and volume. These are continuous rather than categorical measures. *Planimetry* is a subset of morphometry and refers to the analysis of images in two dimensions. *Stereology* refers to the determination of three-dimensional data from two-dimensional images.

## 2.3 Specialist Shape Descriptors

Nuclear pleomorphism is one of the hallmarks of neoplasia and, further, it is incorporated into current grading systems for tumours, for example, carcinoma of the breast (BLOOM and RICHARDSON 1957; SEARS et al. 1982; STENKVIST et al. 1982; FUSTER et al. 1983). In morphometric terms pleomorphism shows as irregularity and variability of nuclear shape. Given that nuclear pleomorphism correlates with the biological behaviour of some tumours (BLACK and SPEER 1957; NEALSON et al. 1981; PARL and DUPONT 1982), quantitation of nuclear shape might be expected to show a similar or even better correlation with tumour behaviour. It has been shown that a quantitative index of nuclear roundness separated a group of prostatic cancer patients with otherwise histologically indistinguishable tumours into long-term survivors and those who died (DIAMOND et al. 1982a, b).

The most commonly used quantitative indicator of nuclear shape is the form factor (F) defined as:

$$F = \frac{C^2}{4\pi A}$$

where C is the nuclear circumference and A is the nuclear profile area. Derivatives of this form factor are in use, such as the nuclear contour index (NCI), first described by SCHREK (1972) and reviewed by CROCKER (1984). The NCI is defined as:

$$NCI = \frac{\text{perimeter}}{\sqrt{\text{area}}}$$

A glance at the two formulae shows that for a perfect circle the form factor (F) is 1 while for the same shape the NCI 3.54 is $(= 2/\sqrt{\pi})$. The two shape descriptors are proportionate to one another and increase as the perimeter of the nucleus becomes less circular. Studies using different shape descriptors based on similar formulae can therefore be regarded as comparable.

Although in common use, form factors are relatively insensitive indicators of minor deviations from circularity (YOUNG et al. 1974; BOWIE and YOUNG 1977; TEICH et al. 1979). This insensitivity led GSCHWIND et al. (1986) to investigate other shape descriptors concerned with ellipticality, concavity and irregularity of contour (PANOZZO and HUERLIMAN 1983).

One approach towards measuring irregularity is assessment of ellipticality of nuclear contour. Thios can be achieved either by measuring the ratio of the short and long axes of the corresponding ("best fit") ellipse or by measuring the difference in area or roundness between actual contour and its corresponding ("best fit") ellipse.

Another type of shape descriptor which is insensitive to "smooth" contours, but very sensitive to sharp angles is the "bending energy" (YOUNG et al. 1974; BOWIE and YOUNG 1977). This can be simply described as, supposing the nuclear perimeter were made of stiff wire, the amount of energy which would be required to bend it back into a circle.

A further class of shape descriptors, reviewed by GSCHWIND et al. (1986), measure the concavity of contours and are hence a way to quantitate irregularity or "crinkliness".

A mathematical approach to the crinkliness of outline is through Fourier spectrum analysis of the frequencies derived from angular changes in direction of the nuclear outline. Comparing this and other shape descriptors using scatter diagrams from the same set of outlines analysed by different methods, GSCHWIND et al. (1986) found poor correlation between different types of shape descriptor. This poor correlation suggests that some of the novel shape descriptors may perform differently from others and thus merit future investigation in nuclear quantitation.

Addressing the quantitation of cleaved nuclei in non-Hodgkin's lymphoma, STEVENS et al. (1987) introduced two new shape descriptors previously employed in a non-medical science (TSUBAKI and JIMBO 1979). The first descriptor, "convexity-concavity" (C-C), is based on the ratio between mean feret diameter (fd) and the perimeter-equivalent diameter:

$$\text{C-C} = \overline{\text{fd}}/\text{Pd},$$

where Pd = perimeter/$\pi$. The mean feret diameter is the mean of a series of readings of diameter taken in, say 16 different directions across the nucleus as though using a pair of calipers (in practice achieved by digital image analysis). The maximum value for a nucleus without any concavity (invagination) of its outline would be 1.

The second descriptor, "slimness" (SL), describes nuclear profiles which are elongated or ellipsoidal. It is defined as the coefficient of variation (CV) of the feret diameter; in other words, it is the standard deviation of the feret diameter divided by the mean feret diameter.

$$\text{SL} = \text{CVfd} = \text{SDfd}/\overline{\text{fd}}$$

Since the SDfd is zero for a circular nuclear profile, the index has a value of zero for this shape.

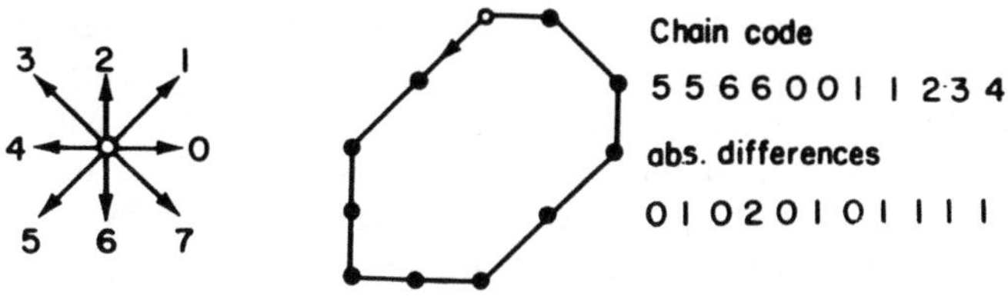


**Fig. 1.** The nuclear chain code is obtained by following the nuclear contour, each segment being listed according to the direction on the star diagram (*left*) with which it corresponds. The smoothness of nuclear contour is obtained by addition of the absolute differences in directional change along the contour. (WIED et al. 1981)

These two shape descriptors, novel to pathology, were shown to be useful in classifying ellipsoidal and cleaved nuclei in non-Hodgkin's lymphoma (STEVENS et al. 1987) and deserve investigation in other fields.

An entirely different approach to the measurement of smoothness of nuclear contours is the chain code method (Fig. 1). This method measures the absolute differences in directional change along the contour.

### *2.3.1 Morphometry of Subnuclear Structures*

While most morphometric studies of the nucleus have concentrated on nuclear overall size, shape and DNA ploidy, some have analysed subnuclear structures. These have included the nucleolus (COUVE 1985; VASQUEZ-NIN et al. 1986; MCLEAN and GAMEL 1988; SOKOL 1989), the chromatin pattern (RENAU-PIQUERAS and CERVERA 1983), including symmetry of its distribution (PANNO 1988), and electron microscopic studies of different nuclear components (BARANSKA et al. 1983; MARALDI et al. 1986 a, b).

## 2.4 Static Cytophotometry

As discussed earlier, due to its advanced mechanisms for adaptation to light levels, the human eye is a very poor judge of absolute light intensity. Cancer cells are characterised by hyperchromatic nuclei, meaning that the chromatin within the nucleus, as stained by routine stains such as haematoxylin, appears excessively dark. In stating this, the pathologist is implying that the nucleus contains more DNA than an ordinary diploid nucleus, in other words that there is DNA aneuploidy. In referring to nuclear ploidy as determined from static and from flow cytometry, the term "DNA ploidy" should be used in preference to the term "ploidy" (HIDDEMANN et al. 1984). This is because true ploidy can be determined only from standard cytogenetic preparations; cytophotometrically determined "DNA ploidy" is only accurate to the order of about ±3 whole chromosomes (LOVETT et al. 1984; QUIRKE and DYSON 1986).

Photometric (densitometric) equipment capable of measuring light intensity was developed from morphometric equipment. The first device, the cytophotometer, was developed by CASPERSSON (1936). The original device worked only with ultraviolet light, but was subsequently developed to work within the visible light spectrum (SWIFT and RASCH 1956). Most descriptions of cytophotometry deal with attempts to estimate the total nuclear DNA content, although there is no theoretical reason why any other compound capable of binding a dye should not be quantitated in this way.

For analysis of DNA content, routine stains such as haematoxylin are suboptimal since they bind non-stoichiometrically to DNA intensity of staining depends on the concentration of the dye used in the staining procedure. Dyes with a stoichiometric relationship to DNA should be used, where it is possible to be confident that the amount of dye bound is proportional to the amount of DNA in the nucleus. Many initial studies employed the fluorescent dye, acridine orange, but ease of operation, including the avoidance of the need to generate ultraviolet light, has led to more extensive use of the Feulgen technique. After controlled hydrolysis in hydrochloric acid, this results in a purple coloured absorptive dye stoichiometrically bound via the periodic acid-Schiff reaction to sugar moieties on DNA (NAORA 1955).

Since the absorption peak of the dye is at 546 nm, ideally the nuclei should be illuminated in monochromatic light at this wavelength so that they will have maximum optical density, giving the highest possible signal-to-noise ratio (ERHARDT et al. 1985). Some studies have used stringent production of monochromatic light by interference filters such as the Schott line filter (ADAMS 1968), while in others specific filtration has not been described (FLINT et al. 1988). In the absence of an interference filter a polychromatic green filter may yield adequate nuclear contrast.

Early static cytophotometric techniques used a fixed aperture to determine the area of sampling. Because the aperture could not precisely fit the nuclear contour, the nuclear areas are intentionally under- or oversampled.

Undersampling the nuclear area was called the "plug" method, since a plug of density information was sampled from the centre of the nucleus. This information was then multiplied by a calculated nuclear area derived from a series of nuclear diameters on the basis that total DNA is proportional to nuclear area multiplied by optical density. Irregularities in nuclear outline would degrade the accuracy of this technique.

The alternative approach of deliberately oversampling the nuclear area carried the risk of inclusion of debris outside the perimeter of the nucleus in the optical density measurement – a significant problem in solid tumours.

Irregularities in the distribution of stain within the nucleus introduce a second type of error. For example, some nuclei such as those of plasma cells and the "Orphan Annie" nuclei of papillary carcinoma of the thyroid (OYAMA 1989) have peripherally dispersed chromatin. Since optical density is a function of the logarithmic ratio of transmitted to incident photon flux, irregularities in distribution produce a positive bias into estimation of total density. This distributional problem can be overcome in two ways. Instruments such as the Deeley scanning microphotometer used a slotted mechanical disk to mask the image, within which several point readings were taken so that logarithmic transformation could be applied before summation (ATKIN 1970). Current methods employ electronic masking of the nuclear contour together with individual analysis of the grey-scale values within the enclosed area (MARCHEVSKY et al. 1987a).

Since even with current computer-assisted image analysis systems the individual nuclei to be assessed have to be selected by a trained observer, the procedure is relatively slow; most studies are restricted to about 100 nuclei per sample (HALL and FU 1985). This leads to a relatively high coefficient of variation in the results with consequent insensitivity to small changes in DNA ploidy (BAAK 1987a,b). For this reason, coupled with the increasing use of flow cytometry with its high sampling rate of 10000–100000 cells per minute (LOVETT et al. 1984) and applicability even to paraffin-embedded tissues (HEDLEY et al. 1983, 1985), the use of static cytophotometry is less widespread than that of flow cytometry.

Nevertheless, the DNA ploidy results from static cytophotometry have been validated by demonstration of high correlation with flow cytometric data (AUER and TRIBUKAIT 1980; CORNELISSE et al. 1984; KREICBERGS et al. 1981; FOSSA et al. 1983). Further, static cytophotometry has a key advantage over one insuperable feature of flow cytometry. The flow cytometer analyses cell suspensions from tissue homogenates. Therefore if a minority cell population is the subject for study the results from that cell population are likely to be swamped by data from more prominent cell populations. This is well exemplified by scirrhous breast carcinoma, where only about 20% of the tissue volume is tumour cells and the rest is stroma (UNDERWOOD 1972). Further, many tumours are surrounded by large numbers of inflammatory cells (UNDERWOOD 1974). With static cytophotometry, the cells for study are actively selected by the observer and unwanted cells are ignored.

Just as in flow cytometry, where cells can be classified by multiple parameters (for example, nuclear DNA ploidy coupled with a particular cytological staining reaction), this may also be possible with static cytophotometry. The development of real-colour computerised image analysis systems will allow simultaneous assessment of several cell variables.

## 2.5 Hardware for Video-Based Image Analysis

The increasing sophistication and reasonable pricing of microcomputers, together with advances in video technology, have meant that most researchers involved in quantitation of the nucleus have turned to video-based image analysis (WIED et al. 1981; BAAK et al. 1982; MARCHEVSKY et al. 1987a). The considerable advances in digital image processing, fostered by commercial incentives to meet the pattern recognition requirements of military surveillance, quality control and robotics, are available in analysis of video images.

Increasing use of video systems is a logical progression since they avoid some of the shortcomings of human vision. The most rudimentary video-based image analysis system includes: a computer with specialised hardware, a video camera and monitor, interactive peripherals and data storage systems.

### *2.5.1 Computers with Specialised Hardware*

A wide range of commercially available desk-top personal computers can form the centre of a successful video-based image analysis system. A popular choice for adaptation to image analysis is the IBM PC AT (MIZE 1985), but cheaper computers such as the BBC microcomputer can be employed in limited applications (LOWE and HARVEY 1986). Commercially available image analysis systems contain dedicated microprocessors for digital image analysis and for interface to digitiser tablets (for example the Kontron MOP Videoplan) or to mouse input (Seescan Solitaire Plus, Seescan Ltd). Computers suitable for digital image analysis must have at least options of hardware for analogue/digital converters (ADC), digital/analogue converters (DAC), frame buffers and ancillary peripheral interfaces.

#### 2.5.1.1 Analogue/Digital and Digital/Analogue Converters

Most video cameras produce an analogue output (typically a voltage) which is a format which cannot be accepted by a computer. It must be digitised (made numerical) before the computer can recognise the data (CASTLEMAN 1979), using an ADC. The converse of this is true when an image returns from the computer to the video system; the DAC converts the digital signal into analogue voltages to drive a video monitor.

Video signals carry information as a series of horizontal parallel lines scanning the cathode ray tube, each with a succession of different light intensities (grey levels). The ADC divides each line into a series of light points called pixels; it assigns a numerical value to the grey scale level for each pixel. In this way the video image is converted into a matrix of numbers suitable for presentation to the computer. Most ADCs will accept either 256 lines of 256 pixels each or twice this number. This is far inferior to human visual acuity which can resolve the equivalent of about 10000 lines of information (SADUN 1985). In the event of a video camera offering higher resolution than this, the extra information becomes redundant.

Analogue/digital converters resolve different numbers of grey scale intensities. Those based on a six-bit converter resolve 64 ($= 2^6$) levels while the increasingly common eight-bit converters resolve 256 ($= 2^8$) levels. Thus the full intensity scale in the latter instance ranges from black (0) to white (255). This number of grey levels far exceeds the capabilities of the human eye, which can resolve only about 64 grey levels in an image. The conversion time for most ADCs and DACs is 1/30th of a second or faster.

2.5.1.2 Frame Buffer

The amount of information in a monochrome video image of average resolution is large even by the standards of modern data storage systems. For example, an eight-bit grey level matrix of 512 × 512 pixels is digitised into 262 144 bytes. Theoretically, this information could be stored in a standard mass storage system such as a hard disk drive, but storage and retrieval times render this impracticable. The frame buffer (temporary video image store) is an additional dedicated memory connected to the output of the ADC which can hold at least one digitised image at a time. Typically, the frame buffer overwrites the previous video image each time the image is scanned (usually every 25th or 50th of a second in the UK). At any time, the user can capture the stored image for processing, preventing its being overwritten.

2.5.1.3 Ancillary Hardware Components

Specialised hardware capable of performing standard image-processing functions has been developed, installation of which lifts the burden of operation from what would otherwise be complex programs with long operation times (PRATT 1978).

Look-up tables (LUTs) are one of the most widely used aids to image analysis. They are electronic boards operating within the ADC which correlate one set of values with another arbitrarily selected set of values. For example, an LUT can map the range of pixel grey intensities from 0 to 255 with another set ranging from 255 to 0, resulting in a negative image. A range of other useful functions is possible. Contrast enhancement can be achieved by redefining which grey shades are mapped to black and to white when the image is displayed. Pseudocolour can be inserted by assigning colours to different ranges of grey shades to highlight features. The pseudocolour image is then stored in a colour LUT, LUTs can perform image averaging where several different images of the same object are averaged to reduce the effect of electronic noise on the signal. Image subtraction can be used to cancel out problems of uneven object illumination. For example, by subtracting the image of a slide with no specimen present from that of the section to be observed, uneven illumination is digitally compensated.

Image division is possible through LUTs and is essential in the measurement of optical density needed in microdensitometry. Since optical density is the logarithmic ratio of original light intensity over that seen when an object (such as a cell nucleus) is placed in the light path, it is necessary to divide images of samples by an image of the background light intensity without the sample present.

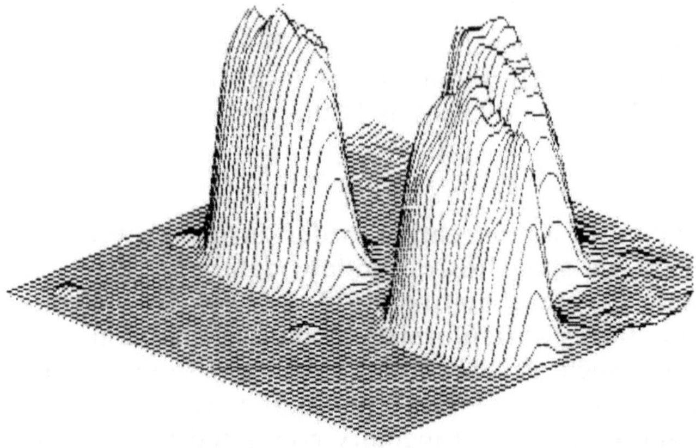

**Fig. 2.** Three-dimensional histogram of optical density (*vertical axis*) of three Feulgen-stained lymphocyte nuclei. The abrupt change in optical density at the edge of each nucleus, shown here as the steep sides of each "plateau", allow recognition of their boundaries by selecting an arbitrary optical density level above which objects are considered to be nuclei

These image manipulations within LUTs are carried out by an arithmetic logic unit (ALU) interfaced to the LUT, ALUs can also support useful functions such as deblurring, filtering and edge detection based on rates of change within the grey scale values or by thresholding (Fig. 2). A useful arithmetic function carried out in some systems is correction for convex or concave distortion – a defect of most optical systems – by shifting the image to positions which have been predetermined by testing the system with the image of a grid of squares.

Histogram extractors are mechanisms for calculating the frequency distribution of grey scale levels in an image stored in the frame buffer. This information is used not only to display histograms but also for edge detection and contrast enhancement.

#### 2.5.1.4 Video Cameras and Monitors

Many nuclear quantitative studies have been performed with photographic prints (BOON et al. 1980, 1984; DIAMOND et al. 1982a, b; TANNENBAUM et al. 1982; BONDESON et al. 1983; EPSTEIN et al. 1984; LESTY et al. 1986). This indirect approach is time-consuming, uses photographic materials and introduces two stages of magnification change and distortions into the measuring system. Real-time (live) video-based systems which allow direct visualisation of specimens are becoming increasingly popular (MARCHEVSKY et al. 1985, 1986, 1987a; MARCHEVSKY and GIL 1986). This trend is likely to continue as video technology advances.

Appropriate choice of video camera is essential for valid video-based image analysis. A checklist of some key features which must be considered in selecting a monochrome video camera is shown in Table 1. The two basic types of camera are electron tube cameras and charge-coupled-device

**Table 1.** Checklist of key features for selection of a monochrome video camera

| |
|---|
| Resolution adequate for experimental requirements |
| Satisfactory performance at experimental illumination intensity |
| Appropriate wavelenght sensitivity |
| Satisfactory signal to noise ratio |
| Ability to withstand high light intensity |
| Satisfactory operation at prevailing temperatures |
| Gamma value of unity for optical density measurement |
| No automatic light circuits for optical density measurement |
| Robust enough for application |
| Absence of aberrations such as blooming and lag |

(CCD) cameras. The electron tube type is the older established technology with presently better performance per unit price than CCD cameras in most applications. They have a higher spatial resolution and a better signal to noise ratio than CCD cameras. CCD cameras are physically smaller than electron tube types and have a more accurate geometrical resolution.

Videcon cameras are the commonest electron tube type. They have poor sensitivity to low light, their response to light is non-linear and they are normally fitted with automatic light control circuits which adjust the black level and gain of the camera. They cannot therefore make valid measurements of absolute grey scale values as are required in optical density measurement. The performance characteristic of most video monitors complements some of the short comings of Videcon cameras. This, coupled with their peak spectral response in green light, means that they give video images which appear most like human vision. They are therefore satisfactory for applications where absolute grey scale measurement is not required. Their sensitivity to damage by excessively bright light must be considered.

The Newvicon camera is a more expensive electron tube type than the Videcon. It offers advantages including: twenty times higher light sensitivity than the Videcon, relative resistance to light damage and a linear response to light intensity (unity gamma), having no automatic light gain control circuitry. It is particularly suited to fluorescent microscopy and to densitometric measurements. For densitometric analyses, a black stripe may be applied to one edge of the camera face plate so that a black signal is included in each line of the video image. This allows the writing of system software to calibrate the video signal for true grey scale values. Newvicons produce higher spatial resolution and a better signal to noise ratio than do Videcons. However, the video image from them appears dissimilar to the human visual image due to apparent lack of contrast and a peak spectral response in the red.

Charge-coupled-device cameras consist of an array of photosensitive cells which collect charge which is then dispersed as they are read by the video signal. They are an emergent technology which at present is limited by the relatively low numbers and varying light response of the individual cells. Because of the rectilinear configuration of the cell array, they have high geometrical accuracy. They have even less potential for light overload damage than the Newvicon.

Colour cameras are of two main types. Single tube cameras produce a video signal with two main elements: full resolution monochrome data and low resolution data concerning colour hue. For real-colour image analysis, this composite signal has to be decoded into individual red, green and blue signals. These signals retain the disadvantage of low spatial resolution for the colour information. Three-tube cameras, which have separate red, green and blue tubes, produce colour data of high spatial resolution which is more suitable for some image analysis applications.

Choice of video monitor is less critical than that of the camera since the monitor serves merely to allow the operator to see the image being processed. In selecting a monitor compatible with the camera and system it should be remembered that the image processing hardware frequently is limited to processing 512 or fewer lines. Any extra resolution inherent in the monitor will not be reflected in system performance. In testing the resolution of video monitors it is worth considering that resolution in the horizontal direction is customarily only about 80% of that in the vertical direction.

#### 2.5.1.5 Interactive Peripherals

Fully automatic image analysis systems are not at present widely applicable to pathology since the images being analysed and the data sought from them are too complex for current technology to interpret reliably. Some form of skilled intervention is usual. A range of peripheral equipment has been devised to allow operator control in computerised interactive morphometry (CIM) (MARCHEVSKY and GIL 1986). A typical compromise within CIM is that the skilled operator recognises suspicious cells of interest, free from overlap with other cells and which are likely to be amenable to quantitation by the image analyser. Such cells are pointed out to the CIM system using one of the interactive peripherals to be described, following which the image analysis system takes over the automated functions.

Light pens generate the coordinates of the spot to which they point on the screen. They are common in other areas of computer technology, particularly for selection from menus on the screen. Due to their low resolution and the need for complex software to support the devices, they are infrequently used in CIM.

Digitising tablets (Fig. 3) are devices containing various types of electromechanical tracking hardware over which the operator moves a stylus.

**Fig. 3.** The digitising tablet, to the right of the microscope, is a commonly used electromagnetic device for transferring stylus coordinates to the computer. It is seen here connected to the MOP-Videoplan, a manual video-based image analysis system (Kontron Electronics, St. Albans, England)

The video system generates a confirmatory image of the stylus position on the screen. The stylus may be used to trace either actual photographs placed on the digitising tablet or may be moved over an empty tablet to trace an image on the video monitor. Spatial correspondence has to be assumed and relied upon between coordinates on the tablet and those in the video image.

The mouse is the familiar and convenient roller-ball pointer used in microcomputers as a means of coordinate input (Fig. 4). Tracking is sufficiently accurate for tracing images with the cursor displayed on the screen, in addition to the familiar role in menu selection. One problem with the mouse is that the operator traces in one plane an image which appears in another. This feat is difficult for some operators.

Touch-sensitive screens are transparent overlays with built-in tracking systems similar to those found in digitising tablets; they are placed over the screens of video monitors (SILAGE and GIL 1984; GIL et al. 1986). Most react to any pressure source, but some require a specialised stylus. They offer the advantage of tracing very close to the actual video image and a white line may be made to appear on the screen to show the areas which have been touched. Disadvantages include a resolution usually lower than the pixel resolution of the video image, lack of sensitivity, the need for calibration and wear with heavy use.

### 2.5.1.6 Storage Systems

As discussed in Sect. 2.5.1.2, whole digitised images take up considerable amounts of computer memory. Many projects can be executed without storage of actual images, only the data which result from their analysis being stored. When it is considered necessary to store whole images (for

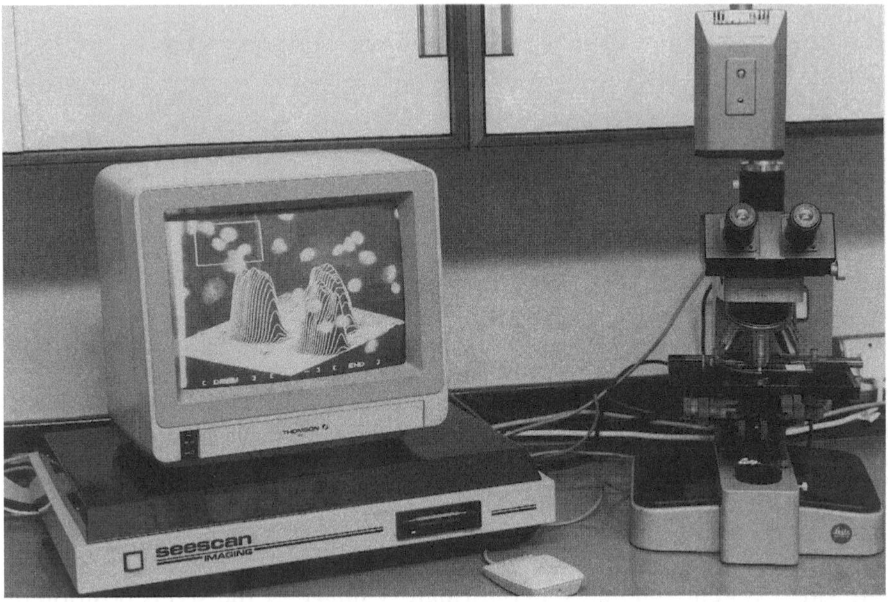


**Fig. 4.** The mouse (*foreground*) is an increasingly popular interactive peripheral for driving the cursor in interactive image analysis. It is seen here as part of the Seescan Solitaire Plus semiautomatic image analysis system (Seescan Ltd, Cambridge, England)

example when the same image is to be processed later by different techniques), the options of floppy disk, hard disk and tape storage are available. In the future, laser disk recorders, with their ability to store large numbers of high resolution images and rapid retrieval, are likely to become widespread.

#### 2.5.1.7 Servo-Driven Microscope Stages

Microscope stages can be manufactured with stepping servomotors driven by a chain of impulses from a computer. The computer can thus move the slide in the X and Y directions for either systematic scanning of the whole slide or for relocating a feature of interest. The coordinates intended and those actually reached by the servo system may not always correspond accurately and most current systems can scan only relatively small areas. The chief use of servo-driven microscope stages so far has been in the systematic zig-zag scanning of cytological preparations (GREENBERG et al. 1986; GARCIA et al. 1987).

### 2.6 Specimen Illumination: Problems and Solutions

With its poor perception of absolute light intensity (SADUN 1985; YATES 1978), the human visual system is often oblivious to unevenness of illumination over either the microscope field or a photograph lying on a copy stand. Video-based systems, with their sensitivity to absolute light intensity, immediately show up uneven illumination. Such unevenness not only invalidates data based on absolute grey scale values, but also hampers image segmentation based on grey scale levels. Before image analysis can commence, the illumination of the system should be made even, perhaps using the video image of the background with the contrast enhanced via one of the LUT functions to make more obvious any unevenness of illumination. With most microscopes, even illumination can be achieved by adjusting the substage condenser and by moving the lamp filament in X, Y and Z directions.

If even illumination still cannot be achieved, the problem can often be reduced by storing an image of the background (an area of the slide without any specimen present) and subtracting this digitally from the specimen image. This also has the advantage of reducing the effect of any dust particles in the light path. An alternative is to subject the image to high-pass filtration, which enhances edge detection while reducing the effect of gradual changes in intensity.

With analysis of absolute grey scale values, including optical density measurements, there is usually a need to divide the background control image by the specimen image. Naturally both images have to be captured under identical conditions of illumination, but this can be problematic where illumination is provided by a filament light source. Since the power dissipated by a resistive filament is proportional to the square of the voltage applied and the light intensity varies exponentially with the voltage over some parts of the voltage range, a constant voltage source is needed. In measuring optical densities, one way round the problem is to measure the background light intensity from the image containing the sample. For example ALLISON et al. (1988) measured background light intensity of a three pixel wide annulus around each nucleus to be quantitated.

In the longer-established method for quantitation of nuclear DNA, nuclei are scanned by a small spot of monochromatic light. The intensity of each spot is measured at each point as it scans across the microscope field. For valid results, the spot size must be chosen carefully (MAYALL and MENDELSOHN 1970; GOLDSTEIN 1970, 1971 a, b; DUIJNDAM 1980 a, b). The incident light must be at the wavelength of peak absorption of the stain and spectrally pure (ALLISON et al. 1981, 1984 a; ALLISON 1985; RASCH 1985). Glare level within the equipment, which reduces the apparent density of the nuclei, must be minimised (GOLDSTEIN 1970). Provided that these factors are considered, the technique can measure DNA content with a precision adequate for most biological purposes (BEDI and GOLDSTEIN 1976; SKLAREW 1982; ALLISON et al. 1984 a) and which betters that available with video-based image analysis (ALLISON et al. 1984 b, 1988).

## 2.7 Common Software Routines in Image Analysis

A number of software routines are in widespread use both in commercial packages and in custom-built systems.

Point-counting routines are an extension of traditional point counting, which used a grid superimposed over a microscope image. Instead of this, a video grid display is superimposed onto a live video image allowing estimates of volume, surface area and numerical particle densities to be made. The user still has to communicate to the computer information on which points overlie which structure. This is time-consuming but may be facilitated by interactive peripherals such as a touch-sensitive screen (SILAGE and GIL 1984).

Tracing perimeters and boundaries is one of the commonest applications of video-based nuclear quantitation (MARCHEVSKY et al. 1986, 1987b, c). Software to support the input of these boundaries from digitising tablets, mice and touch-sensitive screens is available (CASTLEMAN 1979). Attention must be paid to the accuracy and level of control over the tracing line to obtain satisfactory results (GIL et al. 1986). Most software supports computation of boundary lengths, enclosed areas, maximum and minimum diameters, feret diameters and various form factors from the tracings. The tracing techniques must be repeated exactly between different samples and should be tested by the user in various quality control procedures (SLOOTEN et al. 1985; GIL et al. 1986). In particular, observer variability must be monitored closely (BARRY and SHARKEY 1985; DARDICK and CALDWELL 1985; CHAN et al. 1987).

On-screen graphics may be superimposed over live video images to allow estimation of nuclear size. For example, by touching the centre of a nucleus on a digitiser pad, a series of graded circles may be caused to appear (MARCHEVSKY et al. 1985, 1986). The operator selects the best fitting circle, the size of which is known to the computer. Various other simple shapes can be generated in this way (MARCHEVSKY et al. 1987a). The generation of series of enlarging circles is a common routine in automated image analysis systems for recognition of cells and their nuclei (Fig. 5).

Linear distances may be measured by the graphic generation of a line between two chosen points, a common use of which is the calculation of selected nuclear diameters (MARCHEVSKY et al. 1987a).

Most CIM systems allow area measurement from segmentation of the image by grey scale levels from an LUT (MIZE 1985). The user observes the video image (Figs. 6, 7) and selects the various threshold grey scale values at which pseudocolours are added to the image (Fig. 8). Following this the image is edited to remove unwanted structures such as debris and overlapping nuclei. The computer then draws in the boundaries of the segmented objects (Fig. 9) for measurement. In some software packages, routines to assist separation of touching objects are included, such as eroding all objects by one pixel and then dilating them by one pixel without

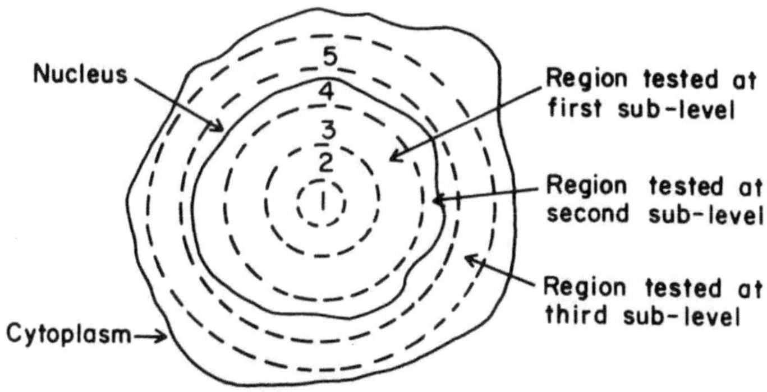


CELLULAR FEATURES ASSESSED
USING ANNULAR SCAN
Nuclear Size and Density
Integrated Nuclear Density
Average Cytoplasmic Density
Clarity of the Nuclear Border
Clarity of the Cytoplasmic Border
Average Background Density

**Fig. 5.** The annular scan method used to detect a cell and its nucleus in the CYBEST, CERVISCAN and TICAS-CDM systems. (WIED et al. 1981)

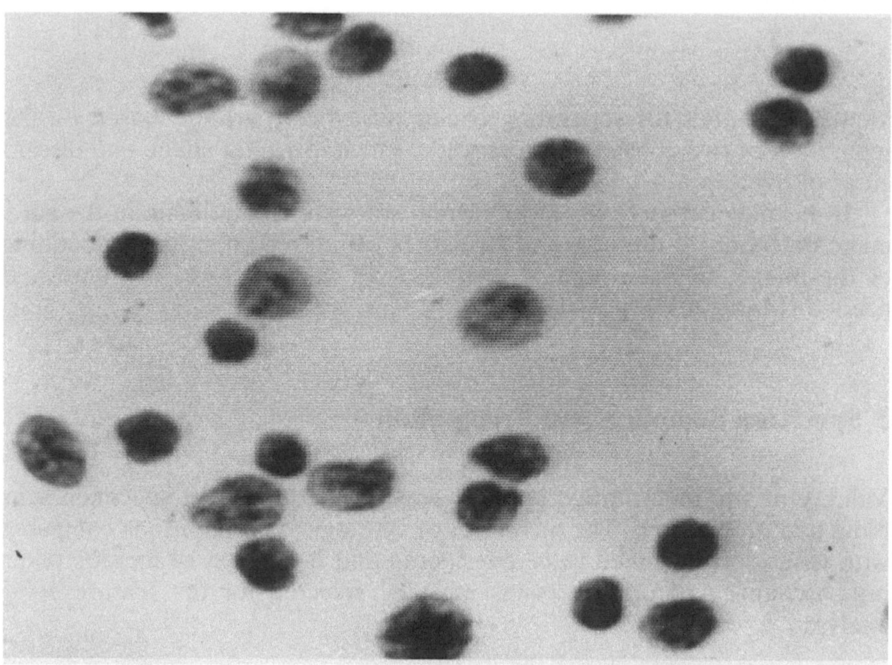

**Fig. 6.** Video image, at 256 × 256 pixels resolution, of a Feulgen-stained fine needle aspirate of a breast tumour

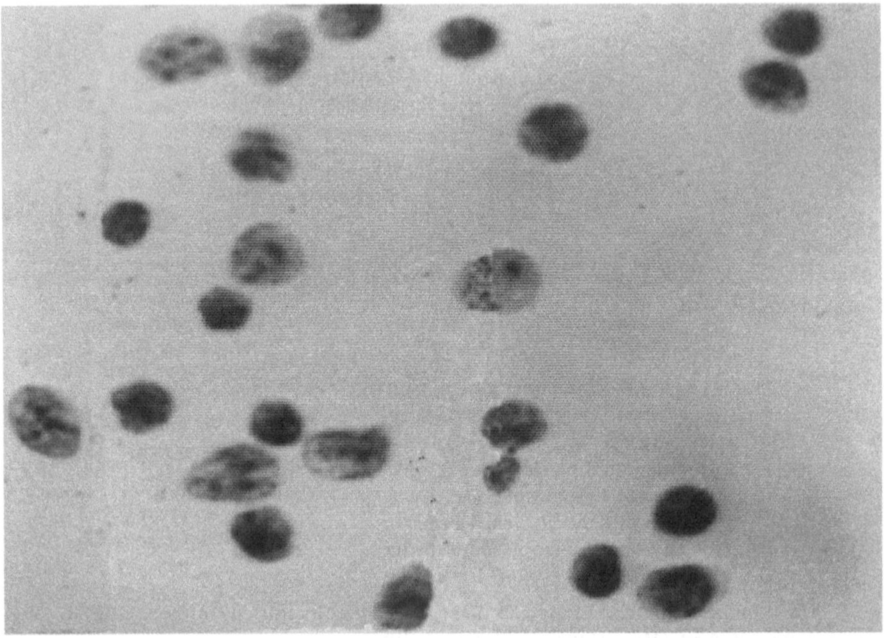

**Fig. 7.** The same video image with its grey scale histogram superimposed. The cell nuclei lie within a comparatively narrow band of grey scale values, making them suitable objects for recognition by grey scale level thresholding

joining. Routines for separating circles have been written, based on the separation of two convex surfaces; these inevitably alter shape and discard areas of overlap.

In histological sections, most current software is unreliable in the automatic detection of the edges of structures of interest; considerable editing of the image, or even manual tracing of all the structures of interest, is needed (MARCHEVSKY et al. 1987 a).

## 3 Specimen Sampling and Preparation

Validity in any quantitative study is dependent on correct specimen sampling and preparation. The attributes of cytological preparations compared with tissue sections need to be considered and the effects of fixation taken into account. Appropriate stains must be selected for the feature being analysed.

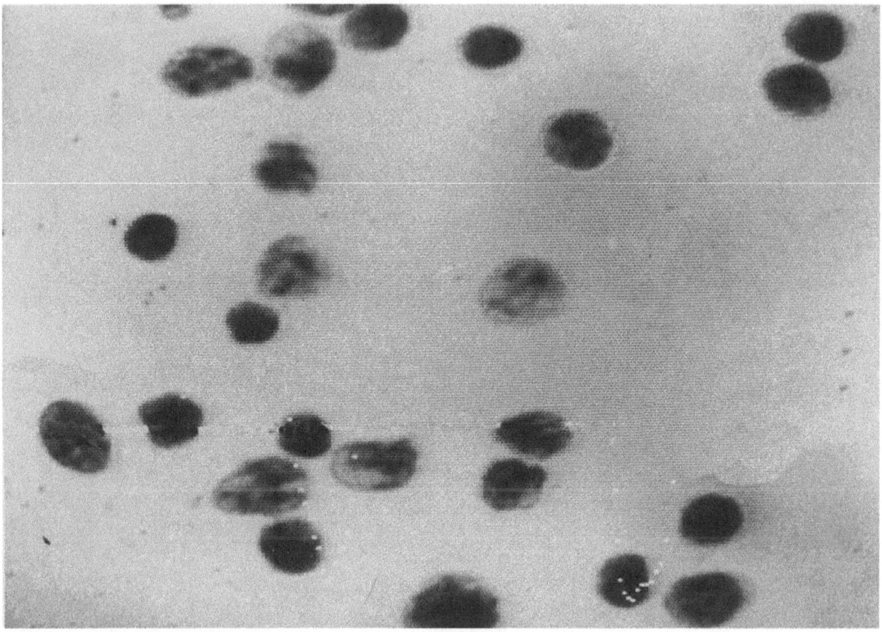

**Fig. 8.** The same video image with a cursor (*horizontal straight white line*) superimposed. This can be moved up and down the image while the grey scale plot of the pixels traversed by it (*curved white line*) is superimposed. This allows selection of threshold values for grey scale segmentation of the image, which is then marked in pseudocolour

## 3.1 Cytological Preparations

Tumour cells may be sampled from mucosal surfaces (ANONYMOUS 1989) or from body fluids such as pleural and peritoneal effusions (MARCHEVSKY et al. 1987b), or the cerebrospinal fluid where they remain viable after exfoliation. In other fluids such as the urine, exfoliated cells rapidly degenerate. Since this occurs out of the experimenter's control, it is likely to have unpredictable effects on quantitative studies.

Fine needle aspiration (FNA) biopsy is an increasingly popular source of material for diagnostic cytology. Smears made from FNA specimens display both single cells and small solid tissue fragments. A method for obtaining and transporting FNA specimens has been described (HARRIS et al. 1987), involving flushing out the FNA syringe into cytological transport medium which is then made into cytospin preparations. In addition to providing excellent nuclear preservation, this method gives good cell dispersion and freedom from background debris, yielding ideal conditions for nuclear quantitative studies.

Mirroring their popularity in diagnostic cytology, FNA specimens have been used chiefly for nuclear quantitation in the breast (SPRENGER et al. 1979; ZAJDELA et al. 1979; AUER et al. 1980; BOON et al. 1982b;

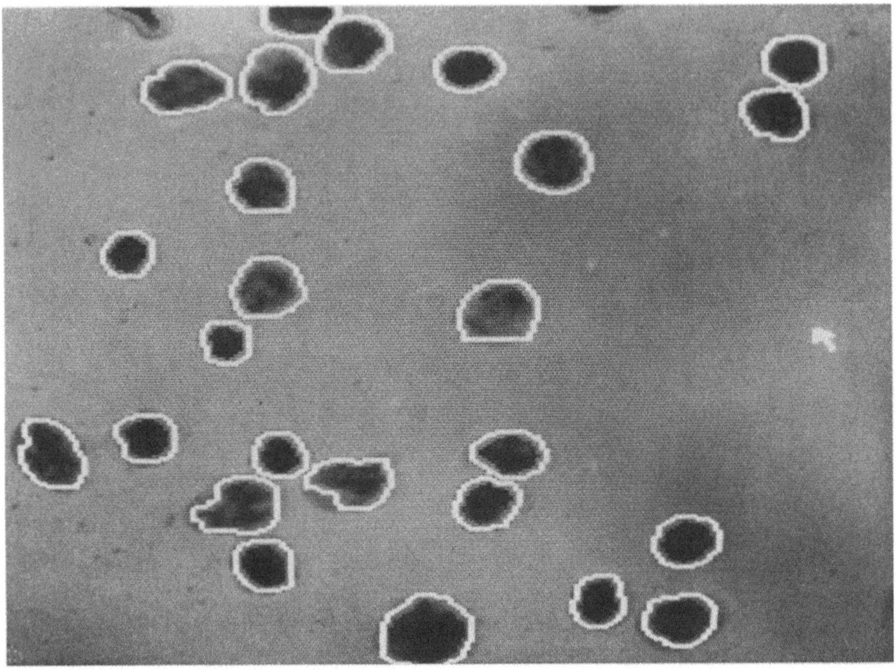

**Fig. 9.** The same image, after segmentation based on grey scale values. The selected areas are dilated by one pixel, following which the optical density enclosed within the frames so created is integrated with the area

STENKVIST et al. 1982), thyroid (SPRENGER et al. 1977; BOON et al. 1980, 1982a; LUCK et al. 1982; BONDESON et al. 1983) and prostate (SPANDER et al. 1982; TRIBUKAIT et al. 1983; BOCKING et al. 1984a, b).

Cytological preparations have the advantage, compared with tissue sections, of including the whole nucleus. For this reason they are particularly suitable in quantitative DNA studies (PEET and SAHOTA 1984). Their disadvantages include: (a) loss of all architectural features, leaving the cells with no histological context, rendering analysis of nuclear polarity impossible; (b) potential difficulty in identification of the cell type; (c) problems of overlapping nuclei in clumps, which are liable to harbour the most atypical cells; and (d) a different nuclear shape compared with that seen in tissue sections.

The technique for preparing single cell suspensions from paraffin-embedded tissue suitable for flow cytometry (HEDLEY et al. 1983) may also provide preparations suitable for cytophotometry, given that age of paraffin-embedded tissues does not affect subsequent Feulgen staining (KREICBERGS and ZETTERBERG 1980).

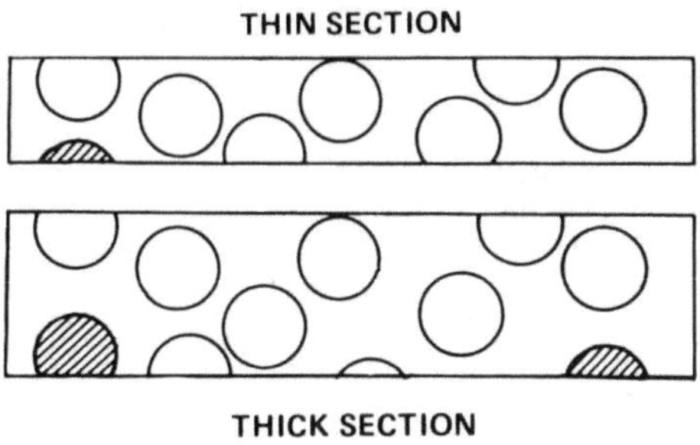


**Fig. 10.** Section thickness affects the content of nuclei and their fragments in the section. Although the number of intact nuclei and their hidden fragments (*shaded*) increases with section thickness, the number of visible fragments remains approximately the same. The difference in count gives the number of whole particles per section thickness

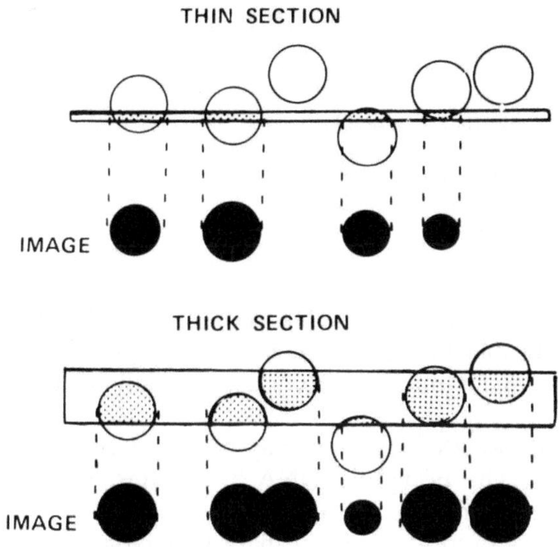


**Fig. 11.** Area proportions are increased if sections are relatively thick. Greater section thickness increases both the number of nuclear profiles and their mean area. (BECK and ANDERSON 1987)

### 3.2 Tissue Sections

In contrast to the intact nuclei typical of cytological preparations, nuclei in tissue sections are frequently sectioned by the microtome (FU and HALL 1985). The proportion of incomplete nuclei in a section increase as the nuclear size approaches or exceeds the section thickness (Figs. 10, 11). In quantitation of nuclear size or cytophotometric DNA analysis, an artefactual skew on the low side is to be expected, resulting from the inclusion of nuclear fragments. In practice, this has been shown to become a greater problem the thinner the tissue section (WRIGHT and ALISON 1984).

Despite skew introduced by the inclusion of nuclear fragments, clinically relevant quantitative studies may still be performed on tissue sections. Grossly aneuploid tumours are still detected because a proportion of the cells are included with intact nuclei. Thus when chondrosarcomas were sectioned at 4, 8 and 12 μm (KREICBERGS and ZETTERBERG 1980),

**Table 2.** Size distribution in a sample of 400 nuclei measured in 5-μm sections

| 1<br>Nuclear diameter (μm) | 2<br>Number of nuclei | 3<br>Cumulative count | 4<br>Calculated distribution |
|---|---|---|---|
| 10 | 15 | 14 | 133 |
| 9 | 37 | 52 | 143 |
| 8 | 89 | 141 | 154 |
| 7 | 65 | 206 | 167 |
| 6 | 68 | 274 | 182 |
| 5 | 65 | 339 | 200 |
| 4 | 23 | 362 | 222 |
| 3 | 27 | 389 | 250 |
| 2 | 11 | 400 | 286 |

Column 2: the observed size distribution.
Column 3: the number of nuclei of or above the corresponding size shown in column 1.
Column 4: values for p calculated from Abercrombie's formula, where A = 40, M = 5 μm and L is the corresponding value shown in column 1.

the proportion of tumour cells with estimated DNA content over the 90th centile of that present in the fibroblast controls was independent of section thickness, although estimated mean nuclear DNA content fell with decreasing section thickness.

More subtle population differences, however, may be obscured by the inclusion of nuclear fragments. The polyploid DNA histograms of some benign cell populations may obscure aneuploid cell populations. Where more than one cell population is present or when there is polyploidy, the stem cell modal DNA content can be measured accurately only by determining the intact nuclear DNA content. An area of study where polyploidy may mask aneuploidy is the interaction between human papillomavirus (HPV) (which induces polyploidy) and dysplasia (typified by aneuploidy) in the cervix (FU et al. 1981, 1983).

Measuring true nuclear diameter is also made difficult by inclusion of nuclear fragments in tissue sections. SIMNETT (1967), discussing the measurement of mitotic incidence, described a method of calculating true nuclear diameter, which has been recommended (WRIGHT and ALISON 1984) over other more cumbersome methods. This method is set out briefly below.

Tissue sections contain nuclear fragments; therefore the true nuclear diameter cannot be obtained from the mean of the diameters of the nuclei (and their fragments) seen in sections. If the mean diameter of nuclei (even if fragmented) which have their midpoints in the section could be taken, the true diameter would be found. ABERCROMBIE (1946) gave the proportion of nuclei with their midpoint in the section as:

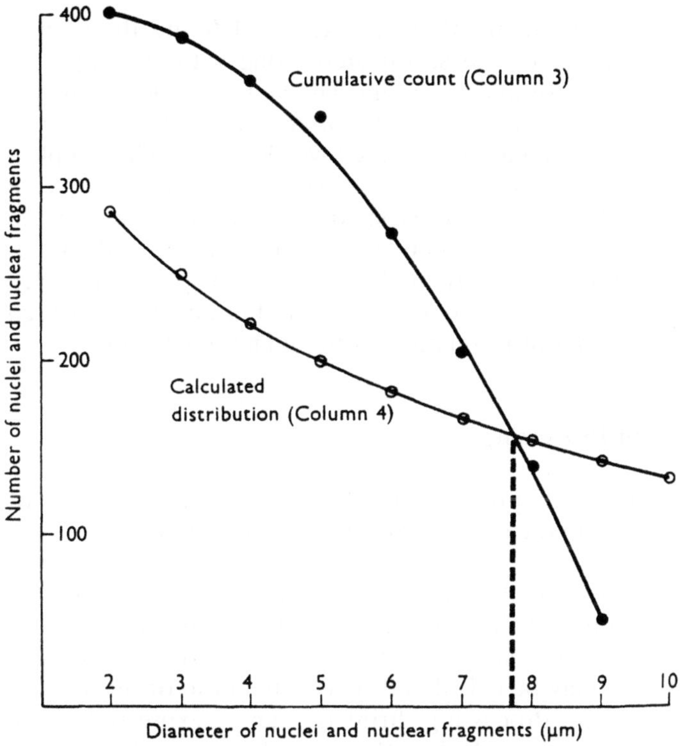


**Fig. 12.** Graphic solution to the data from Table 2. The curves for columns 3 and 4 intersect at 7.75 µm on the x-axis. This value is taken as the true nuclear diameter. (SIMNETT 1968)

$$p = \frac{A \cdot M}{L + M}$$

where p is the number of nuclei with their midpoints in the section, A is the number of both complete and fragmented nuclei in the section, M is the section thickness and L is the nuclear diameter.

If the nuclear diameter were known together with the section thickness, it would then be possible to predict the proportion of nuclei with their midpoints in the section. In practice it is the nuclear diameter which is being sought and if ABERCROMBIE's relationship (1946) is used in a comparison of the expected proportion and observed numbers of nuclear fragments of different sizes, the diameter can be calculated.

Table 2 shows the size distribution, measured by SIMNETT (1967), of 400 fragmented alveolar cell nuclei in 5-µm sections from a previous experiment (SIMNETT and HEPPLESTON 1966). The cumulative count (column 3) is the number of fragments of, or above, each given size. We need to know the level in this column at which nuclei are sectioned in such a way as to project their full diameter (the level at which nuclei have their midpoints in the section). To generate the expected values for p (column 4) the number of nuclei (whole or fragmented) is taken as 400 and the section thickness as 5 μm; these values are substituted along with the apparent nuclear diameter into Abercrombie's formula. The value of nuclear diameter from column 1 at which columns 3 and 4 *approximate* most closely is nearest to the true nuclear diameter (in other words 8 μm). The simplest way to obtain the nuclear diameter at which columns 3 and 4 would be *equal* is a graphical solution (Fig. 12). From this, it can be seen that the curves intersect at 7.75 μm – this is taken as the true diameter of the interphase nucleus. This method is strictly applicable only to spherical nuclei, but nuclei which are not spherical can also be treated in this way provided that their orientation is random in relation to the plane of section.

### 3.3 Tissue Fixation and Processing

Fixation and processing both produce tissue shrinkage, the amount of which depends on the techniques used. Accurate work requires these volume changes to be known. Further, there may be differential shrinkage of tissue elements: for example, collagen fibres often show greater shortening than other cellular elements. If the volume of an organ is recorded before and after fixation (possibly by displacement of water in a tank), the fixation shrinkage factor ($f^3$) may be calculated. This is the ratio of the volume after fixation to that before. Processing shrinkage can be expressed as the processing shrinkage factor ($p^2$), which is the area of the traced outline of a block after processing to that before. The fixation and processing factors are multiplicative, so an overall linear shrinkage factor (F) can be used to correct linear measurements. Similarly, $F^2$ is used to correct area and $F^3$ to correct volume measurements.

Many nuclear quantitative studies are relative, comparing a study group of nuclei with a control group; in these circumstances tissue shrinkage need not be taken into account provided that the methods of fixation and processing remain constant during the study. Absolute values will not be directly comparable between studies using different techniques.

### 3.4 Staining

Standardisation of staining techniques is crucial to the reproducibility of quantitative studies (WITTEKIND 1985). Most planimetric CIM work has used traditional staining methods such as haematoxylin and eosin for sections, Papanicolaou stain for cervical smears and air-dried Giemsa or alcohol-fixed Papanicolaou techniques for FNA specimens. Cytophotometric studies have generally used stoichiometric stains such as the Feulgen technique. Since density of haematoxylin staining shows correlation with Feulgen staining (WIED et al. 1981), with suitable controls and calibration it could possibly be used in some cytophotometric studies.

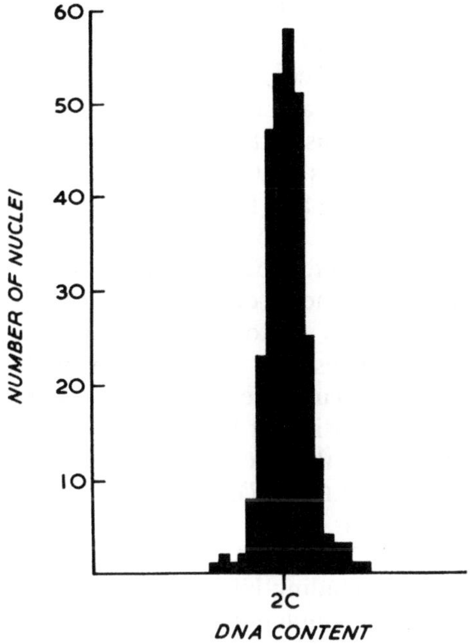


**Fig. 13.** Frequency distribution of 420 lymphocyte nuclei from around the tumour. The mean Feulgen microdensitometrically determined DNA ploidy is taken as the diploid control value. (DIXON and STEAD 1977)

## 4 Analysis of Morphometric and Cytophotometric Data

The software provided in many commercial image analysis systems provides for statistical comparisons of normally distributed sets of data allowing comparisons of cell populations without retrieval of the raw data from the equipment. Failing this, the mean, range and standard deviation are often provided and these should be selected for output since they are the basis for many statistical comparisons. Before selecting an image analysis system which cannot perform statistics internally it is worth investigating the interface capability with other computers to determine whether data can be transferred directly into the files of another statistical package, avoiding the need for tedious re-entering of data.

### 4.1 Analysis of the DNA Histogram

Much of the reported nuclear quantitation literature is concerned with analysis of the DNA histogram. DNA histograms do not have any unique features or problems in analysis which are not present in other histograms. The problems discussed are applicable to other quantitative data on the nucleus.

In static cytophotometry, control cells such as fibroblasts and lymphocytes are commonly used to determine a mean diploid DNA content for calibration (Fig. 13). However, variations in instrument sensitivity can affect whether diploid and aneuploid cells are distinguished and thus change the classification of a tumour ploidy pattern. DNA measurements by static cytophotometry usually produce coefficients of variation (the standard deviation of the data divided by its mean) ranging from 2% to 3%. They cannot detect aneuploid cell populations with abnormal chromosome numbers within $46 \pm 2$. A further problem is that some calibrant cell populations such as fibroblasts may be replicating and have some apparent tetraploid cells in the $G_2$ phase of the cell cycle.

The definition of aneuploidy can be difficult. True aneuploidy is detected from formal chromosomal karyotyping and is defined as the presence of abnormal chromosome numbers. It is customarily taken as a marker of malignant potential. Benign processes may mimic DNA aneuploidy as determined by cytophotometry. Facultative polyploidy is an example of a phenomenon likely to cause confusion; it has been observed in normal hyperplastic states of tissues such as the liver (BOHM and SANDRITTER 1975), thyroid (GILBERT and PFITZER 1977), pancreas (POHL et al. 1981) and endometrium (WAGNER et al. 1968) and is normal in megakaryocytes (PENNINGTON and OLSEN 1970; TROWBRIDGE et al. 1983, 1984). It may also be produced by viruses, including HPV, superimposing polyploid cell populations on the aneuploid populations associated with cervical intraepithelial neoplasia (FU et al. 1981, 1983).

Tetraploidy is encountered when cell proliferation occurs, for example following injury or in the stroma around tumours. This leads to a tetraploid peak representing cells in the $G_2$ phase of the cell cycle. Additionally there may be some cells with DNA ploidy between the diploid and tetraploid peaks (DORMER 1987) which are replicating their DNA (in the S phase of the cycle). Where there are many cells in S phase, the detection of minor aneuploid populations may be rendered impossible.

Many authors have classified DNA histograms by simple visual inspection (Fig. 14). Clinically useful classification of tumour samples has been achieved without more complex techniques (AUER et al. 1980; HALL and FU 1985).

The commonest parameter used to describe a DNA histogram is the stem cell modal value for $G_0/G_1$ cells. ATKIN et al. (1959) defined this as the "average value ... of cells that fall within about 15% of the mode". Useful clinical data can be produced from this simple value (ATKIN 1972). Where there is increased stem cell turnover, formulae for modal correlation based on assumptions regarding the distribution of S phase cells between the diploid and tetraploid peaks are available (OKAGAKI and IZUO 1978).

A comparative review of the various computerised modelling methods available for analysing DNA histogram data has been published (BAISCH 1982). The models which performed best were those using a large mainframe computer. These usually contained a large number of variables for which solutions had to be found. These included the mean of the position of the $G_1$ and $G_2+M$ peaks, their standard deviation, the age distribution

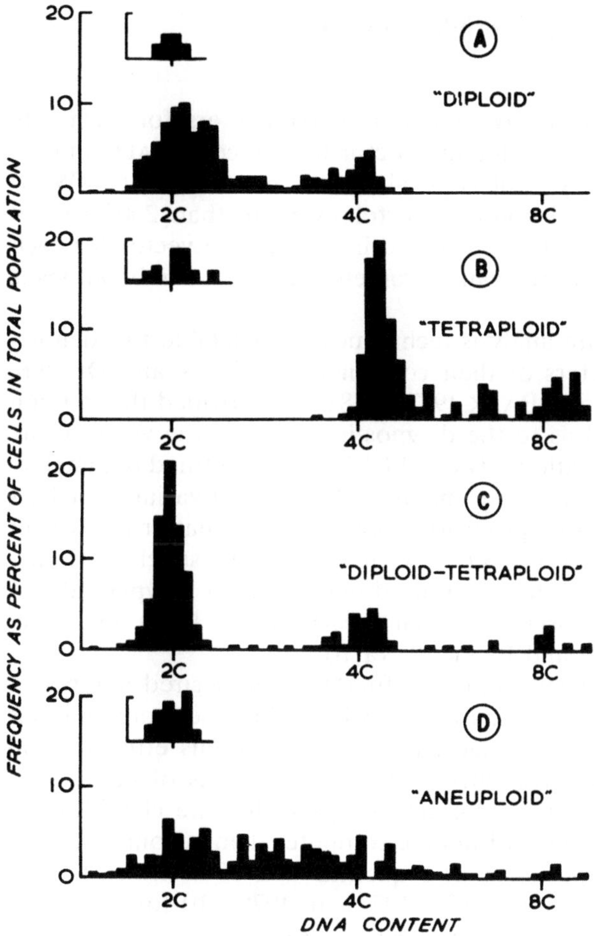


**Fig. 14.** DNA frequency distributions from four cervical tumour biopsies (*A*–*D*). The insets in *A*, *B* and *D* show the distributions for the control population of lymphocyte nuclei. (DIXON and STEAD 1977)

of the cell population (STEEL 1968), the rate of DNA synthesis and the relative durations of the $G_1$, S and $G_2 + M$ phase times together with their standard deviations. WATSON et al. (1987) have suggested an alternative method for modelling which can be performed on a microcomputer with only 28 K of addressable memory, which was reported to be satisfactory in nearly all cases.

An additional problem encountered with some DNA histograms is poor approximation to a normal distribution. In this case analysis is more difficult due to unsuitability of parametric statistics.

### 4.2 Classification of Nuclei by Multiple Parameters and Multivariate Analysis

Modern CIM can measure a large number of parameters for each cell; these are sometimes termed the feature vector for the cell (HALL and FU 1985). For example in analysis of cervical smears more than 100 (WIED et al. 1981) and in bronchial epithelial cytology more than 200 (SWANK et al. 1983) parameters have been included in the feature vector for each cell. Extracting the clinically relevant parameters or their combinations is a complex task.

A variety of multivariate analysis techniques are available for identifying the important parameters or their combinations (BAAK and DIEGENBACH 1977; DIEGENBACH and BAAK 1977, 1978 a, b). Provided that a rigorous set of tests is satisfied as to the diagnostic or prognostic value of the parameters or their combinations (BAAK 1987 b), the identified parameters can be used in clinical decision making. Most multivariate analysis techniques compute composite predictors based on combinations of weight ed features. Feature weightings are selected to minimise classification errors. This type of approach has been used in calculating "atypia indices" for cervical (WIED et al. 1981), bronchial (SWANK et al. 1983) and endometrial cells (DIEGENBACH and BAAK 1978 a, b).

Selection of features becomes more difficult as the desired number of classification groups increases. This is true when cytological preparations contain several different cell lineages, for example: serous effusions and cervical or bronchial cytological specimens. Different sets of parameters become relevant for separating different groups within the classification. The procedure for classification within the computer then becomes sequential rather than simultaneous. Use of sequential sets of decisions is illustrated in the work of KOSS et al. (1975, 1977 a, b, 1978 a, b) for classifying cells in urine. Their classification provides for degenerate cells, benign, atypical and malignant transitional epithelial cells, HPV-infected cells and other cell populations such as renal tubular cells. In classifying cells in cervical exfoliative cytology, TAYLOR et al. (1978) used a series of eight binary decisions.

### 4.3 Specimen Classification

When any binary decision process classifies biological samples, two types of error occur. Classification of a benign sample as malignant is called a false positive and classification of a malignant sample as benign is called a false negative. Sensitivity of the test is given as:

$$\text{Sensitivity} = \frac{\text{true positives}}{\text{true positives} + \text{false negatives}},$$

while specificity is:

$$\text{Sensitivity} = \frac{\text{true negatives}}{\text{true negatives} + \text{false positives}} .$$

The thresholds for specimen classification can be arbitrarily altered to favour the type of error made. Thus if nuclear quantitation were used as a screening method with cytological preparations it might be decided to err on the side of false positives. This was the case when CASTLEMAN and WHITE (1980), studying cervical smears, used a two-stage classification where cell classification was followed by specimen classification.

### 4.4 Criteria for Cell and Sample Classification

When human observers are trained to classify nuclei on the basis of the traditional subjective criteria of size, hyperchromaticism and pleomorphism, they are trained to minic the behaviour of an "expert" pathologist or group of already trained pathologists. Since the "expert" pathologist (being human) is partially unreliable, perfect agreement between the trainee and the expert cannot be expected.

When decision-making alogarithms are built up, or "trained", the training process may differ in that no expert is present at the outset. The data are screened, using multivariate analysis, for similar patterns and correlations between features. KOSS et al. (1977 a, b) found similarities between the results of each type of training process.

## 5 Some Problems of Statistical Analysis

Apart from the general considerations in the statistical analysis of data from biological samples, special problems in analysis of nuclear quantitative data arise from non-normal distribution of data.

With the current trend in some journals towards the presentation of confidence intervals around the mean of a population (GARDNER and ALTMAN 1986; MORRIS and GARDNER 1988), specific problems arise when confidence intervals have to be calculated from non-parametric data (BROWN and BECK 1988 c). Distribution-free methods have to be used for this and the intervals may more appropriately be expressed about a median than the mean (CAMPBELL and GARDNER 1988).

A series of review articles has been published which includes advice on current statistics packages for microcomputers and their compatibility (BROWN and BECK 1988 a), confidence intervals and significance tests (BROWN and BECK 1988 b), distribution-free methods (BROWN and BECK 1988 c) and other problem areas (BROWN and BECK 1989 a–c). All of these contain criticisms of the various graphic options available for presenting data for publication.

## 5.1 Sample Size

One of the commonest practical problems encountered in quantitation of the nucleus is selection of the appropriate sample size. Since many of the procedures in CIM are time-consuming, no observer wishes to study an excessive sample size where perhaps a smaller sample could have been used in statistically valid comparisons. On the other hand, choice of an inadequately sized sample may lead to a large standard deviation or to wide confidence intervals. This would prevent detection of any differences between the means or medians of the populations.

A simple and popular test valuable in planning an experiment is the method of cumulative means. In this graphical analysis, one nuclear parameter is measured and its cumulative mean plotted against sample size (say, 10, 20, 30 ... 100 cells). Initially the line representing the cumulative mean fluctuates widely, but it eventually becomes nearly straight at the true mean value. If this value is marked on the graph and lines ±5% of the measured value from this are drawn in, the sample size at which the cumulative mean ceases to fluctuate by more than this can be visualised. This simple exercise performed for all the parameters being investigated gives a rough idea of the minimum appropriate sample size.

## 5.2 Observer Variability

Observer variability in histopathological diagnosis and grading of tumours and of dysplastic states (partly dependent on nuclear features) is well known (ISMAIL et al. 1989; ROBERTSON et al. 1989). If we look to nuclear quantitation to inject science and objectivity into observation of nuclei, it is essential to ask how reproducible are the results of nuclear quantitation. A basic rule in morphometry is that the feature being studied must be unambiguously recognisable (HAUG 1980). This may be the case with *recognition* of nuclei, but different observers may *interpret* and trace the nuclear boundaries differently.

To test this, BARRY and SHARKEY (1985) evaluated intra- and interobserver variability in planimetry of area, perimeter and form factors in non-Hodgkin's lymphomas using linear regression analysis and *t*-tests. They found good correlation for area measurement, but poor correlation for measurement of perimeters and form factors due to difficulty in following complex nuclear contours. This study has been criticised on the grounds of the intrinsically complicated shape of the nuclei being quantitated, the low nuclear magnification at the surface of the digitising tablet, and the use of a stylus rather than a cross-hair cursor (DARDICK and CALDWELL 1985). The latter group of authors found better levels of intra- and interobserver agreement, but again the nuclear contour indices were the area of poorest correlation. Their better agreement is possibly explicable by their use of more highly magnified images on the digitiser tablet. Where complex

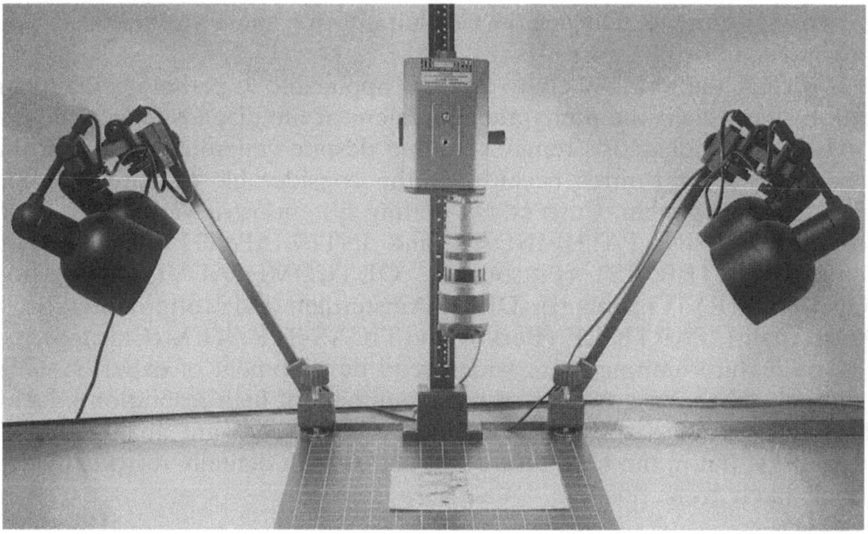

**Fig. 15.** Where nuclear size is small or when subnuclear components are to be quantitated, micrographs are imaged using a video camera with a zoom lens on a copy stand

boundaries of small nuclei must be traced, increased accuracy may be achieved through the higher magnification possible when photomicrographs are imaged by a video camera on a copy stand (Fig. 15) compared with that available from a video camera directly fitted to the microscope.

Again in a study of lymphoma, CHAN et al. (1987) measured intra- and interobserver variability for various planimetric parameters. In this study there was good intra- but poor interobserver correlation for form factors. It is therefore possible that these variations result from different observers' interpretations of the level of care with which the nuclear perimeter needs to be traced.

Whatever the cause of these discrepancies, it is clear that descriptions of nuclear quantitative studies should include a statement about observer variation and the degree to which the different observers were trained.

## 6 Expert Systems

Expert systems involve the application of artificial intelligence (AI) to quantitative pathology (BAAK 1988). Definition of AI is difficult, just as it is difficult to define human intelligence. A satisfactory definition of AI for practical purposes is: "computer programs that can perform a task which normally requires an intelligent person". This was the definition used by the The Second International Conference on Artificial Intelligence System

(Expert Systems) as Diagnostic Consultants for the Cytologic and Histologic Diagnosis of Cancer (BAAK 1988).

It is clear that AI is likely to become applicable to pathology via quantitation since this is the numerate equivalent of the diagnostic art. The use of AI in cancer diagnosis remains limited despite generous funding of the research in some centres because of the considerable time required to build an expert system. Currently operating systems reviewed (BAAK 1988) included: MYCIN, PATHFINDER and INTELLIPATH (developed at Stanford), INTERNIST (Pittsburgh), OUTCOME ADVISOR(R) and CONSULT-I(R) (Cincinnati), DEW (Amsterdam and Rotterdam), PECS (Amsterdam), PARTICLE (Berlin) and TICAS-STRATEX (Chicago).

One feature limiting more widespread development of expert systems is difficulty with mass storage of large numbers of high resolution images. The fastest storage and retrieval systems at present are analogue laser video disks, but in the near future fast access with digitally recorded video disks is likely to be possible.

### 6.1 Special Languages for Expert Systems

The special software and languages required to build expert systems are called "high end tools". Most are complex and difficult to understand. Some, such as Gold Words, can run on personal computers. Most, such as KEE and ART, require large computers. These languages are not restricted to representing knowledge in the IF ... THEN format, but support more flexible logic such as "multiple inheritance frame system" and "integrated forward and backward chaining" (BAAK 1988).

## 7 Nuclear Quantitation in Specific Tissues

Nuclear quantitation has been used widely throughout the organ systems in analysis of dysplastic and neoplastic states, both to assist diagnosis and for assigning prognosis. Non-neoplastic states have been studied to a much lesser extent. The following review of nuclear quantitation in specific tissues does not attempt to be comprehensive, but merely to highlight the main findings and their correlation with other diagnostic or prognostic aids and to show the gaps in our present knowledge.

### 7.1 Mammary Gland

While most nuclear quantitative studies have been concerned with malignant neoplasms, a few studies have concentrated on physiological conditions and benign neoplasms. For example, predominantly diploid DNA histograms with variable tetraploidy have been reported in lactation (IZUO et al. 1971 a), apocrine metaplasia (IZUO et al. 1971 b), sclerosing adenosis (STENGER et al. 1980) and fibroadenomas (ZAJICEK et al. 1970; SPRENGER et al. 1979). BHATTACHARJEE et al. (1985) advocated nuclear quantitation as at least a partly sensitive means of distinguishing epitheliosis from intraduct carcinoma.

Studying DNA ploidy using Feulgen cytophotometry and its potential for diagnosis, AUER and TRIBUKAIT (1980) had difficulty distinguishing invasive carcinomas from some benign conditions due to normal DNA ploidy in the malignancies. BOQUOI et al. (1975) had previously detected aneuploidy in 65% of carcinomas. TAVARES (1968) found a modal tetraploid peak in most carcinomas, but with medullary types tending to have lower ploidy values.

Planimetric studies of nuclear size in breast carcinomas have found considerable variation in nuclear size within carcinomas and overlap with some benign conditions, limiting their potential diagnostic use (WALLGREN and ZAJICEK 1976; ZAJDELA et al. 1979; CORNELISSE et al. 1981; BOON et al. 1982 b; WITTEKIND and SCHULTZE 1987). This problem may arise because some carcinomas have small nuclei. Some mammary carcinomas, for example lobular carcinoma, have uniform small nuclei similar to those in benign states. It is therefore interesting to consider whether bringing in optical density information through Feulgen cytophotometry can detect differences between in situ and invasive lobular carcinoma. SACHS (1970) found higher stem cell modal DNA values in the diploid-tetraploid range in carcinomas and in "type B" (HAAGENSEN 1971) in situ neoplasms than in the "type A" small cell in situ neoplasm. It is not known whether the aneuploid type B in situ neoplasms have a higher malignant potential.

Cytophotometric studies have documented better survival in patients with diploid than with aneuploid tumours after correction for stage (ATKIN 1972; AUER et al. 1980).

Computerised interactive morphometry has yielded prognostic information in invasive carcinomas. Nuclear density and area variances have been correlated with mitotic activity and recurrence rates (STENKVIST et al. 1981, 1982). BAAK (1985) reported an additional contribution from nucleolar morphometry over the standard assessment of prognosis in invasive carcinoma (PARL and DUPONT 1982). Discriminant analysis of several morphometric features besides nuclear quantitation has been reported to be a better predictor of 6-year survival (BAAK 1985) than TNM staging (NEALSON et al. 1981; FUSTER et al. 1983).

There is current interest in oestrogen receptor (ER) status of carcinomas as a predictor of recurrence (KNIGHT 1977; WESTERBERG et al. 1980; UNDERWOOD et al. 1983) and sensitivity to endocrine therapy (HEUSON et al. 1977). Correlations between nuclear morphometry and ER status have been reported. ER has been found to be present more commonly in diploid than in DNA aneuploid tumours (BICHEL et al. 1982; CORNELISSE et al. 1984) as determined by cytophotometry. Purely planimetric studies, too, have shown correlation. ER positive tumours have been shown to have a smaller mean nuclear diameter (ANTONIADES and SPECTOR 1979) and area (MOSSLER et al. 1982) than ER negative tumours.

## 7.2 Female Genital Tract

With its range of dysplastic states, in situ carcinomas and tumours of borderline malignancy, all posing problems of recognition, grading and classification, the female genital tract has proved an interesting area for the application of quantitative studies.

### *7.2.1 Uterine Cervix*

Exfoliative cytological and histological specimens are easily obtained from the uterine cervix. Consequently, this site has been subject to numerous quantitative studies. Early karyotyping and DNA measurements confirmed the unifying concept of cervical intraepithelial neoplasia embracing both dysplasias and carcinoma in situ. They showed aneuploid patterns in these lesions indistinguishable from invasive carcinoma (WILBANKS et al. 1967; SPRIGGS et al. 1971; ATKIN 1976a). The changes of cervical intraepithelial neoplasia (CIN) may regress, persist or progress to invasive carcinoma. FU et al. (1981) have shown that lesions which regress are largely diploid or polyploid, while all the lesions progressing to invasive carcinoma are DNA aneuploid.

Progression from CIN to invasive carcinoma may be associated with evolution of new stem cell lines. An increase in aneuploid stem cells with DNA aneuploidies generally lower than triploid has been observed in surface epithelium adjacent to microinvasive carcinoma (FU et al. 1980) while the actual tongue of invasive carcinoma contains both high and low ploidy DNA aneuploid stem cells (ATKIN 1976a, b).

ATKIN (1976a) noted that tumours with a high DNA aneuploid pattern were more radiosensitive than those with only slight aneuploidy. The prognostic importance of ploidy was found to exceed that of histological grade in those tumours treated with radiotherapy.

Recently HPV infection has been linked to the formation of flat cervical condylomata in the case of HPV types 6 and 11 and to CIN with HPV types 16, 18 and 31 (CRUM et al. 1984; BURNS et al. 1987; COLLINS et al. 1988). The characteristic DNA ploidy abnormality associated with HPV infection is polyploidy (FU et al. 1983). In transition from warty atypia to CIN, DNA aneuploid cells become demonstrable (FU et al. 1983).

The majority of invasive carcinomas of the cervix are of the squamous cell type, but relative reduction in their incidence possibly following cervical cytological screening now means that 10%–20% of cervical carcinomas are adenocarcinomas (REAGAN and NG 1973; WELLS and BROWN 1986). In cervical adenocarcinomas, degree of differentiation correlates with Feulgen cytophotometrically determined DNA ploidy (FU et al. 1982). Poorly differentiated carcinomas tend to have the higher DNA aneuploid levels. High-degree DNA aneuploidy is a predictor of aggressive behaviour in cervical adenocarcinoma (ATKIN 1976b; FU et al. 1982).

In the rare clear cell metaplasias of the cervix (and vagina) in the progeny of diethylstilbestrol-exposed mothers, measurement of DNA ploidy has shown that most cases have a diploid pattern and that abnormal mitotic figures remain one of the most important indicators of malignant potential (FU et al. 1978, 1979).

Automated techniques for screening cervical exfoliative cytological specimens are currently subject to intensive investigation, partly due to potential commercial advantages from automation of this substantial, labour-intensive workload. The problems of cellular aggregates (not suitable for video-based image analysis, yet possibly containing the most significantly abnormal cells), nuclear overlap and debris currently hamper attempts at automated screening. An early planimetric study showed progressive reduction in nuclear area from dysplasia with increasing grade of CIN and progression to invasive carcinoma (REAGAN et al. 1957).

More recently, experimental automated systems analysing parameters such as nuclear size, nuclear-cytoplasmic ratio, nuclear optical density, shape and chromatin texture show promise for prescreening of cervical smears. Examples of such systems include: CYBEST (TANAKA et al. 1979, 1982), CERVISCAN (TUCKER and HUSSAIN 1981) and bioPEPR (ZAHNISER et al. 1979, 1980). The most promising approaches to automation may lie in contextual analysis where information about a whole smear preparation is collected by pattern recognition software (GARCIA et al. 1987).

A completely different approach to analysis of cervical smears relies on linear discriminant analysis of multispectral texture features (sometimes called intermediate squamous cell markers) which are undetected by the human eye (WIED et al. 1981, 1984). The histochemical basis of intermediate markers is unknown, although it has been suggested that they result from alterations in the isoelectric point of nuclear histone proteins (WIED et al. 1984).

### *7.2.2 Endometrium*

Early detection of endometrial adenocarcinoma is made difficult by small numbers of malignant endometrial cells in routine cervical smears and by difficulty in their recognition. Even if these are correctly recognised, the ensuing endometrial curettings may show various atypical hyperplasias which may be difficult to distinguish from adenocarcinoma (FOX and BUCKLEY 1982). Studies by DIEGENBACH and BAAK (1977, 1978a, b),

BAAK and OORT (1983) and BAAK et al. (1981 a, b) have shown that morphometric analysis of glands in the atypical hyperplasias is useful in their classification. In contrast to this, classification by nuclear morphological features is less predictable when combined with architectural features (BAAK et al. 1988).

In predicting which cases of atypical hyperplasia will progress to carcinoma, nuclear morphometric features have been shown to be predictive (COLGAN et al. 1983; BAAK 1986), especially when combined with architectural features (BAAK et al. 1988).

Fewer studies have attempted to distinguish hyperplasias from carcinomas on cytological preparations. SKAARLAND (1985) found the nuclei of epithelial fragments from both the Isaacs cell sampler and from the Endoscann instrument to be suitable for morphometric analysis. In cytological preparations, REAGAN and NG (1973) found only a slight increase in nuclear size in simple hyperplasias compared with normal endometrium. In the same study, the nuclei of adenocarcinomas could be distinguished by size and nucleolar content from normal cells, although distinction between atypical hyperplasias and well-differentiated adenocarcinomas could not be achieved.

In Feulgen cytophotometric studies, the DNA ploidy of normal cyclical endometrium lies within the diploid-tetraploid range and a clear-cut diploid distribution is found during the late secretory phase of the cycle (WAGNER et al. 1968). Distinction of hyperplasias from proliferative phase endometrium is therefore difficult (WAGNER et al. 1967). In endometrial adenocarcinomas, DNA ploidy correlates with histological grade and prognosis (ATKIN 1976 a, b; MOBERGER et al. 1984). In the latter study, DNA ploidy proved a better predictor of survival than either clinical stage or histological grade.

### *7.2.3 Ovary*

ATKIN (1971), in the first substantial series of ovarian carcinomas assessed for DNA ploidy by Feulgen cytophotometry, showed that tumours could be classified broadly into those with near-diploid DNA ploidy and those with higher degrees of aneuploidy. The latter group tended to be in a more advanced clinical stage at presentation and to have a worse prognosis. The results of this study have been confirmed in malignant common epithelial tumours of the ovary. DNA ploidy has been correlated with histological grade (ERHARDT et al. 1985), with general prognosis (ERHARDT et al. 1984; BAAK et al. 1987) and with response to cisplatin treatment (BAAK et al. 1988 b).

Borderline common epithelial tumours have been studied intensively by nuclear and general morphometric techniques to determine whether prognosis can be predicted and in an attempt to define borderline tumours as a distinct histological entity. They have a heterogeneous DNA content

(WEISS et al. 1969; FRIEDLANDER et al. 1983), with about one-third exhibiting DNA aneuploidy (WEISS et al. 1969). This study also found that DNA aneuploidy correlated with a more advanced clinical stage at presentation. Combining nuclear quantitation with other morphometric features in discriminant analysis, BAAK et al. (1981 a) and BAAK and OORT (1983) have been able to distinguish borderline from malignant tumours and this approach has been illustrated in clinical decision making (BAAK and VAN DER LEY 1984).

Most nuclear quantitative studies have been concerned with common epithelial tumours, but one study has correlated nuclear quantitative features with prognosis in granulosa tumours (SASSEN and BAAK 1986). This showed useful prognostic correlations in this tumour with otherwise unpredictable behaviour.

## 7.3 Urogenital System

Being susceptible to dysplastic field changes, multifocal tumours and multiple tumour recurrences, the urogenital system has been a fruitful area for nuclear quantitation. This has been facilitated by the ready availability of cytological preparations.

### *7.3.1 Bladder*

Grading transitional cell carcinomas is important in prognosis and in planning treatment. Grade II and III tumours are managed more aggressively than grade I tumours. Any quantitative procedure capable of refining grading or yielding extra prognostic data has potential clinical value. In cytophotometric studies of DNA ploidy, grade I tumours have been shown to be almost uniformly diploid, while grade III tumours are aneuploid (TAVARES et al. 1966; FOSSA 1977; BJELKENKRANTZ et al. 1982). Grade II tumours vary between diploid and DNA aneuploid (HERDER et al. 1982).

Transitional cell carcinomas can be detected and followed up for recurrence by urinary cytology. Studying exfoliated transitional cells in grade I and II tumours, BOON et al. (1984) observed good correlation between the nuclear-cytoplasmic ratio of the voided cells and tumour grade.

KOSS et al. (1975, 1977 a, b, 1978 a, b) used CIM to measure nuclear optical density, shape and chromatin texture. Overlap was found in the features of benign and atypical cells. Within the atypical cell group two distinct subgroups, those with slight and those with marked atypia, could be distinguished. This impression has since been confirmed by others (TRIBUKAIT and GUSTAFSON 1980; HERDER et al. 1982).

### 7.3.2 Kidney

While there have been several flow cytometric analyses of renal tumours, Feulgen cytophotometry of these lesions has been less popular. BENNINGTON and MAYALL (1983) found correlation between nuclear morphometric features, including DNA cytophotometry, and tumour grade of renal cell carcinoma. Although both techniques distinguished the nuclei of carcinoma from those of normal tubules, neither reliably separated renal cell carcinoma from renal cortical "adenoma". TOSI et al. (1986 a) reported correlation between nuclear morphometric features and prognosis in stage I renal cell carcinomas.

### 7.3.3 Prostate

Establishing quantitative criteria to detect and grade carcinoma in cytological preparations from the prostate is likely to be increasingly important in view of the trend towards use of FNA for initial diagnosis of this increasingly common tumour. Recognition of those malignancies which are low grade is important since these may be managed conservatively.

Tumour grade has been correlated with DNA ploidy in Feulgen cytophotometric analyses (ZETTERBERG and EPISOTI 1980), but this technique alone is not useful in diagnosis since even moderately differentiated tumours may have near-diploid DNA ploidy (RONSTROM et al. 1981).

Morphometric nuclear analysis has yielded additional diagnostic information. Nuclei and the nucleoli within them tend to be enlarged in carcinoma (BOCKING et al. 1984 b). Nuclear form factors (roundness factors) have been consistently shown to correlate with prognosis (CANTRELL et al. 1981; DIAMOND et al. 1982 a, b). DIAMOND et al. (1982 a) found this index useful in distinguishing benign from malignant prostatic cells. Nucleolar enlargement quantitated at scanning electron micrography has been shown useful in prediction of metastasis (TANNENBAUM et al. 1982).

### 7.3.4 Testes

The incidence of malignant germ cell tumours is increased in the maldescended testis (ANDERSON 1988), for unknown reasons. Testicular biopsies from maldescended testes have revealed increased nuclear size and DNA content of germ cells (MULLER and SKAKKEBAEK 1984).

Both spermatocytic and classical seminomas display aneuploidy (ATKIN 1973; MULLER et al. 1981; TALERMAN et al. 1984). The spermatocytic variant did not contain any haploid cells (TALERMAN et al. 1984), with the implication that this tumour is not composed of neoplastic spermatocytes capable of meiotic division. Ploidy in relation to meiosis in germ cell neoplasms has been discussed by FOX (1987).

## 7.4 Gastrointestinal Tract

The gastrointestinal tract offers challenges to nuclear quantitation, including investigation into diagnosis, grading and prognosis of dysplastic lesions and analysis of the "adenoma-carcinoma" sequence.

### *7.4.1 Salivary Gland*

A wide range of histological types of tumour occur within the salivary glands, each with characteristic biological behaviour. Within a given tumour type, such as mucoepidermoid carcinoma, behaviour and histological degree of differentiation have been correlated with Feulgen cytophotometrically determined DNA values (KINO et al. 1973). In pleomorphic adenomas and their malignant counterpart DNA ploidy has been shown to be of some prognostic value (THORUD et al. 1980).

### *7.4.2 Oesophagus*

Examining severe oesophageal squamous epithelial dysplasias and carcinomas, MAKUDA et al. (1978) found that these were commonly distinguished from normal samples by DNA tetraploidy. In a study correlating prognosis with Feulgen cytophotometrically determined DNA ploidy, postsurgical recurrence was strongly associated with gross DNA aneuploidy (SUGIMACHI et al. 1984).

### *7.4.3 Stomach*

With increasing interest in dysplastic states and early carcinomas of the stomach, nuclear quantitation has been used in dysplasia classification and in prediction of cancer prognosis. CIM has been applied successfully to gastric brushing cytology (BOON et al. 1981) in an attempt to reduce subjective errors. In a study of a gastric dysplasias, nuclear size proved the most useful single morphometric feature for their classification (JARVIS and WHITEHEAD 1985). With their known malignant potential, a variety of gastric polyps has been studied by Feulgen cytophotometry (WIENDL et al. 1974; SPRENGER and WITTE 1978), although overlap between the DNA histograms of carcinomas and benign conditions was found. In a study of early gastric cancer, those with the superficially spreading growth pattern were generally diploid, while those with penetrating growth tended to be aneuploid (INOKUCHI et al. 1983).

### *7.4.4 Colon*

Although the adenoma-carcinoma sequence in the large intestine has been studied by general morphometry (KAYSER et al. 1985; JARVIS et al. 1987), there has been a relative paucity of quantitative microscopic studies of the nucleus. GRAHAM et al. (1988) studied nuclear placement within the epithelium together with nuclear area and its variance within normal colon, adenomas and adenocarcinomas. Both proved useful in distinguishing the three states. In view of the current difficulty in classifying dysplastic states of the colon (DUNDAS et al. 1987), particularly in relation to ulcerative colitis, there is potential for nuclear quantitation in this area.

## 7.5 Respiratory Tract

Nuclear quantitative microscopy has been used extensively in the classification of cytological preparations and of solid tumours from the respiratory tract in addition to the study of dysplastic states.

### *7.5.1 Nasal Cavity and Larynx*

Dysplastic states in the upper respiratory tract, some with specific occupational risk factors, have been studied by morphometric analysis of the epithelial basal layer nuclei (BOYSEN and REITH 1983; BOYSEN et al. 1983; RIGAUT et al. 1985). In thin plastic sections, nuclear and nucleolar size distinguished normal epithelium from squamous metaplasia and dysplasia.

The management and follow-up of dysplasias in laryngeal squamous metaplastic epithelium is difficult. A useful finding is the association of high-degree DNA aneuploidy with recurrence and progression to invasion (BJELKENKRANTZ et al. 1983). In established squamous cell carcinomas of the head and neck, nuclear morphometry and DNA ploidy measurement correlate with prognosis (HOLM 1982; DAVIS et al. 1987).

### *7.5.2 Lung*

Early detection of lung cancer, possibly through improvements in objectivity of pulmonary cytology, is one way in which the currently poor prognosis may eventually be improved. In one cytophotometric study of sputum cytology, DNA measurement was found to correlate with nuclear atypia (NASIELL et al. 1978). PAK et al. (1981), examining bronchial smears by morphometric and cytophotometric methods, were able to distinguish different degrees of epithelial atypia. Automated image analysis used as a screening procedure for sputum samples distinguished metaplasias, dysplasias and carcinomas (GREENBERG et al. 1982, 1986; SWANK et al. 1983).

Within the often difficult classification of lung cancer type, nuclear quantitation has so far had limited application. BROERS et al. (1988) described the successful distinction of small cell and small cell variant types, which has clinical importance.

## 7.6 Haematological Specimens

While there have been some nuclear quantitative studies on non-neoplastic haematological cells (MAYALL and MENDELSOHN 1970; PENNINGTON and OLSEN 1970; TROWBRIDGE et al. 1983, 1984; DARDICK 1984; DARDICK and DARDICK 1984; RICCO et al. 1985; PAYNE et al. 1985, 1987; GIROUD et al. 1988), most studies have concentrated on differences between normal cells and those of lymphomas and leukaemias (TOSI et al. 1984; DARDICK et al. 1987; MARCHEVSKY et al. 1987c; STEVENS et al. 1988). The use of morphometric techniques to classify nuclei in lymphomas has been well reviewed (CROCKER 1984).

With their complicated nuclear shapes, T-cell lymphomas and leukaemias have been a fertile subject for nuclear morphometry (BARANSKA et al. 1983; TOSI et al. 1988). These studies have included examination of nuclear shape and chromatin texture at the ultrastructural level (SHUM et al. 1986; SIMON 1987). In mycosis fungoides, indices of nuclear shape have been found useful in diagnosis (MEIJER et al. 1980), while nuclear DNA ploidy correlates with response to therapy (VAN VLOTEN et al. 1979).

Few studies have specifically addressed classification or diagnosis of Hodgkin's lymphoma, perhaps due to its heterogeneous histological types. There have been comparative nuclear morphometric studies with non-Hodgkin's lymphomas (DARDICK et al. 1987), including a study of circulating monocytes in these conditions (SOKOL 1989).

In B-cell non-Hodgkin's lymphomas, quantitation of nuclear size (CROCKER and CURRAN 1979), shape (TOSI et al. 1983; VAN DER VALK et al. 1983; DARDICK et al. 1984; RAPHAEL et al. 1985; PELSTRING and SWERDLOW 1987), nucleolar features (VAN DER VALK et al. 1982, 1983) and ultrastructural features (DARDICK et al. 1985a, b) has been studied extensively in diagnosis and classification. Due to the relatively easy disaggregation of lymphoid tissues to form good cell suspensions, cytometric analyses of lymphomas have been more widespread than cytophotometry. DNA ploidy and S-phase fraction are very important prognostic features (QUIRKE and DYSON 1986).

## 7.7 Skin

The measurement of invasion depth is a well-known important method for assessing prognosis in malignant melanoma. Morphometric studies have also been performed on the nucleus both to distinguish benign naevi from malignant melanoma (BRUNGGER and CRUZ ORIVE 1987) and as a prognostic indicator (TAN and BAAK 1984; BAAK and TAN 1986).

The extensive studies on nuclear shape in the diagnosis and prognosis of mycosis fungoides have been reviewed in Sect. 7.6.

## 7.8 Endocrine Glands

The behaviour of many endocrine tumours is difficult to predict (for example, APUDomas and adrenal tumours). This may partly be due to the often pleomorphic appearance of nuclei in endocrine tissues. The pancreatic islet cell nuclei, for example, vary in size and may be facultatively tetraploid (POHL et al. 1981). This is also the case in the thyroid (GILBERT and PFITZER 1977). Morphometric techniques may therefore be of value in the endocrine system.

The thyroid gland, being accessible to FNA cytology, has been most extensively studied. Examining air-dried FNA smears from follicular tumours, BOON et al. (1980, 1982 a) reported correlation between nuclear area and nuclear-cytoplasmic ratio with the conventional histological criteria in differential diagnosis between adenoma and carcinoma. These findings have not been confirmed by another group (LUCK et al. 1982), who were unable to distinguish benign from malignant neoplasms by nuclear morphometry although neoplasms generally showed larger nuclei compared with simple goitre.

In Hürthle cell tumours of the thyroid, a partial association between nuclear size and biological behaviour has been reported (BONDESON et al. 1983). Using cytophotometric analysis of DNA ploidy in tissue sections, FLINT et al. (1988) showed a significant association between DNA ploidy and tumour invasion.

SLOOTEN et al. (1985) reported morphometric differences between the nuclei of benign and malignant adrenal cortical neoplasms, a classic problem area in conventional histological prediction of malignancy.

## 7.9 Nervous System Including Eye

There have been several flow cytometric analyses of primary brain tumours correlating DNA ploidy with prognosis, but there are few morphometric studies of the nucleus in the nervous system. GIANGASPERO et al. (1984) reported a morphometric comparison of the nuclei in glioblastoma multiforme with those of metastatic carcinoma to the central nervous system (CNS).

Uveal tract melanomas are the CNS neoplasms most widely studied by nuclear morphometry. GAMEL et al. (1982) and MCLEAN and GAMEL (1988) compared cytophotometrically determined DNA ploidy and nuclear morphometric features with survival of similar stage ophthalmic melanomas. Morphometric features of the nucleolus proved a better predictor of survival than simple measurement of nuclear size (GAMEL and MCLEAN 1983).

## 7.10 Bones and Connective Tissues

Nuclear morphometric features have been studied both on normal bone (POLIG et al. 1984) and in bone tumours (BOCKING et al. 1980). Osteosarcomas tend to exhibit relatively high DNA aneuploidy (KREICBERGS 1980, 1981; BOCKING et al. 1984a). The bone cells around aneurysmal bone cysts have been distinguished from those of telangiectatic osteosarcoma by quantitation of nuclear size (RUITER et al. 1977).

Histological differentiation of benign cartilaginous proliferations from chondrosarcoma is difficult. Feulgen cytophotometry has not assisted in this distinction since a high proportion of low-grade chondrosarcomas have been reported to be diploid (KREICBERGS et al. 1980). However, the same study showed that those low-grade chondrosarcomas displaying hyperdiploidy were associated with an aggressive clinical course.

In smooth muscle tumours, especially those of the uterus, the value of mitotic counts in behaviour prediction is well known (TAYLOR and NORRIS 1966). Diploid stem cell lines have been reported in uterine leiomyomas (BOHM and SANDRITTER 1975), but so far leiomyosarcomas have been studied by flow cytometry only.

## 7.11 Inflammatory Disorders

Most microscopic nuclear quantitation has been directed towards tumour investigation. Only one nuclear quantitative study of an inflammatory disorder is known to the author. This examined differences between the nuclear profiles in the granulomas of tuberculosis and of sarcoidosis (TOSI et al. 1986b).

# 8 Conclusions

There is currently an explosive increase in interest in morphometric techniques as an aid to diagnosis and stating prognosis in pathology. With this interest concentrating on tumours, most studies have concerned nuclear features. Observer variability studies (DUNDAS et al. 1987; ISMAIL et al. 1989; ROBERTSON 1989) and the onset external quality assurance in histopathology (LEE and BURNETT 1987) – still the only non-numerate speciality in pathology (BECK and ANDERSON 1987) – have shown up the fallibility of subjective diagnosis. Statements like "the nuclei show hyperchromaticism, pleomorphism and enlargement" and "the mitotic rate did not appear unduly high" are becoming questionable without objective studies.

With proper quality control, quantitative microscopy is reproducible. Advances in video and computer technology, including image storage, are increasing the degree of automation possible. This facilitation of quantitative microscopy and the advent of expert systems must never lead to premature commissioning of automated procedures in clinical diagnosis. BAAK (1987b) has reviewed five levels of validation which must be satisfied before any morphometric rule is applied to clinical use. Adherence to these criteria should prevent the unjustified delegation of the decision-making process from the histo- or cytopathologist to machines. The head of a laboratory remains medicolegally responsible for diagnoses issuing from automatic machines under his control. Before basing clinical reports on their results, he must be fully confident in their reliability.

## References


Abercrombie M (1946) Estimation of nuclear population from microtome sections. Anat Rec 94:239–247

Adams LR (1968) A photographic cytometric method which avoids distributional error. Acta Cytol 12:3–8

Aherne WA, Dunnill MS (1982) Morphometry. Edward Arnold, London

Aherne WA, Al-Wiswazy M, Ford D, Kellerer AM (1977) Assessment of inherent fluctuation of mitotic and labelling indices in human tumours. Br J Cancer 36:577–591

Allison DC (1985) Refinements in absorption-cytometric measurements of cellular DNA content. In: Cowden RR, Harrison FW (eds) Advances in microscopy. Alan R Liss, New York, pp 167–185

Allison DC, Ridolpho PF, Rasch RW, Lohnson TS (1981) Increased accuracy of absorption cytometric DNA values by control of stain intensity. J Histochem Cytochem 29:1219–1223

Allison DC, Lawrence GN, Ridolpho PF, O'Grady BJ, Rasch RW, Rasch EM (1984a) Increased accuracy and speed of absorption cytometric DNA measurements by automatic corrections for nuclear darkness. Cytometry 5:217–227

Allison DC, Meyne J, Ridolpho P, Bose K, Chakerian M, Robertson J (1984b) Computerized measurement of the DNA content, areas, and autoradiographic grains of the same nuclei: demonstration that lightly ($^3$H) thymidine-labelled bone marrow cells are predominantly in $G_0/G_1$ and $G_2$. J Histochem Cytochem 32:1197–1201

Allison DC, Mayall BH, Levin J (1988) Comparison of absorption measurements of DNA stain content by utilising video and scanning image cytometers. Cytometry 9:573–578

Anderson CK (1988) Carcinoma in situ of the testis. J Pathol 155:3–5

Anonymous (1989) Brush cytology for colorectal cancer. Lancet I:477–478

Antoniades K, Spector H (1979) Correlation of estrogen receptor levels with histology and cytomorphology in mammary cancer. Am J Clin Pathol 71:497–503

Atkin NB (1970) Principles and application of the Deeley integrating microdensitometer. In: Wied GL, Bahr GF (eds) Introduction to quantitative cytochemistry, vol II. Academic Press, New York, pp 1–26

Atkin NB (1971) Modal DNA value and chromosome number in ovarian neoplasia: a clinical and histopathologic assessment. Cancer 27:1064–1073

Atkin NB (1972) Modal deoxyribonucleic acid value and survival in carcinoma of the breast. Br Med J I:271–272

Atkin NB (1973) High chromosome numbers of seminomata and malignant teratomata of the testis: a review of data from 103 tumours. Br J Cancer 28:275–280

Atkin NB (1976a) Cytogenetic aspects of malignant transformation. S. Karger, Basel

Atkin NB (1976b) Prognostic significance of ploidy level in human tumours. I. Carcinoma of the uterus. J Natl Cancer Inst 56:909–1004

Atkin NB, Richards BM, Ross AJ (1959) The deoxyribonucleic acid content of carcinoma of the uterus: an assessment of its possible significance in relation to histopathology and clinical course based on data from 165 cases. Br J Cancer 13:773–779

Auer G, Tribukait B (1980) Comparative single cell and flow DNA analysis in aspiration biopsies from breast carcinomas. Acta Pathol Microbiol Scand 88:355–360

Auer G, Caspersson T, Wallgren A (1980) DNA content and survival in mammary carcinoma. Anal Quant Cytol 2:161–165

Baak JPA (1985) The relative prognostic significance of nucleolar morphometry in invasive ductal breast cancer. Histopathology. 9:437–444

Baak JPA (1986) Further evaluation of the practical applicability of nuclear morphometry for the prediction of the outcome of atypical endometrial hyperplasia. Anal Quant Cytol Histol 8:46–48

Baak JPA (1987a) Quantitative pathology today – a technical view. Pathol Res Pract 182:396–400

Baak JPA (1987b) The principal advances of quantitative pathology. Anal Quant Cytol Histol 9:89–95

Baak JPA (1988) Artificial intelligence systems (expert systems) as diagnostic consultants for the cytologic and histologic diagnosis of cancer. J Cancer Res Clin Oncol 114:325–334

Baak JPA, Diegenbach PC (1977) Quantitative nuclear image analysis: differentiation between normal, hyperplastic and malignant appearing uterine glands in a paraffin section. I. Elementary features for differentiation. Eur J Obstet Gynecol Reprod Biol 7:33–42

Baak JPA, Oort J (1983) A manual of morphometry in diagnostic pathology. Springer, Berlin Heidelberg New York

Baak JPA, Tan GJ (1986) The adjuvant prognostic value of nuclear morphometry in stage I malignant melanoma of the skin. A multivariate analysis. Anal Quant Cytol Histol 8:241–244

Baak JPA, Van Der Ley G (1984) Borderline or malignant ovarian tumour? A case report of decision making with morphometry. J Clin Pathol 37:1110–1113

Baak JPA, Blanco AA, Kurver PHJ et al. (1981a) Quantitation of borderline and malignant mucinous ovarian tumours. Histopathology 5:353–360

Baak JPA, Kurver PHJ, Diegenbach PC, Delemarre JFM, Brekelmans ECM, Niewlaat JE (1981b) Discrimination of carcinoma of the endometrium by quantitative microscopy – a feasibility study. Histopathology 5:61–68

Baak JPA, Kurver PHJ, Boon ME (1982) Computer-aided application of quantitative microscopy in diagnostic pathology. Pathol Annu 17:287–306

Baak JPA, Wisse-Brekelmans ECM, Uyterlinde AM, Schipper NW (1987) Evaluation of the prognostic value of morphometric features and cellular DNA content in FIGO I ovarian cancer patients. Anal Quant Cytol Histol 9:287–290

Baak JPA, Nauta JJP, Wisse-Brekelmans ECM, Bezemer PD (1988a) Architectural and nuclear morphometrical features together are more important prognosticators in endometrial hyperplasias than nuclear morphometrical features alone. J Pathol 154:335–341

Baak JPA, Schipper NW, Wisse-Brekelmans ECM, Ceelan T, Bosman FT, Van Geuns H, Wils J (1988b) The prognostic value of morphometrical features and cellular DNA content in cis-platin treated late ovarian cancer patients. Br J Cancer 57:503–508

Bacus JW, Wiley EL, Galbraith W, Marshall PN, Wilbanks GD, Weinstein RS (1984) Malignant cell detection and cervical cancer screening. Anal Quant Cytol 6:121–130

Baisch H, Beck H-P, Christinses IJ et al. (1982) A comparison of mathematical methods for the analysis of DNA histograms obtained by flow cytometry. Cell Tissue Kinet 15:235–249

Baranska W, Kujawa M, Ochocka M (1983) Morphometric ultrastructural analysis of nuclear components in T-cell acute lymphoblastic leukaemias. Folia Haematol (Leipz) 110:481–489

Barry JD, Sharkey FE (1985) Observer reproducibility during computer-assisted planimetric measurements of nuclear features. Hum Pathol 16:225–227

Beck JS, Anderson LM (1987) Quantitative methods as an aid to diagnosis in histopathology. In: Anthony PP, MacSween RNM (eds) Recent advances in histopathology, vol 13. Churchill Livingstone, Edinburgh, pp 255–269

Beck JS, Morley SM, Gibbs JH, Potts RC, Ilias MI, Kardjito T (1986) The cellular responses of tuberculosis and leprosy patients and of healthy controls in skin tests to 'New Tuberculin' and Leprosin A. Clin Exp Immunol 64:484–494

Bedi KS, Goldstein DJ (1976) Apparent anomalies in nuclear Feulgen-DNA contents. J Cell Biol 71:60–64

Bennington JL, Mayall BH (1983) DNA cytometry on four-micrometer sections of paraffin-embedded human renal adenocarcinomas and adenomas. Cytometry 4:31–39

Bhattacharjee DK, Harris M, Faragher EB (1985) Nuclear morphometry of epitheliosis and intraduct carcinoma of the breast. Histopathology 9:511–516

Bichel P, Poulsen HS, Andersen J (1982) Estrogen receptor content and ploidy of human mammary carcinoma. Cancer 40:1771–1774

Bjelkenkrantz K, Herder A, Grontoft O, Stal O (1982) Cytophotometric characterisation of the WHO grades of transitional cell neoplasms. Pathol Res Pract 174:68–77

Bjelkenkrantz K, Lundgren J, Olofsson J (1983) Single cell DNA measurements in hyperplastic, dysplastic and carcinomatous laryngeal epithelia, with special reference to the occurrence of hypertetraploid cell nuclei. Anal Quant Cytol 5:184–188

Black MM, Speer FD (1957) Nuclear structure in cancer tissues. Surg Gynecol Obstet 105:97–102

Bloom HJG, Richardson WW (1957) Histologic grading and prognosis in breast cancer. Br J Cancer 11:359–377

Bocking A, Adler C-P, Common HH, Hilgarth M, Granzen B, Auffermann W (1984a) Algorithm for a DNA-cytophotometric diagnosis and grading of malignancy. Anal Quant Cytol 6:1–8

Bocking A, Auffermann W, Schwarz H, Bammert J, Dorrer G, Vucicuja S (1984b) Cytology of prostatic carcinoma: quantification and validation of diagnostic criteria. Anal Quant Cytol 6:74–88

Bohm N, Sandritter W (1975) DNA in human tumors: a cytophotometric study. In: Grundmann E, Kirsten WH (eds) Current topics in pathology, vol 60. Springer, Berlin Heidelberg New York, pp 152–219

Bondeson L, Bondeson A-G, Lindholm K (1983) Morphometric studies on nuclei in smears of fine needle aspirates from oxyphilic tumors of the thyroid. Acta Cytol 27:437–440

Boon ME, Lowhagen T, Willems JS (1980) Planimetric studies on fine needle aspirates from follicular adenoma and follicular carcinoma of the thyroid. Acta Cytol 24:145–148

Boon ME, Kurver PJH, Baak JPA, Thompson HT (1981) The application of morphometry in gastric cytological diagnosis. Virchows Arch [A] 393:159–164

Boon ME, Lowhagen T, Cordoso PL, Blonk DI, Kurver PJH, Baak JPA (1982a) Computation of preoperative diagnosis probability for follicular adenoma and carcinoma of the thyroid on aspiration smears. Anal Quant Cytol 4:1–5

Boon ME, Trott PA, van Kaam H, Kurver PJH, Leach A, Baak JPA (1982b) Morphometry and cytodiagnosis of breast lesions. Virchows Arch [A] 396:9–18

Boon ME, Kurver PJH, Baak JPA, Ooms ECM (1984) Morphometric differences between urothelial cells in voided urine of patients with grade I and grade II bladder tumors. J Clin Pathol 34:612–615

Boquoi E, Krebs S, Kruezer G (1975) Feulgen-DNA-cytophotometry on mammary tumor cells from aspiration biopsy smears. Acta Cytol 19:326–331

Bowie JE, Young IT (1977) An analysis technique for biological shape. II. Acta Cytol (Baltimore) 21:455–464

Boysen M, Reith A (1983) Discrimination of various epithelia by simple morphometric evaluation of the basal cell layer. Virchows Arch [Cell Pathol] 42:173–184

Boysen M, Marton PF, Pilstrom L, Solberg LA, Torp T (1983) A simple and efficient method for objective discrimination between pseudostratified metaplastic and dysplastic nasal epithelium. J Microsc 130:99–106

Broers JLV, Pahlplatz MMM, Katzko MW, Oud PS, Ramaekers FCS, Carney DN, Vooijs GP (1988) Quantitative description of classic and small cell lung cancer lines by nuclear image cytometry. Cytometry 9:426–431

Brown RA, Swanson Beck J (1988a) Statistics on microcomputers. A non-algebraic guide to

their appropriate use in biomedical research and pathology laboratory practice. A series of six articles. 1. Data handling and preliminary analysis. J Clin Pathol 41:1033–1038
Brown RA, Swanson Beck J (1988b) Statistics on microcomputers. A non-algebraic guide to their appropriate use in biomedical research and pathology laboratory practice. 2. Confidence intervals and significance tests. J Clin Pathol 41:1148–1154
Brown RA, Swanson Beck J (1988c) Statistics on microcomputers. A non-algebraic guide to their appropriate use in biomedical research and pathology laboratory practice. 3. Analysis of variance and distribution-free methods. J Clin Pathol 41:1256–1262
Brown RA, Swanson Beck J (1989a) Statistics on microcomputers. A non-algebraic guide to their appropriate use in biomedical research and pathology laboratory practice. 4. Correlation and regression. J Clin Pathol 42:4–12
Brown RA, Swanson Beck J (1989b) Statistics on microcomputers. A non-algebraic guide to their appropriate use in biomedical research and pathology laboratory practice. 5. Analysis of categorical data. J Clin Pathol 42:117–122
Brown RA, Swanson Beck J (1989c) Statistics on microcomputers. A non-algebraic guide to their appropriate use in biomedical research and pathology laboratory practice. 6. Statistical methods for diagnostic tests. J Clin Pathol 42:225–230
Brown RA, Al Moussa M, Beck JS (1986) Histometry of the normal thyroid in man. J Clin Pathol 39:475–482
Brungger A, Cruz-Orive LM (1987) Nuclear morphometry of nodular malignant melanomas and benign melanocytic nevi. Arch Dermatol Res 279:412–414
Bullough WS (1950) Mitotic activity in tissues of dead mice, and tissues kept in physiological salt solutions. Exp Cell Res 1:410–420
Burns J, Graham AK, Frank C, Fleming KA, Evans MF, McGee JO'D (1987) Detection of low copy human papillomavirus DNA and mRNA in routine paraffin sections of cervix by non-isotopic in situ hybridisation. J Clin Pathol 40:858–864
Campbell MJ, Gardner MJ (1988) Calculating confidence intervals for some non-parametric analyses. Br Med J 296:1454–1456
Cantrell BB, de Klerk DP, Eggleston JC, Boitnett JK, Walsh PC (1981) Pathological factors that influence prognosis in stage A prostatic carcinoma: the influence of extent versus grade. J Urol 125:516–520
Caspersson T (1936) Über den chemischen Aufbau der Strukturen des Zellkerns. Scand Arch Physiol 73 [Suppl 8]:1–12
Castleman KR (1979) Digital image processing. Prentice Hall, Englewood Cliffs (New Jersey)
Castleman KR, White BS (1980) The trade-off of cell classifier error rates. Cytometry 1:156–160
Chan KW, Chiu KY, Fu KH, Ling ML (1987) Observer variability in microcomputer-assisted morphometric study of nuclear parameters. Pathology 19:407–409
Christopherson WM, Williamson EO, Gray LA (1972) Leiomyosarcoma of the uterus. Cancer 29:1512–1517
Cocker J, Fox H, Langley FA (1968) Consistency in the histological diagnosis of abnormalities of the cervix uteri. J Clin Pathol 21:67–70
Colgan TJ, Norris HJ, Foster W, Kurman RJ, Fox CH (1983) Predicting the outcome of endometrial hyperplasia by quantitative analysis of nuclear features using a linear discriminant function. Int J Gynecol Pathol 1:347–352
Collins JE, Jenkins D, McCance DJ (1988) Detection of human papillomavirus DNA sequences by in situ DNA-DNA hybridisation in cervical intraepithelial neoplasia and invasive carcinoma: a retrospective study. J Clin Pathol 41:289–295
Cornelisse CJ, de Koning HR, Arentz PW, Roatgener JW, van Heerde P (1981) Quantitative analysis of nuclear area variations in benign and malignant breast cytology specimens. Anal Quant Cytol 3:128–138
Cornelisse CJ, de Koning HR, Moolenar AJ, van de Velde CJ, Ploem JS (1984) Image and flow cytometric analysis of DNA content in breast cancer: relation to estrogen receptor content and lymph node involvement. Anal Quant Cytol 6:9–13
Couve E (1985) Morphometric analysis of the nucleolus during the life cycle of human odontoblasts. Anat Rec 213:215–224

Crocker J (1984) Morphometric and related quantitative techniques in the study of lymphoid neoplasms. A review. J Pathol 143:69–80

Crocker J, Curran RC (1979) A study of nuclear diameters in lymph node imprints using the Zeiss Microvideomat. J Clin Pathol 32:670–674

Crum CP, Ikenberg H, Richart RM, Gissman L (1984) Human papillomavirus type 16 and early cervical neoplasia. N Engl J Med 310:880–883

Dardick I (1984) Morphometry of normal human lymphoid tissues. Nuclear invaginations and clefts. Arch Pathol Lab Med 108:197–201

Dardick I, Caldwell D (1985) Reproducibility of morphometric image analysis. Hum Pathol 16:1178

Dardick I, Dardick AM (1984) Morphometry of normal human lymphoid tissues. Nuclear parameters for comparative studies of lymphoma. Arch Pathol Lab Med 108:190–196

Dardick I, Caldwell DR, McCaughey WT, Al-Jabi M (1984) Nuclear morphologic and morphometric analyses of large noncleaved cell and immunoblastic non-Hodgkin's lymphomas. Hum Pathol 15:965–972

Dardick I, Caldwell DR, Bailey DB, Dardick AM, Jeans MT (1985 a) Nuclear morphologic and morphometric analyses of nodular poorly differentiated lymphocytic lymphoma: assessment of small cleaved nuclei. Hum Pathol 16:1187–1199

Dardick I, Dardick AM, Caldwell AR, Jeans MT, Bladon T, Setterfield G (1985 b) Non-Hodgkin's lymphoma classification: ultrastructural morphometric studies for the quantification of nuclear compartents in situ. Hum Pathol 16:1046–1060

Dardick I, Caldwell DR, Silver SS, Tubbs RR (1987) Lymphocyte nuclear morphometry in diffuse well differentiated lymphocytic lymphoma. Comparative morphometry of normal lymphoid tissues, non-Hodgkin's lymphoma and Hodgkin's disease. Arch Pathol Lab Med 111:130–138

Davis RK, Fox C, Heffner DK (1987) Computerised nuclear morphometry: a reproducible cytopathologic marker of head and neck cancer. Otolaryngol Head Neck Surg 96:15–21

De Hoff RT, Rhines FN (1961) Determination of the number of particles per unit volume from measurements in random plane sections: the general cylinder and ellipsoid. Transactions of the American Institute of Mining and Metallurgical Engineers 221:975–982

Denekamp J, Kallman RF (1973) In vivo and in vitro labelling of animal tumours with tritiated thymidine. Cell Tissue Kinet 6:217–227

Derman H, Koss LG, Hyman MP, Penner DW, Soule E, Hicklin MD (1981) Cervical cytopathology. I. Peers compare performance. Pathologist 36:317–321

Diamond DA, Berry SJ, Jewett HJ, Coffey DS (1982 a) A new method to assess metastatic potential of human prostatic cancer. J Urol 128:729–734

Diamond DA, Berry SJ, Umbricht C, Jewett HJ, Coffey DS (1982 b) Computerized image analysis of nuclear shape as a prognostic factor for prostatic cancer. Prostate 3:321–332

Diegenbach PC, Baak JPA (1977) Quantitative nuclear image analysis: differentiation between normal, hyperplastic and malignant appearing uterine glands in a paraffin section. II. Computer assisted recognition by discriminant analysis. Eur J Obstet Gynecol Reprod Biol 7:389–394

Diegenbach PC, Baak JPA (1978 a) Quantitative nuclear image analysis: differentiation between normal, hyperplastic and malignant appearing uterine glands in a paraffin section. III. The use of texture features for differentiation. Eur J Obstet Gynecol Reprod Biol 8:109–116

Diegenbach PC, Baak JPA (1978 b) Quantitative nuclear image analysis: differentiation between normal, hyperplastic, and malignant appearing uterine glands in a paraffin section. IV. The use of Markov chain texture features in discriminant analysis. Eur J Obstet Gynecol Reprod Pathol 8:157–162

Dixon B, Stead RH (1977) Feulgen microdensitometry and analysis of S-phase cells in cervical tumour biopsies. J Clin Pathol 30:907–913

Dormer P (1987) Morphometry and cytometry correlated to quantitative autoradiography. Anal Quant Cytol Histol 9:115–122

Duijndam WAL, Smeulders AWM, Van Duijn P, Verweig AC (1980 a) Optical errors in scanning stage absorbance cytophotometry. I. Procedures for correcting apparent integrated observance values for distributional, glare, and diffraction errors. J Histochem Cytochem 28:388–394

Duijndam WAL, Van Duijn P, Riddersma SH (1980b) Optical errors in scanning stage absorbance cytophotometry. II. Application of correction factors for residual distributional error, glare, and diffraction error in practical cytophotometry. J Histochem Cytochem 28:395–400

Dundas SAC, Kay R, Beck S, Cotton DWK, Coup AJ, Slater DN, Underwood JCE (1987) Can histopathologists reliably assess dysplasias in chronic inflammatory bowel disease? J Clin Pathol 40:1282–1286

Epstein JI, Berry SJ, Eggleston JC (1984) Nuclear roundness factor: a predictor of progression in untreated stage A2 prostatic cancer. Cancer 54:1666–1671

Erhardt K, Auer G, Bjorkholm et al. (1984) Prognostic significance of nuclear DNA content in serous ovarian tumors. Cancer Res 44:2198–2202

Erhardt K, Auer G, Bjorkholm E et al. (1985) Combined morphologic and cytochemical grading of serous ovarian tumors. Am J Obstet Gynecol 151:356–361

Evans DMD, Shelley G (1982) Respiratory cytodiagnosis: study in observer variation and its relation to quality of material. Thorax 37:259–263

Evensen A (1965) Significance of mitotic duration in evaluating kinetics of cellular proliferation. Nature 195:718–719

Flint A, Davenport RD, Lloyd RV, Beckwith AL, Thompson NW (1988) Cytophotometric measurements of Hurthle cell tumors of the thyroid gland. Cancer 61:110–113

Fossa SD (1977) DNA-values in transitional cell carcinoma of the urinary bladder. Eur J Cancer 13:1155–1161

Fossa SD, Thorud E, Vaage S, Shoaib MC (1983) DNA cytometry of primary breast cancer. Acta Pathol Microbiol Scand [A] 91:235–243

Fox H (1987) Biology of teratomas. In: Anthony PP, MacSween RNM (eds) Recent advances in histopathology, vol 13. Churchill Livingstone, Edinburgh, pp 33–43

Fox H, Buckley CH (1982) The endometrial hyperplasias and their relationship to endometrial neoplasia. Histopathology 6:493–510

Friedlander ML, Taylor IW, Russell P, Musgrove EA, Hedley DH, Tattersall MHN (1983) Ploidy as a prognostic factor in ovarian cancer. Int J Gynecol Pathol 2:55–63

Fu YS, Hall TL (1985) DNA measurement in tissue sections. Anal Quant Cytol 7:90–96

Fu YS, Robboy SJ, Prat J (1978) Nuclear DNA study of vaginal and cervical squamous cell abnormalities in DES-exposed progeny. Obstet Gynecol 52:129–137

Fu YS, Reagan JW, Richart RM, Townsend DE (1979) Nuclear DNA and histopathologic studies of genital lesions in DES-exposed progeny. I. Intraepithelial squamous abnormalities. Am J Clin Pathol 72:503–514

Fu YS, Temmin L, Olaizola YM, Reagan JW (1980) Nuclear DNA characteristics of microinvasive squamous carcinoma of the uterine cervix. In: Fenoglio CM, Wolff MW (eds) Progress in surgical pathology, vol 1. Masson, New York, pp 233–244

Fu YS, Reagan JW, Richart RM (1981) Definition of cervical precursors. Gynecol Oncol 12:S220–223

Fu YS, Reagan JW, Fu AS, Janiga KE (1982) Adenocarcinoma and mixed carcinoma of the uterine cervix. II. Prognostic value of nuclear DNA analysis. Cancer 49:2571–2577

Fu YS, Braun L, Shah KV, Lawrence WD, Robboy SJ (1983) Histologic, nuclear DNA and human papillomavirus studies of cervical condylomas. Cancer 52:1705–1711

Fuster E, Garcia-Vilanova A, Narbona B, Romero R, Llombart-Bosch A (1983) A statistical approach to an individualized prognostic index (IPI) for breast cancer survivability. Cancer 52:728–736

Gamel JW, McLean IW (1983) Computerised histologic assessment of malignant potential. II. A practical method for predicting survival following enucleation of uveal melanoma. Cancer 52:1032–1038

Gamel WG, McLean IW, Greenberg RA, Zimmerman LE, Lichtenstein SJ (1982) Computerised histologic assessment of malignant potential: a method for detecting the prognosis of uveal melanomas. Hum Pathol 13:893–897

Garcia GL, Kuklinski WS, Zahniser DJ, Oud PS, Vooys PG, Brenner JF (1987) Evaluation of contextual analysis for computer classification of cervical smears. Cytometry 8:210–216

Gardner MJ, Altman DG (1986) Confidence intervals rather than P values: estimation rather than hypothesis testing. Br Med J 292:746–750

Giangaspero F, Muhlbaier LH, Burger PC (1984) The glioblastoma multiforme and the metastatic carcinoma: a morphometric study of nuclear size and shape. Appl Pathol 2:160–167

Gil J, Marchevsky AM, Silage DA (1986) Applications of computerized morphometry in pathology. I. Tracings and generation of graphic standards. Lab Invest 54:222–227

Gilbert P, Pfitzer P (1977) Facultative polyploidy in endocrine tissue. Virchows Arch [Cell Pathol] 25:233–242

Giroud F, Gauvain C, Seigneurin D, von Hagen V (1988) Chromatin texture changes related to proliferation and maturation in erythrocytes. Cytometry 9:339–348

Goldstein DJ (1970) Aspects of scanning microdensitometry I. Stray light (glare). J Microsc 92:1–16

Goldstein DJ (1971 a) Aspects of scanning microdensitometry II. Spot size, focus, and illumination. J Microsc 93:15–21

Goldstein DJ (1971 b) Aspects of scanning microdensitometry III. The monochromator system. J Microsc 105:33–38

Gonzales RC, Thomason MG (1978) Syntactic pattern recognition. An introduction. Addison Wesley, Reading (Mass.)

Goudie RB (1987) Immunohistochemistry in diagnostic histopathology. In: Anthony PP, MacSween RNM (eds) Recent advances in histopathology, vol 13. Churchill Livingstone, Edinburgh, pp 233–254

Graem N (1979) Mitotic activity and delay of fixation of tumour tissue. Acta Pathol Microbiol Scand [A] 87:375–378

Graham AR, Paplanus SH, Bartels PH (1988) Microphotometry of colonic lesions. Lab Invest 59:397–402

Greenberg SD, Smith S, Swank PR et al. (1982) Visual cell profiles for quantitation of premalignant cells in sputum: a preliminary report. Acta Cytol 26:809–813

Greenberg SD, Hunter NR, Taylor GR (1986) Application of cell image analysis to the diagnosis of cellular atypia in sputum: a review. Diagn Cytopathol 2:168–174

Gschwind R, Umbricht CB, Torhorst J, Oberholzer M (1986) Evaluation of shape descriptors for the morphometric analysis of cell nuclei. Pathol Res Pract 181:213–222

Haagensen CD (1971) Diseases of the breast, 2nd ed. W.B. Saunders, Philadelphia

Hall TL, Fu YS (1985) Applications of quantitative microscopy in tumour pathology. Lab Invest 53:5–21

Harris SC, Currie A, Anderson G, Howat AJ (1987) Transport media for fine needle aspiration cytology. J Clin Pathol 40:1263

Haug H (1980) The significance of quantitative stereologic experimental procedures in pathology. Pathol Res Pract 166:144–164

Hedley DW, Friedlander ML, Taylor IW, Rugg CA, Musgrove EA (1983) Method for analysis of cellular DNA content of paraffin-embedded pathological material using flow cytometry. J Histochem Cytochem 31:1333–1335

Hedley DW, Friedlander ML, Taylor IW (1985) Application of DNA flow cytometry to paraffin-embedded archival material for the study of aneuploidy and its clinical significance. Cytometry 6:327–333

Herder A, Bjelkenkrantz K, Grontoft O (1982) Histopathological subgrouping of WHO grade II urothelial neoplasms by cytophotometric measurements of nuclear atypia. Acta Pathol Microbiol Immunol Scand [A] 90:405–408

Hermens AF, Barendsen GW (1969) Changes of cell proliferation characteristics in a rat rhabdomyosarcoma before and after irradiation. Eur J Cancer 5:173–189

Heuson JC, Longeval E, Mattheiem WH, Deboel MC, Sylvester RJ, Leclerq G (1977) Significance of quantitative assessment of estrogen receptors for endocrine therapy in advanced breast cancer. Cancer 39:1971–1977

Hiddemann W, Schumann J, Andreeff M (1984) Convention on nomenclature for DNA cytometry. Cytometry 5:445–446

Holm LE (1982) Cellular DNA amount of squamous cell carcinomas of the head and neck region in relation to prognosis. Laryngoscope 92:1064–1069

Inokuchi K, Kodama Y, Sasaki O, Kamegawa T, Okamura T (1983) Differentiation of growth patterns of early gastric cancer determined by cytophotometric DNA analysis. Cancer 51:1138–1141

Ismail SM, Colclough AB, Dinnen JS et al. (1989) Observer variation in histopathological diagnosis and grading of cervical intraepithelial neoplasia. Br Med J 298:707–710

Izuo M, Okagaki T, Richart RM, Lattes R (1971 a) Nuclear DNA content of acinar cells of the human breast during lactation. Am J Clin Pathol 56:443–447

Izuo M, Okagaki T, Richart RM, Lattes R (1971 b) DNA content in "apocrine metaplasia" of fibrocystic disease of the breast. Cancer 27:643–650

Jarvis LR, Whitehead R (1985) Morphometric analysis of gastric dysplasia. J Pathol 147:133–138

Jarvis LR, Graff PS, Whitehead R (1987) Correlation of nuclear ploidy with histology in adenomatous polyps of colon. J Clin Pathol 40:26–33

Kayser KK, Modlinger F, Postl K (1985) Quantitative low resolution analysis of colon mucosa. Anal Quant Cytol Histol 7:205–212

Kino I, Richart RM, Lattes R (1973) DNA in salivary gland tumors. I. Warthin tumors, benign mixed tumors, and mucoepidermoid carcinomas. Arch Pathol 95:245–251

Knight WA, Livingston RB, Gregory EJ, McGuire WL (1977) Estrogen receptor as an independent prognostic factor for early recurrence in breast cancer. Cancer Res 37:4669–4671

Koss LG, Bartels PH, Bibbo M, Freed SZ, Taylor J, Wied GL (1975) Computer discrimination between benign and malignant urothelial cells. Acta Cytol 19:378–391

Koss LG, Bartels PH, Bibbo M, Freed SZ, Sychra JJ, Taylor J, Wied GL (1977 a) Computer analysis of atypical urothelial cells. I. Classification by supervised learning alogarithms. Acta Cytol 21:247–260

Koss LG, Bartels PH, Bibbo M, Synchra JJ, Wied GL (1977 b) Computer analysis of atypical urothelial cells. II. Classification by unsupervised learning alogarithms. Acta Cytol 21:261–265

Koss LG, Bartels PH, Synchra JJ, Wied GL (1978 a) Computer discriminant analysis of atypical urothelial cells. Acta Cytol 22:382–386

Koss LG, Bartels PH, Synchra JJ, Wied GL (1978 b) Diagnostic cytologic sample profiles in patients with bladder cancer using TICAS system. Acta Cytol 22:392–397

Kreicbergs A, Zetterberg A (1980) Cytophotometric DNA measurements in tissue sections from old paraffin-embedded specimens. Anal Quant Cytol 2:84–92

Kreicbergs A, Zetterberg A, Soderberg G (1980) The prognostic significance of nuclear DNA content in chondrosarcoma. Anal Quant Cytol 2:271–279

Kreicbergs A, Cewrien G, Tribukait B, Zetterberg A (1981) Comparative single cell and flow DNA analysis of bone sarcoma. Anal Quant Cytol 3:121–127

Laskey RA (1987) Basic molecular and cell biology: the cell nucleus. Br Med J 295:1121–1123

Lee FD, Burnett RA (1987) Quality assurance in histopathology. J Pathol 152:247–251

Lesty C, Raphael M, Nonnenmacher RL, Leblond-Missenard V, Delcourt A, Howard A, Binet JL (1986) An application of mathematical morphology to analysis of the size and shape of nuclei in tissue sections of non-Hodgkin's lymphoma. Cytometry 7:117–131

Lovett EJ, Schmitzer B, Keren DF, Flint A, Hudson JL, McClatchey KD (1984) Application of flow cytometry to diagnostic pathology. Lab Invest 50:115–140

Lowe JS, Harvey L (1986) Low cost image digitisation using a BBC microcomputer. J Pathol 148:109 A

Luck JB, Mamaw VR, Frable WJ (1982) Fine needle aspiration biopsy of the thyroid: differential diagnosis by videoplan image analysis. Acta Cytol 26:793–796

Makuda T, Sasano N, Sato E (1978) Evaluation of esophageal dysplasia by cytofluorometric analysis. Cancer 41:1399–1404

Maraldi NM, Marinelli F, Cocco L, Papa S, Santi P, Manzoli FA (1986 a) Morphometric analysis and topographical organisation of nuclear matrix in freeze-fractured electron microscopy. Exp Cell Res 163:349–362

Maraldi NM, Marinelli F, Ranaldi R, Papa S, Mariuzzi GM, Manzoli FA (1986 b) Morphometric and topologic analysis of freeze-fractured interphase nuclei. Anal Quant Cytol Histol 8:343–348

Marchevsky AM, Gil J (1986) Applications of computerised interactive morphometry in pathology. II. A model for computer generated 'diagnosis'. Lab Invest 54:708–716
Marchevsky AM, Gil J, Caccami D (1985) Computerised interactive morphometry. A study of malignant mesothelioma in pleural biopsy specimens. Arch Pathol Lab Med 109:1102–1105
Marchevsky AM, Gil J, Silage DA (1986) Computerised interactive morphometry as a potentially useful tool for the classification of non-Hodgkin's lymphomas. Cancer 57:1544–1549
Marchevsky AM, Gil J, Jeanty H (1987a) Computerised interactive morphometry in pathology: current instrumentation and methods. Hum Pathol 18:320–331
Marchevsky AM, Hauptman E, Gil J, Watson C (1987b) Computerized interactive morphometry as an aid for the diagnosis of pleural effusions. Acta Cytol 31:131–136
Marchevsky AM, Klapper E, Gil J (1987c) Computerized interactive morphometry: automatic classification of nuclear profiles in non-Hodgkin's lymphoma. Am J Clin Pathol 87:561–568
Marr D (1982) Vision: a computational investigation into the human representation and processing of visual information. W.H. Freemann, San Francisco
Mayall BH, Mendelsohn ML (1970) Deoxyribonucleic acid cytophotometry of stained human leucocytes. II. The mechanical scanner of CYDAC, the theory of scanning photometry, and the magnitude of residual errors. J Histochem Cytochem 18:383–407
McLean IW, Gamel JW (1988) Prediction of metastasis of uveal melanoma: comparison of morphometric determination of nucleolar size and spectrophotometric determination of DNA. Invest Ophthalmol Vis Sci 29:507–511
Meijer CJLM, van der Loo EM, van Vloten WA, van der Velde EA, Scheffer R, Cornelisse CJ (1980) Early diagnosis of mycosis fungoides and Sézary's syndrome by morphometric analysis of lymphoid cells in the skin. Cancer 45:2864–2871
Mize RR (1985) The microcomputer in cell and neurobiology research. Elsevier, New York
Moberger B, Forsslun AG, Moberger G (1984) The prognostic significance of DNA measurements in endometrial carcinoma. Cytometry 5:430–438
Moller U, Keiding N (1982) Circadian variations in influx and efflux of the S phase in a partially synchronised system. Methodological problems concerning pulse labelling of epithelial cells in the hamster cheek pouch. Cell Tissue Kinet 11:405–413
Morris JA, Gardner MJ (1988) Calculating confidence intervals for relative risks (odds ratios) and standardised ratios and rates. Br Med J 296:1313–1316
Mossler JA, McCarty KS, Woodard BH, Mitchener LM, Johnston WW (1982) Correlation of mean nuclear area with estrogen receptor content in aspiration cytology of breast carcinoma. Acta Cytol 26:417–421
Muller J, Skakkebaek NE (1984) Abnormal germ cells in maldescended testes: a study of cell density, nuclear size, and deoxyribonucleic acid content in testicular biopsies from 50 boys. J Urol 131:730–733
Muller J, Skakkebaek NE, Lundsteen C (1981) Aneuploidy as a marker for carcinoma-in-situ of the testis. Acta Pathol Microbiol Scand [A] 89:67–68
Naora H (1955) Microspectrophotometry of cell nuclei stained by the Feulgen reaction. Exp Cell Res 8:259–265
Nasiell M, Kato H, Auer G, Zetterberg A, Roger V, Karlen L (1978) Cytomorphological grading and Feulgen DNA-analysis of metaplastic and neoplastic bronchial cells. Cancer 41:1511–1521
Nealson TF, Nkongho A, Grossi CE, Ward R, Nealon C, Gillooley JF (1981) Treatment of early cancer of the breast (T1NOMO and T2NOMO) on the basis of histologic characteristics. Surgery 89:279–289
Okagaki T, Izuo M (1978) Correction of the modal DNA values obtained by microspectrophotometry and test for their shifts. J Natl Cancer Inst 60:1251–1258
Ooms ECM, Anderson WAD, Alons CL, Boon ME, Veldhuizen RW (1983a) Analysis of the performance of pathologists in the grading of bladder tumours. Hum Pathol 14:140–143
Ooms ECM, Anderson WAD, Alons CL, Boon ME, Veldhuizen RW (1983b) Morphometric grading of bladder tumors in comparison with histologic grading by pathologists. Hum Pathol 14:144–150

Oyama T (1989) A histopathological, immunohistochemical and ultrastructural study of intranuclear cytoplasmic inclusions in thyroid papillary carcinoma. Virchows Arch [A] 414:91–104
Pak HY, Ashdjian V, Yokoto SB, Teplitz RL (1981) Quantitative DNA determinations by image analysis. I. Application of human pulmonary cytology. Anal Quant Cytol 4:95–104
Panno JP (1988) Symmetry analysis of cell nuclei. Cytometry 9:195–200
Panozzo R, Huerliman H (1983) A simple method for the quantitative discrimination of convex and convex-concave lines. Microscopica Acta 87:169–176
Parl FF, Dupont WD (1982) A retrospective cohort study of histologic risk factors in breast cancer patients. Cancer 50:2410–2416
Payne CM, Hicks MJ, Kim A (1985) Ultrastructural morphometric analysis of normal human lymphocytes stimulated in vitro with mitogens and antigens. Am J Pathol 120:263–275
Payne CM, Hicks MJ, Bjore CG Jr, Kibler R (1987) Ultrastructural morphometric analysis of nuclear contour irregularity in normal cord and adult blood: correlation with distinct lymphocyte subpopulations. Diagn Clin Immunol 5:41–53
Peet FG, Sahota TS (1984) A computer-assisted cell identification system. Anal Quant Cytol 6:59–65
Pelstring RJ, Swerdlow SH (1987) Improved nuclear contour indices for lymphoid morphometry. Anal Quant Cytol Histol 9:469–474
Pennington DG, Olsen TE (1970) Megakaryocytes in states of altered platelet production: cell numbers, size and DNA content. Br J Haematol 18:447–463
Pohl MN, Swartz FJ, Carstens PHB (1981) Polyploidy in islets of normal and diabetic humans. Hum Pathol 12:184–186
Polig E, Kimmel DB, Jee WS (1984) Morphometry of bone cell nuclei and their location relative to bone surfaces. Phys Med Biol 29:939–952
Pratt WK (1978) Digital image processing. Wiley Interscience, New York
Quirke P, Dyson JED (1986) Flow cytometry: methodology and applications in pathology. J Pathol 149:79–88
Rajewsky MF (1965) In vitro studies of cell proliferation in tumours. Eur J Cancer 1:281–287
Raphael M, Lesty C, Nonnenmacher L, Delcourt A, Missenard-Leblond V, Binet JL (1985) Morphometric characterisation of nuclei in non-Hodgkin's malignant lymphoma. Anal Quant Cytol Histol 7:283–287
Rasch EM (1985) Measurement of DNA 'standards' and the range of accurate DNA estimates by Feulgen absorption microspectrophotometry. In: Cowden RR, Harrison FW (eds) Advances in microscopy. Alan R Liss, New York, pp 137–166
Reagan JW, Ng ABP (1973) The cells of uterine adenocarcinoma, 2nd edn. S. Karger, Basel
Reagan JW, Hamonic MJ, Wentz WB (1957) Analvtical study of the cells in cervical squamous cell cancer. Lab Invest 6:241–248
Renau-Piqueras J, Cervera J (1983) Chromatin pattern of isolated human small thymocytes. A morphometric and stereological study. Virchows Arch [Cell Pathol] 42:315–325
Ricco R, De Benedictis G, Giardina C, Bufo P, Resta L, Pesce-Delfino V (1985) Morphometric analytical evaluators of lymphoid populations in nonneoplastic lymph nodes. Anal Quant Cytol Histol 7:288–293
Rigaut JP, Boysen M, Reith A (1985) Karyometry of pseudostratified, metaplastic and dysplastic nasal epithelium by morphometry and stereology. 2. Automated image analysis (IBAS) of the basal layer of nickel workers. Pathol Res Pract 180:151–160
Robertson AJ, Anderson JM, Beck JS et al. (1989) Observer variability in histopathological reporting of cervical biopsy specimens. J Clin Pathol 42:231–238
Ronstrom L, Tribukait B, Eposti PL (1981) DNA pattern and cytological findings in fine-needle aspirates of untreated prostatic tumors: a flow cytometric study. Prostate 2:79–88
Ruiter DI, Cornelisse CJ, van Rijssel TG, van der Velde EA (1977) Aneurysmal bone cysts and telangiectatic osteogenic sarcoma: histopathological and morphometric study. Virchows Arch [A] 373:311–325
Sachs H (1970) Zytophotometrische Untersuchungen bei Präkanzerosen der Mamma. Beitr Pathol 143:360–377
Sadun AA (1985) Vision: a multimodal sense. Bull Clin Neurosci 50:61–68

Sassen RJ, Baak JPA (1986) Morphometry in the differential diagnosis of granulosa cell tumors of the ovary. Anal Quant Cytol Histol 8:245–249

Schrek R (1971) Ultrastructure of blood lymphocytes from chronic lymphocytic and lymphosarcoma cell leukaemia. J Natl Cancer Inst 48:51–64

Sears HF, Janus C, Levy W, Hopsen R, Creech R, Grotzinger P (1982) Breast cancer without axillary metastasis. Are there high risk biologic subpopulations? Cancer 50:1820–1827

Seybolt JF, Johnson WD (1971) Cervical cytodiagnosis problems: a survey. Am J Obstet Gynecol 109:1089–1103

Sherman AB, Koss LG, Adams SE (1984) Intraobserver and interobserver differences in the diagnosis of urothelial cells: comparison with classification by computer. Anal Quant Cytol 6:112–120

Shum DT, Roberts JT, Smout MS, Wells GA, Simon GT (1986) The value of nuclear contour index in the diagnosis of mycosis fungoides. An assessment of current ultrastructural morphometric diagnostic criteria. Cancer 57:298–304

Siegler EE (1956) Microdiagnosis of carcinoma in situ of the uterine cervix: a comparative study of pathologists' diagnoses. Cancer 9:463–466

Silage DA, Gil J (1984) Use of a touch-sensitive screen in interactive morphometry. J Microsc 134:315–321

Simnett JD (1967) The measurement of mitotic incidence and radioautographic labelling index from tissue sections: some mathematical considerations. J R Microsc Soc 88:371–382

Simnett JD, Heppleston AG (1966) Cell renewal in the mouse lung. The influence of sex, strain and age. Lab Invest 15:1793–1801

Simon GT (1987) The value of morphometry in the ultrastructural diagnosis of mycosis fungoides. Ultrastruct Pathol 11:687–691

Skaarland E (1985) Morphometric analysis of nuclei in epithelial structures from normal and neoplastic endometrium: a study using the Isaacs cell sampler and Endoscann instruments. J Clin Pathol 38:496–501

Sklarew RJ (1982) Simultaneous Feulgen densitometry and autoradiographic grain counting with the Quantimet 720 D image-analysis system. I. Estimation of nuclear DNA content in ($^3$H) thymidine-labeled cells. J Histochem Cytochem 30:35–48

Slooten HV, Schaberg A, Smeenk SD (1985) Morphologic characteristics of benign and malignant adrenocortical tumors. Cancer 55:766–773

Sokol RJ (1989) Nuçleoli of blood monocytes in malignant lymphoma: a morphometric study. Am J Clin Pathol 91:60–63

Spander PJ, Ruiter DJ, Hermans J, de Vogt HJ, Brusse JAM, Boon ME (1982) The implications of subjective recognition of malignant cells in aspirations for grading of prostatic cancer using cell image analysis. Anal Quant Cytol 2:123–127

Sprenger E, Witte S (1978) The diagnostic significance of nuclear DNA measurement in cytologic smears of benign and malignant gastric lesions. Pathol Res Pract 163:148–157

Sprenger E, Lowhagen T, Vogt-Schoden M (1977) Differential diagnosis between follicular adenoma and follicular carcinoma of the thyroid by nuclear DNA determination. Acta Cytol 21:528–530

Sprenger E, Ulrich H, Schondorf H (1979) The diagnostic value of nuclear DNA determination in aspiration cytology of benign and malignant lesions of the breast. Anal Quant Cytol 1:29–36

Spriggs AI, Bowey CE, Cowdell RH (1971) Chromosomes of precancerous lesions of the cervix uteri: new data and a review. Cancer 27:1239–1254

Stansfeld AG (1985) Lymph node biopsy interpretation. Churchill Livingstone, Edinburgh

Steel GG (1968) Cell loss from experimental tumours. Cell Tissue Res 1:193–198

Stenger HE, Bahnsen J, Hinz B (1980) Cytophotometric analysis of nuclear DNA content in so-called obliterating mastopathy with epithelial hyperproliferation. Pathol Res Pract 170:146–151

Stenkvist B, Bengtsson E, Eriksson O, Jarkrans T, Nordin B, Westman-Naeser S (1981) Correlation between cytometric features and mitotic frequency in human breast carcinoma. Cytometry 1:287–291

Stenkvist B, Bengtsson E, Dahlqvist B, Eklund G, Eriksson O, Jarkrans T, Nordin B (1982) Predicting breast cancer recurrence. Cancer 50:2884–2893

Stevens MW, Fazzalari NL, Crisp DJ (1987) Quantitation of nuclear shape in non-Hodgkin's lymphoma: alternate shape descriptors. Anal Quant Cytol Histol 9:459–468
Stevens MW, Crowley KS, Fazzalari NL, Woods AE (1988) Use of morphometry in cytological preparations for diagnosing follicular non-Hodgkin's lymphomas. J Clin Pathol 41:370–377
Sugimachi K, Ide H, Okamura T, Matsuura H, Endo M, Inokuchi K (1984) Cytophotometric DNA analysis of mucosal and submucosal carcinoma of the oesophagus. Cancer 53:2683–2687
Swank PR, Greenberg SD, Montalvo J et. al. (1983) Classification of bronchial epithelial atypias by the atypia status index. Anal Quant Cytol 5:255–262
Swift HH, Rasch EM (1956) Microphotometry with visible light. Phys Tech Biol Res 3:353–358
Talerman A, Fu YS, Okagaki T (1984) Spermatocytic seminoma: ultrastructural and microspectrophotometric observations. Lab Invest 51:343–349
Tan GJKH, Baak JPA (1984) Evaluation of prognostic characteristics of stage 1 cutaneous malignant melanoma. Anal Quant Cytol 6:147–154
Tanaka N, Akida H, Ueno T, Mukawa A, Kamitsuma K (1979) Field test and experimental use of CYBEST Model 2 for practical gynaecologic mass screening. Anal Quant Cytol 1:122–128
Tanaka N, Ueno T, Ikeda H et. al. (1982) CYBEST Model 3: automated cytologic screen system for uterine cancer utilising image analysis processing. Anal Quant Cytol 4:297–285
Tannenbaum M, Tannenbaum S, De Sanctis PH (1982) Prognostic significance of nucleolar surface area in prostatic cancer. Urology 19:546–551
Tannock IF (1967) A comparison of the relative efficiencies of various metaphase arrest agents. Exp Cell Res 47:345–356
Tannock IF (1968) The relation between cell proliferation and the vascular system in a transplanted mouse mammary tumour. Br J Cancer 22:258–273
Tannock IF (1970) Population kinetics of carcinoma cells, capillary endothelial cells, and fibroblasts in a transplanted mouse mammary tumour. Cancer Res 30:2470–2476
Tavares AS (1968) Ploidy and histological types of mammary carcinomas. Eur J Cancer 3:449–455
Tavares AS, Costa J, de Carvalho A, Reis M (1966) Tumor ploidy and prognosis in carcinomas of the bladder and prostate. Br J Cancer 20:438–441
Taylor HB, Norris HJ (1966) Mesenchymal tumours of the uterus. IV. Diagnosis and prognosis of leiomyosarcomas. Arch Pathol 81:40–44
Taylor J, Puls J, Synchra JJ, Bartels PH, Bibbo M, Wied GL (1978) A system for scanning biological cells in three colors. Acta Cytol 22:29–35
Teich JM, Young IT, Sher SE, Lee JS (1979) Transformation of nuclear morphology during cellular maturation. J Histochem Cytochem 27:193–198
Thorud E, Rolstad A, Lexow P (1980) Malignancy in pleomorphic adenomas of the parotid gland. Flow Cytometry 4:458–465
Tosi P, Cintorino M, Luzi P, Leoncini L, Spina D (1983) A morphometric semiautomated method for analysing cell nuclei in lymph node sections from non-Hodgkin's lymphomas. Significance of data. Exp Pathol 24:237–241
Tosi P, Leoncini L, Spina D, del Vecchio MT (1984) Morphometric nuclear analysis of lymphoid cells in center cell lymphomas and in reactive germinal centers. Am J Pathol 117:-12–17
Tosi P, Lusi P, Baak JPA et. al. (1986a) Nuclear morphometry as a prognostic factor in stage I renal cell carcinoma. Cancer 58:2512–2518
Tosi P, Miracco C, Luzi P, Cintorino M, Kraft R, Cottier H (1986b) Morphometric distinction of granulomas in tuberculosis and sarcoidosis. Difference in nuclear profiles. Anal Quant Cytol Histol 8:233–240
Tosi P, Leoncini L, Baak JPA, Luzi P, Cintorino M, Barbini P (1988) Comparative morphometric analysis of nuclear area and shape in peripheral T-zone lymphomas and in paracortical areas of normal and reactive lymph nodes. Anal Quant Cytol Histol 10:285–293
Tribukait B, Gustafson H (1980) Impulscytophotometrische DNS-Untersuchungen bei Blasenkarcinomen. Onkologie 6:278–288

Tribukait B, Ronstrom L, Episoti PL (1983) Quantitative and qualitative aspects of flow DNA measurements related to the cytologic grade in prostatic carcinoma. Anal Quant Cytol 5:107–111

Trowbridge EA, Martin JF, Slater DN, Kishk YT, Warren CW (1983) Platelet production: a computer based biological interpretation. Thromb Res 31:329–350

Trowbridge EA, Martin JF, Slater DN, Kishk YT, Warren CW, Harley PJ, Woodcock B (1984) The origin of platelet count and volume. Clin Phys Physiol Meas 5:145–170

Tsubaki, J, Jimbo G (1979) A proposed new characterisation of particle shape and its application. Powder Technolgoy 22:161–169

Tucker JH, Hussain OAN (1981) Trials with the CERVISCAN experimental prescreening device on polylysine-prepared slides. Anal Quant Cytol 3:117–120

Tutton PJM (1973) Variation in crypt cell cycle time and mitotic time in the small intestine of the rat. Virchows Archiv [Cell Pathol] 13:68–75

Underwood EE (1970) Quantitative stereology. Addison-Wesley, Reading (Mass.)

Underwood JCE (1972) A morphometric analysis of human breast carcinoma. Br J Cancer 26:234–237

Underwood JCE (1974) Lymphoreticular infiltration in human tumours: prognostic and biological implications. Br J Cancer 30:538–548

Underwood JCE (1987) Introduction to biopsy interpretation and surgical pathology, 2nd edn. Springer, Berlin Heidelberg New York

Underwood JCE, Dangerfield VJM, Parsons MA (1983) Oestrogen receptor assay of cryostat sections of human breast carcinomas with simultaneous quantitative histology. J Clin Pathol 36:399–405

van der Valk P, Hermans J, Brand R, Cornelisse CJ, Spaander PJ, Meijer CJLM (1982) Morphometric characterisation of diffuse large-cell (histiocytic) lymphomas. Am J Pathol 107:327–335

van der Valk P, Mosch A, Kurver PJ, Meijer CJLM (1983) Morphometric characterisation of 52 B-cell non-Hodgkin's lymphomas. J Clin Pathol 36:289–297

van Vloten WA, Scheffer E, Meijer CJLM (1979) DNA-cytophotometry of lymph node imprints from patients with mycosis fungoides. J Invest Dermatol 73:275–277

Vasquez-Nin GH, Echeverria OM, Zavala G, Jimenez-Garcia LF, Gonzales MA, Parra R (1986) Relations between nucleolar morphometric parameters and pre-rRNA synthesis in animal and plant cells. Acta Anat (Basel) 126:141–146

Wagner D, Richart RM, Terner JY (1967) Deoxyribonucleic acid content of presumed precursors of endometrial carcinoma. Cancer 20:2067–2077

Wagner D, Richart RM, Terner JY (1968) DNA content of human endometrial cells during the menstrual cycle. Am J Obstet Gynecol 100:90–97

Wallgren A, Zajicek J (1976) The prognostic value of aspiration biopsy smear in mammary carcinoma. Acta Cytol 20:479–485

Watson JV, Chambers SH, Smith PJ (1987) A pragmatic approach to the analysis of DNA histograms with a definable G1 peak. Cytometry 8:1–8

Weibel ER (1963) Principles and methods for the morphometric study of the lungs and other organs. Lab Invest 12:131–155

Weibel ER (1979) Stereological methods, vol 1, Academic Press, London

Weibel ER (1980) Stereological methods, vol 2, Academic Press, London

Weibel ER, Gomez DM (1962) A principle for counting tissue structures on random sections. J Appl Physiol 17:343–348

Weiss RR, Richart RM, Okagaki T (1969) DNA content of mucinous tumors of the ovary. Am J Obstet Gynecol 103:409–415

Wells M, Brown LJR (1986) Glandular lesions of the uterine cervix: the present state of our knowledge. Histophathology 10:777–792

Westerberg H, Gustafson SA, Nordenskjold B, Silversward C, Wallgren A (1980) Estrogen receptor level and other factors in early recurrence of breast cancer. Int J Cancer 26:429–435

Wied GL, Bibbo M, Bartels PH (1981) Computer analysis of microscopic images: application in cytopathology. In: Sommers SC, Rosen PP (eds) Pathol Annu 16:367–408

Wied GL, Bibbo M, Pishotto FT, Bartels PH (1984) Intermediate cell markers for malignancy: consistency of expression. Anal Quant Cytol 6:243–246

Wiendl HJ, Schwabe M, Becker G, Kowatsch J (1974) Feulgen cytophotometric studies of gastric mucosal smears in malignant and benign diseases of the stomach. Acta Cytol 18:222–230

Wilbanks GD, Richart RM, Terner JY (1967) DNA content of cervical intraepithelial neoplasia studied by two-wavelength Feulgen cytophotometry. Am J Obstet Gynecol 98:792–799

Wittekind D (1985) Dyes and staining standards. Anal Quant Cytol 7:6–30

Wittekind C, Schultze E (1987) Computerised morphometric image analysis of cytologic nuclear parameters in breast cancer. Anal Quant Cytol Histol 9:480–484

Wright NA, Alison M (1984) The biology of epithelial cell populations. Oxford University Press, Oxford, pp 179–186

Yates JT (1978) Contrast sensitivity: characteristics of a large, young, adult population. Am J Optom Physiol Opt 64:519–527

Young IT, Walker JE, Bowie JE (1974) An analysis technique for biological shape. 1. Inform Control 25:357–370

Zahniser DJ, Oud PS, Raaijmakers MCT, Vooys GP, Van de Wall RT (1979) BioPEPR: a system for the automatic prescreening of cervical smears. J Histochem Cytochem 27:635–639

Zahniser DJ, Oud PS, Raaijmakers MCT, Vooys GP, Van de Walle RT (1980) Field test results using the BioPEPR cervical smear prescreening system. Cytometry 1:200–203

Zajdela A, De La Riva L, Ghossein NA (1979) The relation of prognosis to the nuclear diameter of breast cancer cells obtained by cytologic aspiration. Acta Cytol 23:75–80

Zajicek J, Casperson T, Jacobsson P, Kudynowski J, Linsk J, Us-Krasovek M (1970) Cytologic diagnosis of mammary tumors from aspiration biopsy smears: comparison of cytologic and histologic findings in 2111 lesions and diagnostic use of cytophotometry. Acta Cytol 14:370–378

Zetterberg A, Episoti PL (1980) Prognostic significance of nuclear DNA levels in prostatic carcinoma. Scand J Urol Nephrol 55 [Suppl]:53–58

# Flow Cytometry in the Quantitation of DNA Aneuploidy and Cell Proliferation in Human Disease

P. QUIRKE

## 1 Introduction

The use of the technique of flow cytometry has rapidly expanded in the last 5 years and in certain areas such as immunology and the pharmaceutical industry flow cytometers are routine instruments. Concomitant with this expansion there has been an explosion in the volume of the literature, especially within the cancer journals. In this chapter the principles of flow cytometry will be explained, its present and future applications reviewed and examples of new techniques which may be of diagnostic and prognostic importance introduced. There are several reviews on this subject (BARLOGIE et al. 1983; FRIEDLANDER et al. 1984a; QUIRKE and DYSON 1986; RYAN et al. 1988) along with a few books (MELAMED et al. 1979; SHAPIRO 1985) from which further information can be obtained.

Flow cytometry is concerned with the analysis and sorting of populations of cells or particles. The parameters assessed may be physical or optical, such as cell size, light scatter, polarisation measurements or the quantitation of fluorescence over a variety of excitation or emission wavelengths; these variables depend on the type of light source (mercury arc lamp or laser) and the filters used for the selection of emitted wavelengths. The largest use of these instruments is within the field of immunology, where flow cytometry is replacing manual counting techniques for the quantitation of subpopulations of white cells by cell surface bound fluorescence conjugated monoclonal antibodies. This review, however, will be restricted to applications relevant to the nucleus and the measurement of cell proliferation.

## 2 Technique

A single cell suspension is a prerequisite of the method. Blood or semen provides natural single cell suspensions; lymphoid tissue requires only minimal disruption by mincing and syringing whereas most epithelial tissues and other solid organs require careful disaggregation, usually by enzymatic methods. The tissue can be disaggregated fresh or recovered from liquid nitrogen or paraffin-embedded material (HEDLEY et al. 1983, 1985; STONE et al. 1985; SCHUTTE et al. 1985). Postmortem material can also be successfully measured (RABER et al. 1984; QUIRKE, unpublished observations; see Fig. 8).

As shown in Fig. 1, the stained cell suspension (*1*) is injected into the flow cell under gas pressure where it is hydrodynamically focussed (*2*) into a single-file cell stream for passage through the interrogation point, excitation taking place either in an enclosed quartz chamber or in air (*3*). The cell physically interacts with the exciting beam, scattering it in all directions. Light scattered in the forward direction is related to the size of the cell and that scattered at 90° is related to the level of refraction of its inter-

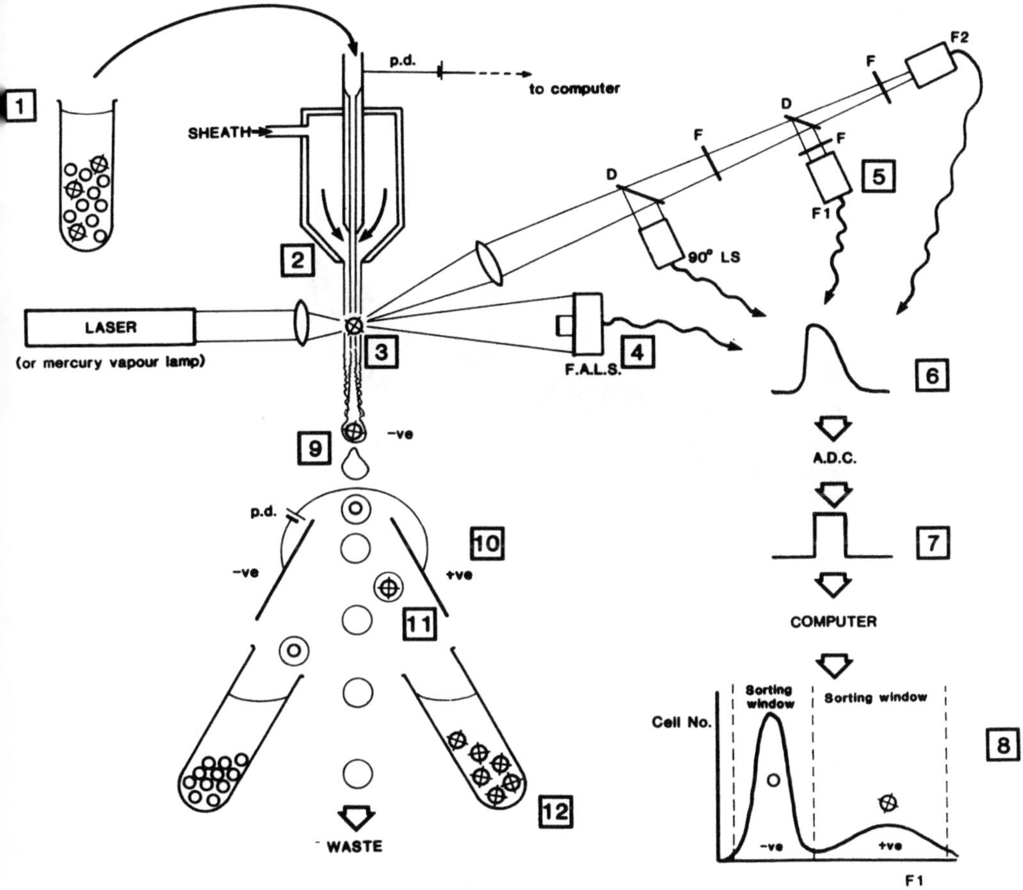


**Fig. 1.** Schematic representation of flow cytometry. For explanation of the diagram see the technique section of the text. The *crossed circles* represent fluorescence positive cells and the *open circles*, fluorescence negative cells. *F.A.L.S.*, forward angle light scatter detector; *F1* and *F2* fluorescence photomultiplier tube detectors; 90° LS, 90° light scatter detector; *D*, dichroic mirror; *F*, filter; *p.d.*, potential difference; *A.D.C.*, analogue to digital converter. (Reproduced by kind permission of the Editor of the Journal of Pathology)

nal structure (*4*). Volume and polarisation measurements are also possible if an appropriate sizing device or polarising filters are available. Excitation of the fluorochrome(s) occurs at the interrogation point (*3*) with the emission of light of a longer wavelength. This is also collected at 90° with selected wavelengths directed by dichroic mirrors to photomultiplier detectors (*5*); unwanted fluorescence and scattered light are blocked by optical filters. Analogue electrical signals are generated for each particle (*6*), converted into digital signals (*7*) for processing by computer software in order to generate one (*8*), two or three parameter-correlated histograms (Fig. 2). The number of quantitated parameters is dependent on the excitation

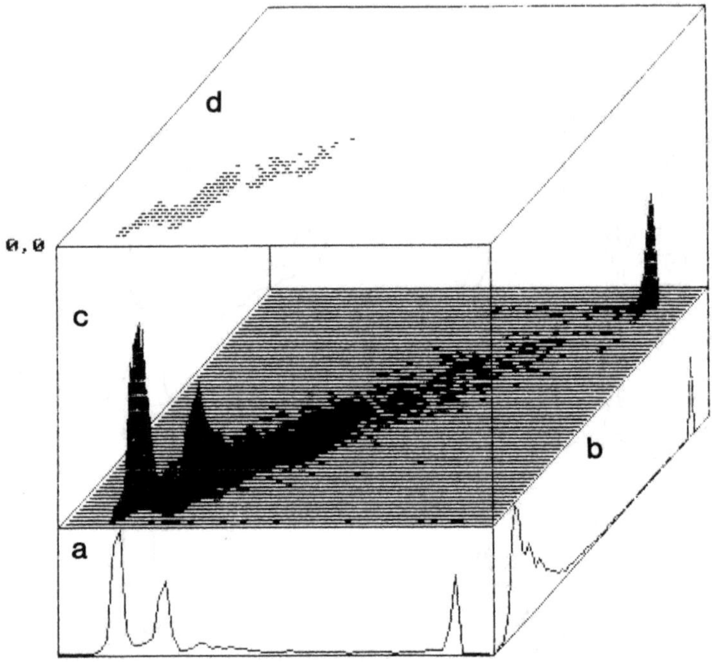


**Fig. 2.** One parameter histograms (*a* and *b*), a two parameter scattergram (*d*) and an isometric plot of DNA vs Coulter volume of HCT 18 colorectal cancer cells (*c*). *a*: DNA histogram of a DNA aneuploid population with a DNA index of 1.7. *b*: Coulter volume measurements of the nuclear volume. *c*: Isometric plot of DNA vs volume. *d*: Scatter plot of DNA vs volume. The DNA aneuploid $G_0/G_1$ population can be seen to have a higher volume than the diploid $G_2/M$ population and is larger than the $G_0/G_1$ cells from which it arose

wavelengths available, photomultiplier detectors installed, analogue-to-digital converters and the power of the computer hardware and software. If cell sorting is required then the properties of the cells to be selected are programmed into the computer (*8*). After leaving the interrogation point the stream breaks up into droplets due to the vibrations of a piezo-electric crystal. If a cell with the selected properties is sensed by the computer, either a positive or a negative charge is applied to the stream at the time of droplet beak-off, thus charging the droplet containing the cell (*9*). This falls between the two high-voltage plates (*10*) and is deflected according to its charge (*11*) into the collecting vessel (*12*), allowing further investigation of its properties.

## 3 Staining

A large variety of stains are available for flow cytometry (see review by SHAPIRO 1985 and Table 1). The physical limitations on their usage are the ability of the light source to excite the stain efficiently and the emission of

**Table 1.** Useful stains in flow cytometry (reproduced by kind permission of the Editor of the Journal of Pathology)

| Stain | Reaction agent | Excitation wavelength (nm) | References |
|---|---|---|---|
| Propidium iodide | DNA | 488 | TAYLOR 1980; HAMILTON et al. 1980; TAYLOR and MILTHORPE 1980; WALLEN et al. 1982 |
| Ethidium bromide | DNA | 488 | TAYLOR 1980; TAYLOR and MILTHORPE 1980; RÖHMER and ELLWART 1981 |
| Acridine orange | DNA/RNA | 488 | TRAGANOS et al. 1977; TAYLOR and MILTHORPE 1980; WALLEN et al. 1982; DYSON et al. 1984 NICOLINI et al. 1979 |
| Mithramycin | DNA | 450 | TAYLOR 1980; TAYLOR and MILTHORPE 1980; CRISSMAN and TOBEY 1974 |
| DAPI | DNA | 361 | TAYLOR 1980; HAMADA and FUJITA 1983 |
| Hoescht 33258 | DNA | 361 | MULLER and GAUTIER 1975; LATT and STETTEN 1976; ARNDT-JOVIN and TOVIN 1977 |
| Hoescht 33342 | DNA | 361 | MILLER et al. 1984; LALANDE and MILLER 1979; SHAPIRO 1981 |
| Pyronin Y | RNA/mitochondria | 545 | SHAPIRO 1981; POLLACK et al. 1982; OWDEN and CURTIS 1983 |
| Fluorescein isothiocyanate | Protein/antibody labelling | 488 | CRISSMAN et al. 1981; ROTI-ROTI et al. 1982; POLLACK et al. 1984 |
| Fluorescein diacetate | Cleaved to fluorescein in cell to form cytoplasmic solution | 488 | MEISINGSET and STEEN 1981 |
| Phycoerythrin | Antibody labelling | 488 | VERNON et al. 1982 |
| Rhodamine isothiocyanate | Protein/antibody labelling | 568 | CRISSMAN and STEINKAMP 1982 |
| Rhodamine 123 | Mitochondria | 488 | DARZYNKIEWICZ et al. 1982 |
| Fluorescamine | Plasma membrane | 390 | HAWKES and BARTHOLOMEW 1977; POCCIA et al. 1979 |
| Methylum-belliferones | Enzyme activities | 361 | WATSON 1980 |
| Nitroheterocycles | Hypoxic cells | 388 | OLIVE and DURAND 1983 |

a quantifiable amount of light. The biological limitations are that the stain must specifically and stoichiometrically react with the substance or physical property to be quantitated. Quantitation of fluorescence is relative and usually performed by comparison of the sample fluorescence to a control; in the case of DNA measurements this can be chicken or trout erythrocytes or more frequently peripheral blood lymphocytes or disaggregated tonsil. In the latter two, however, small differences in the access of stain may be found between quiescent and stimulated lymphocytes (WOLLEY et al. 1982a). Stains such as Hoechst 33258, Hoechst 33342 and acridine orange will enter viable cells but other stains such as propidium iodide or ethidium bromide are dependent on fixation of the cell, membrane permeablisation or stripping techniques for entry (KRISHAN 1975; TRAGANOS et al. 1977; TAYLOR 1980).

## 4 Measurement of DNA Content and Cell Proliferation

The fundamental importance of DNA in cellular processes has been proved in a number of seminal papers: for example, the reporting of a method for stoichiometric DNA staining (FEULGEN and ROSSENBECK 1924), the doubling of DNA content during the cell cycle (CASPERSSON and SCHULTZ 1938), that DNA contained the genetic code (AVERY et al. 1944), its double helical structure (WATSON and CRICK 1953) and the number of chromosomes (TIJO and LEVAN 1956). The recognition of the central role of DNA in neoplasia stems from the description of the Philadelphia chromosome in chronic myeloid leukaemia (NOWELL and HUNGERFORD 1960) and subsequent chromosomal studies (see reviews by SANDBERG 1980; SANDBERG and TURC-CAREL 1987) and more recently from molecular biological investigations. Microdensitometry using Feulgen staining of DNA became popular in the late 1950s and 1960s (see PEARSE 1968; WIED and BAHR 1970; CASPERSSON 1979) but lost favour in the 1970s due to the tedium of the technique and apparent lack of clinical relevance. Recent workers, especially Auer's group at the Karolinska Hospital, have resurrected this technique and with the development of automated methods of DNA analysis for use on disaggregated cell populations, such as the MIAMED-CYTO instrument, cytophotometry may yet find a niche in the routine diagnostic laboratory. Tissue cytophotometric studies have the advantage of specificity as they permit selection of cell populations for study but their major disadvantage is the small number of cells assessed. Whilst the presence of DNA aneuploidy can be detected, the technique is not as sensitive in the detection of small populations of abnormal cells, and reliable estimates of cell proliferation are not obtainable. Cytological cytophotometry as practised using the LEYTAS instrument yields results more akin to flow cytometry since it measures large numbers of cells (i.e. > 10000) and does not discriminate between neoplastic and non-neoplastic tissue.

The major advantages of flow cytometry are the ease of performance of measurements, the rapidity of collection of large numbers of events giving a high degree of statistical safety and most importantly the quantitation and correlation of multiple parameters on single cells.

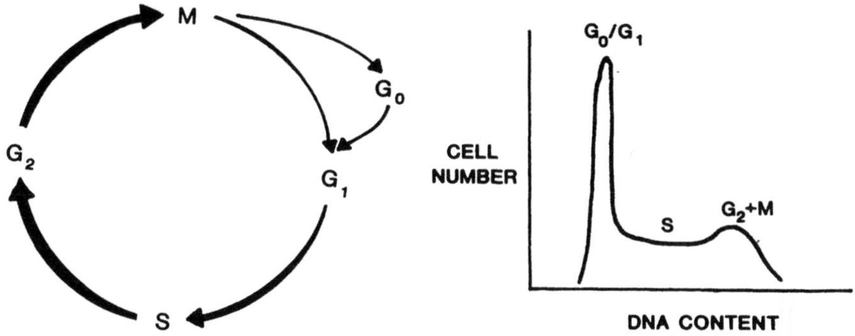


**Fig. 3.** Diagramatic representation of cell cycle and a diploid DNA histogram. See text for details. (Reproduced by kind permission of the Editors of the Oxford Textbook of Pathology)

As stated above, DNA measurements rely upon the use of a fluorescent stain that binds to DNA in a stoichiometric fashion; therefore as the DNA content of the cell increases during S phase, the fluorescence increases until it reaches double the normal level at $G_2$. At mitosis the cell divides, leaving two daughter cells with diploid amounts of DNA (see Fig. 3). The flow cytometer produces a frequency distribution histogram of fluorescence intensity (proportional to DNA content) versus cell number upon which cell cycle computer analysis can be performed. Methods of cell cycle analysis have been reviewed and broadly categorised into four patterns (BAISCH et al. 1982); the first three types use similar parametric statistics in that they assume a Gaussian distribution for the $G_0/G_1$ and $G_2/M$ phases and fit a variety of distributions to the S phase (see Fig. 4). Type I models use several Gaussian distributions, type II a single Gaussian distribution (Fig. 4b) and type III either a broadened polynominal curve (Fig. 4c) or a single rectangle (Fig. 4a) for fitting the S phase area. Type IV models match sequences of computed histograms to the experimental histogram. All models were reported to be relatively accurate though differences were apparent between them. Estimates tended to be least accurate when small populations were present in any phase of the cell cycle and all models appear relatively independent of the coefficient of variation (c.v.). In our experience the latter has not been the case for PARA 1 (Coulter Electronics, Hialeh, Florida), where marked variation is apparent with changes in the c.v.; this program, however, is less stringent in its requirement for a perfect histogram and yields reasonably accurate cell proliferation values up to 14% if the %S phase is added to the %$G_2$/M (QUIRKE 1987) but not if the %S phase alone is used. No "gold standard" currently exists for the estimation of the S phase fraction as all methods of determination of cell proliferation are open to criticism. Such values should be treated as for normal ranges in biochemistry laboratories in that an abnormal value is one outside the defined normal range for that laboratory.

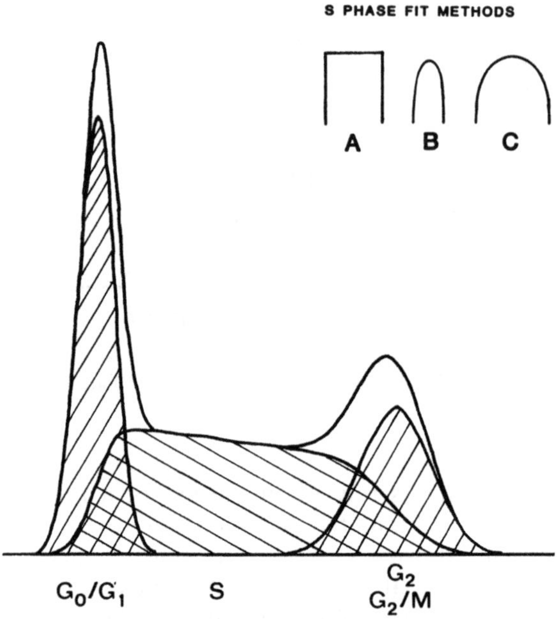


**Fig. 4.** Schematic representation of a DNA histogram broken down into its constituent parts. Gaussian distributions are computed for the $G_0/G_1$ and $G_2/M$ peaks and the S phase content is derived by one of a variety of cell cycle analysis models. *A*, rectangular/multirectangular; *B*, Gaussian; *C*, broadened polynomial. (Reproduced by kind permission of the Editors of the Oxford Textbook of Pathology)

Abnormalities of chromosome number are well described. The great majority of malignant tumours are chromosomally aneuploid, that is they have an increase or less often a decrease in the number of chromosomes. Flow cytometry can detect this abnormality when the aberration exceeds a loss or gain of three chromosomes, i.e. $< 43$ or $> 49$. Clinical studies have established that between 50% and 85% of malignant tumours demonstrate this abnormality and are termed DNA aneuploid. DNA aneuploidy is defined as the presence of an abnormal DNA stem line recognised by the appearance of at least two separate $G_0/G_1$ peaks on the DNA histogram. The term DNA aneuploidy should not be used in the absence of a second peak when dealing with human solid tumour material as an increased access of stain to the DNA may sometimes occur, leading to a spurious elevation of the $G_0/G_1$ peak when compared to an external standard (WOLLEY et al. 1982a; SHACKNEY et al. 1984).

The prevalence of DNA aneuploidy varies with the sensitivity of the measurements and the criteria applied for DNA aneuploidy. Low c.v.s, acceptance of less than 5% of cells in an abnormal peak, and a failure to require a second DNA aneuploid $G_2/M$ peak can lead to a higher incidence of DNA aneuploidy. While the significance of small DNA aneuploid peaks remains to be established, it has been suggested that the higher the number of DNA aneuploid cells present within a tumour the poorer the prognosis (JONES et al. 1988 a). Reproducibility is one of the major advantages of DNA measurement over histological assessment; this has been confirmed in a recent quality control multicentre study performed on the same paraffin-embedded bladder tumour material (COON et al. 1988).

## 5 Flow Cytometric Measurements and Neoplastic Disease

Numerous papers have appeared recently on the incidence and relationship to survival of DNA aneuploidy and cell proliferation in a variety of neoplastic conditions, though few reports on normal tissue have appeared. The predictive value of DNA aneuploidy in these studies of malignant disease is variable, and depends both on instrinsic properties of the tumour itself and on the effectiveness of treatment regimens. Thus in certain tumours, response to chemotherapeutic treatment can result in an apparent reversal of the general rule that DNA aneuploidy or high cell proliferation confers a poor prognosis (see below). The published work will be discussed in four parts: first, the importance of DNA aneuploidy in malignant disease; second, its development in premalignant disease; third, its presence in benign endocrine tumours and lastly the effects of treatment on DNA aneuploidy.

### 5.1 Malignant Tumours

More than 30 types of tumour have now been investigated. Most work to date has been performed on gastrointestinal, genitourinary, breast and haematological malignancies, which will be discussed in the greatest detail.

#### *5.1.1 Gastrointestinal Tract*

##### 5.1.1.1 Colorectal Carcinoma – DNA Aneuploidy

The incidence of DNA aneuploidy ranges from 50% to 71% (PETERSEN et al. 1980; ROGNUM et al. 1980, 1982; TRIBUKAIT et al. 1983 a; ARMITAGE et al. 1985; BANNER et al. 1985; FINAN et al. 1986; KOKAL et al. 1986; HAMMARBERG et al. 1986; QUIRKE et al. 1987; etc) when using definitions close to those of a recent international agreement (HIDDEMANN et al. 1984). Heterogeneity exists as determined by mixed diploid/DNA aneuploid tumours, or more than one DNA aneuploid stem line, the incidence ranging from 30% to 61% (PETERSEN et al. 1980; TRIBUKAIT et al. 1983 a; QIRKE et al. 1985) depending on the number of samples taken and the technical methods utilised. The presence of different stem lines in a tumour bears no relationship to the area the sample is taken from and does not parallel histopathological heterogeneity (QUIRKE et al. 1985; QUIRKE 1987). Diploid and DNA aneuploid primary tumours give rise to predominantly diploid or DNA aneuploid metastases respectively while heterogeneous tumours yield a mixed pattern of metastases (QUIRKE 1987; ARENDS et al. 1987).

The first report of a relationship between ploidy and prognosis came from WOLLEY et al. (1982b). This often quoted study on 30 patients, however, used very different criteria to those defined by HIDDEMANN et al. (1984). Tumours were divided into diploid and non-diploid; tumours with a dominant DNA aneuploid peak were placed in the non-diploid group and tumours containing a small DNA aneuploid peak were placed in the diploid category. However, subsequent reports have confirmed the relationship with prognosis using more standard definitions of DNA aneuploidy though the association now appears less striking than originally claimed (ARMITAGE et al. 1985; KOKAL et al. 1986; STREFFER et al. 1986; QUIRKE et al. 1987; SCHUTTE et al. 1987; SCOTT et al. 1987; EDMIN et al. 1987; WIRSCHING et al. 1987; GOH et al. 1987). In certain studies ploidy appeared prognostic in Dukes' stage B tumours (JONES et al. 1988b) or stage C tumours only (ROGNUM et al. 1987). The strong relationship of ploidy to survival seen in the study of KOKAL et al. (1986) may be explained by the stringent entry requirements of their study which selected curative resections only. Failure of ploidy to predict survival in patients with metastatic liver disease (FINAN et al. 1986) supports the usefulness of this measurement only in tumours from patients undergoing curative operations.

In the majority of studies there is no relationship between either the Dukes' stage or the grade of the tumour and DNA aneuploidy though a few dissenting reports exist (BANNER et al. 1985; KOKAL et al. 1986).

Fewer studies have appeared utilising measurements of cell proliferation in this condition (see Table 2). Using autoradiography MEYER and PRIOLEAU (1981) reported a median S phase value of 17% in 90 tumours. Flow cytometric studies vary depending on the computer model used and the cell cycle phases quoted. In rectal carcinoma two groups have reported a better prognosis in diploid tumours with a low cell proliferation as opposed to a high level of cell proliferation (see Fig. 5; STREFFER et al. 1986; QUIRKE et al. 1987). In the latter study cell proliferation appeared to provide additional prognostic information over ploidy in the Cox's regression model. A further feature of this study was the demonstration of a relationship between the proliferative index ($\%S + G_2$) and the invasive pattern of the border. Tumours with a low cell proliferation invaded more frequently with a pushing as opposed to an infiltrating border.

Flow cytometric assessments in most studies provide useful prognostic information in colorectal carcinoma. Such quantitative measurements may well replace more subjective histopathological assessments. Larger scale, prospective and carefully staged studies are now required in this condition to assess its true predictive value.

**Table 2.** Cell kinetic measurements on colorectal carcinomas

| Authors | Measurement and method | No. of cases | Description | Value (%) |
|---|---|---|---|---|
| MEYER and PRIOLEAU (1981) | S phase fraction autoradiography | 90 | All tumours | 17.8 |
| TRIBUKAIT et al. (1983 a) | S phase flow cytometry | 66 | Diploid | 13.8 |
| | | | DNA aneuploid-DNA index 1.1–1.2 | 17.4 |
| | | | DNA aneuploid-DNA index 1.4–2.5 | 19.4 |
| STREFFER et al. (1986) | S phase flow cytometry | 129 | All tumours | 19.7 |
| | | | Diploid | 18.3 |
| | | | DNA aneuploid | 25.9 |
| HAMMARBERG et al. (1986) | $S + G_2$ phases flow cytometry | 28 | Diploid | 25.7 |
| | | | DNA aneuploid-DNA index 1.1–1.2 | 23.6 |
| | | | DNA aneuploid-DNA index 1.3–2.5 | 32.4 |

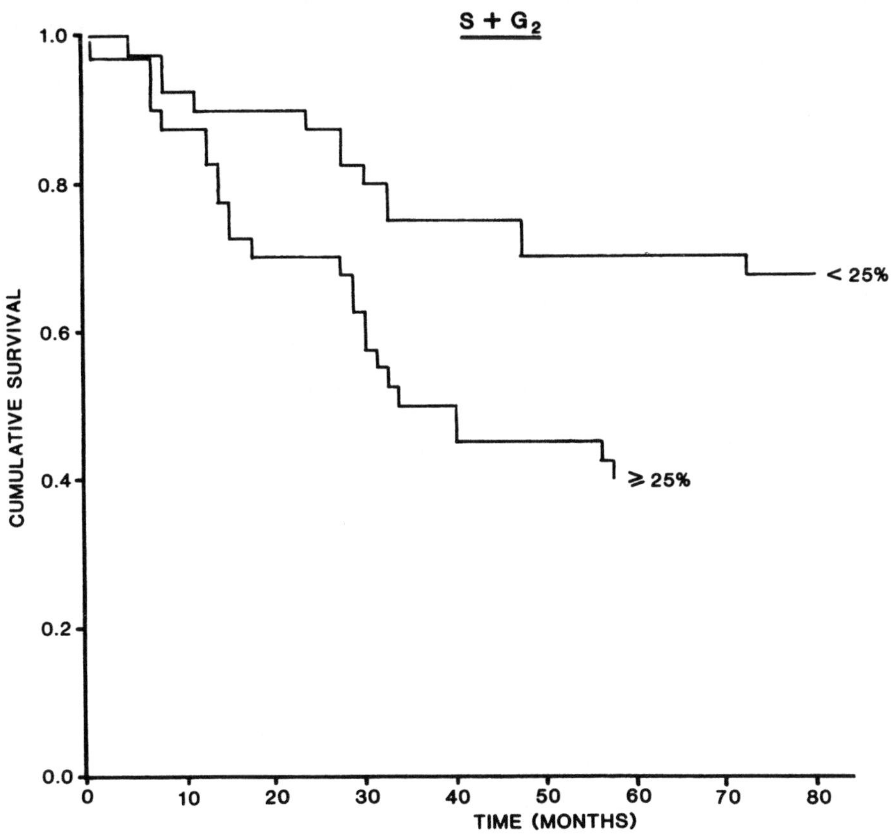


**Fig. 5.** Survival curves for patients with diploid rectal adenocarcinoma with a low cell proliferation ($\%S + G_2 < 25\%$) and a high cell proliferation ($\%S + G_2 > 25\%$), showing the poorer prognosis in the latter group. (Reproduced by kind permission of the Editor of the Journal of Pathology)

#### 5.1.1.2 Gastric Carcinoma

Fewer flow cytometric studies have been performed on gastric as opposed to colorectal adenocarcinoma (MACARTNEY et al. 1986a; BALLANTYNE et al. 1987; WYATT et al. 1989) though a number of DNA cytophotometric papers on early gastric cancer have appeared from Japan showing pathological differences between diploid and non-diploid tumours and with a significant survival advantage to the former (INOKUCHI et al. 1983; KORENAGA et al. 1985; HATTORI et al. 1986; KAMEGAWA et al. 1986; OKAMURA et al. 1987). Using flow cytometry, BALLANTYNE (1987) in a study of 44 curative resections in patients with gastric carcinoma found no relationship between DNA aneuploidy and prognosis whereas WYATT et al. (1989) in 76 patients undergoing curative resection for gastric cancer demonstrated a significantly improved survival in patients with diploid tumours over those with DNA aneuploid tumours. However, on multiple regression analysis the presence of lymph node metastases and resection margin involvement were the only factors of independent prognostic significance, overriding the importance of ploidy in this model. Interestingly, on subdividing the tumours into intestinal and diffuse types (LAUREN 1965), ploidy was only significantly related to prognosis in the intestinal group and not the diffuse tumours.

### *5.1.2 Genitourinary Malignancies*

#### 5.1.2.1 Bladder Carcinoma

Early investigators of bladder carcinomas demonstrated a strong association between ploidy and the grade of differentiation (TRIBUKAIT et al. 1982; GUSTAFSON et al. 1982b, 1982c; FARSUND et al. 1984; CHIN et al. 1985; BLOMJOUS et al. 1988), between ploidy and stage (TRIBUKAIT et al. 1982; CHIN et al. 1985; BLOMJOUS et al. 1988), and between cell proliferation and grade (FARSUND et al. 1984). Flow cytometric measurements have also been found useful in the early prediction of recurrence (KLEIN et al. 1982) and progression of early stage tumours (GUSTAFSON et al. 1982b, 1982c; BLOMJOUS et al. 1988). In a preliminary study of 61 untreated transitional cell carcinomas the presence of DNA aneuploidy was an important prognostic indicator, with 2 of 22 patients with diploid tumours dead from this disease within 5 years as compared to 11 of 22 (50%) with DNA aneuploid tumours (BLOMJOUS et al. 1988).

The bladder is particularly suitable for cytological assessment. A recent study (BADALAMENT et al. 1987) has suggested that bladder-wash flow cytometric assessments are more accurate (83%) in predicting the presence of a tumour than bladder-wash cytology (61%) and that both were more sensitive than voided urinary cytology (41%) though the latter improves with three specimens (60%). A larger study of 249 bladder irrigation specimens found that in the presence of a DNA aneuploid cell population 85% of patients harboured a bladder tumour. The improved sensitivity over cytology was demonstrated in the six cystoscopically and cytologically negative patients with DNA aneuploid stem lines, four of whom subsequently developed tumour (TETU et al. 1987). These authors, however, also reported a reduction in accuracy of flow cytometry with prior intravesical chemotherapy. ORIHUELA et al. (1987) found that DNA aneuploid tumours were more prone to recurrence following transurethral resection, and that the response to treatment with intravesical BCG could be predicted from a combination of flow cytometry and ABH blood group markers.

5.1.2.2 Prostatic Carcinoma

Three studies to date are unanimous in their findings of a significant survival advantage in patients with diploid prostatic tumours (FORDHAM et al. 1986; STEPHENSON et al. 1987; LUDBERG et al. 1987) though none have investigated measurements of cell proliferation. STEPHENSON et al. (1987) in a study of 82 patients reported 5-year survivals of 87.9% for diploid tumours and 49.5% for DNA aneuploid tumours with median suvivals of 8.8 and 5.0 years respectively.

5.1.2.3 Testicular Tumours

A preliminary study on testicular tumours revealed DNA aneuploidy in 63% of seminomas, 68% of teratomas and 50% of combined seminoma/teratoma. Teratomas had a unimodal distribution with DNA indices around 1.4 whereas seminomas had a bimodal distribution around 1.5 and 2.0 (QUIRKE et al. 1986a). No relationship to survival was found but this is not surprising owing to the small number of tumours analysed (63) when compared to the good survival in this tumour type.

### *5.1.3 Breast Carcinoma*

Breast cancer has been extensively studied and has given rise to a complex literature. DNA cytophotometric studies have consistently suggested a strong relationship between DNA content and survival (ATKIN 1972; AUER et al. 1984). Early flow cytometric studies on paraffin-embedded material found a significantly improved short-term survival and time to disease recurrence in diploid tumours, but there was no influence on long-term survival (HEDLEY et al. 1984; DOWLE et al. 1987; OWAINATI et al. 1987). Work on larger numbers of patients has consistently shown a long-term survival advantage in patients with diploid tumours compared to those with DNA aneuploid tumours, and in low over high cell proliferation carcinomas. KALLIONEMI et al. (1987) found ploidy, nodal status, tumour size and content of progesterone receptors to be independent predictors of survival. CORNELISSE et al. (1987), in a study involving 565 patients, found overall survival and distant disease-free interval longer in diploid as opposed to DNA aneuploid carcinomas though ploidy was an independcent predictor of prognosis only in postmenopausal patients. HEDLEY et al.

(1987) also reported a long-term survival advantage associated with diploid status, but this was not an independent predictor of survival. Less information is available on the importance of cell proliferation in breast cancer as determined by flow cytometry. HEDLEY et al. (1987) have reported a significant association between S phase content, prognosis and tumour grade. It is likely that such measurements will play an important role in carcinoma of the breast since prospectively conducted autoradiographic studies of cell proliferation have demonstrated its predictive capacity in this disease (HERY et al. 1987; SILVESTRINI et al.; TUBIANA and KOSCIENLY 1988).

Reports of the relationship of DNA index and cell proliferation to oestrogen receptors have appeared; these include suggestions of a lower oestrogen receptor positivity in patients with a high DNA index and high cell proliferation (e.g. MORAN et al. 1984) but if the DNA index is in the tetraploid region, oestrogen receptor positivity may be preserved (BAILDAM et al. 1987).

### *5.1.4 Haematological Malignancies*

#### 5.1.4.1 Lymphomas

Nearly all reports to date have concentrated on non-Hodgkin's lymphomas. These have a lower incidence of DNA aneuploidy, approximately 20%–30%, compared with the 50%–80% found in carcinomas. DNA indices cluster around the peridiploid (1.1–1.2) and tetraploid (1.8–2.0) regions in lymphoma as opposed to the triploid level (D.I. 1.5.) seen in carcinomas. A strong association is seen between the grade of the tumour and flow cytometric assessments with high-grade tumours demonstrating an increased incidence of DNA aneuploidy (≅ 40%) and a higher median cell proliferation than low-grade tumours (BRAYLAN et al. 1980; DIAMOND et al. 1982; SHACKNEY et al. 1984; ROOS et al. 1985; MACARTNEY et al. 1986b; MORGAN et al. 1986; BAUER et al. 1986; YOUNG et al. 1987; GRIFFIN et al. 1988).

In several recent series, a high cell proliferation has been associated with a poorer prognosis (BAUER et al. 1986; YOUNG et al. 1987; LEHTINEN et al. 1987; GRIFFIN et al. 1988; O'BRIAN et al. 1988a). In this respect, flow cytometric assessments predict outcome more successfully than histological grade (GRIFFIN et al. 1988; O'BRIEN et al. 1988a) or immunophenotype (O'BRIEN et al. 1988b). A novel DNA-malignancy grading method has been reported by BOCKING et al. (1986a) which takes into account the degree of cell proliferation, presence of a DNA aneuploid stemline and the DNA index of the stem line. This grading system strongly correlated with a variety of clinical parameters such as the stage of the tumour and response to treatment and provided information above that derived from the Kiel classification (BOCKING et al. 1986b). Unlike carcinomas, the presence of DNA aneuploidy has not been shown to be of prognostic significance in non-Hodgkin's lymphomas, despite its association with histologically high-grade lesions.

It is of great interest that a simple assessment of cell proliferation, mitosis counting, has been recently reported to be of prognostic value in non-Hodgkin's Lymphomas (BAUER et al. 1986; AKERMAN et al. 1987; GRIFFIN et al. 1988). However, when mitotic counts were compared with flow cytometric assessment of cell proliferation, no relationship was demonstrable. Indeed mitotic activity, like cell proliferation, was found to be an independent prognostic variable (GRIFFIN et al. 1988). At first sight this might appear surprising but the number of mitoses seen in a section are dependent on the level of cell proliferation together with the duration of mitosis.

In a preliminary study of Hodgkin's disease assessments of DNA aneuploidy and cell proliferation failed to predict outcome (MORGAN et al. 1988).

#### 5.1.4.2 Leukaemias

The incidence of DNA aneuploidy in acute myeloid leukaemia (AML) has been shown to vary with type; M1 having the lowest rate (25%) and M2, M4 and M5 a higher incidence (45%) (HIDDEMANN et al. 1986). No clear relationship to survival has yet emerged in AML (HIDDEMANN et al. 1986; BARLOGIE et al. 1987).

The most extensively studied group are the childhood acute lymphoblastic leukaemias (ALL), in which the presence of DNA aneuploidy was associated with a better outcome in two series of patients with treated ALL (LOOK et al. 1985; SMETS et al. 1987). This was confirmed by the report of the Third International Workshop on Chromosomes in Leukaemia (1983), which found a better response to therapy in patients with > 50 chromosomes (see Sect. 5.4).

The presence of an abnormal stemline in AML and myeloma has been found useful in monitoring and the detection of relapse (WALLE and NIEDERMAYER 1985; BUNN et al. 1982).

### *5.1.5 Gynaecological Malignancies*

Ovarian adenocarcinomas have been extensively studied, with the presence of DNA aneuploidy predicting a poor prognosis (FRIEDLANDER et al. 1983, 1984b; VOLM et al. 1985a; IVERSEN and LAERUM 1985; BLUMENFELD et al. 1987; RODENBURG et al. 1987). In 84 patients with stage II–IV disease, BLUMENFELD et al. (1987) reported median survivals of 48 months for patients with diploid as opposed to 19 months for patients with DNA aneuploid tumours; DNA tetraploid tumours behaved as diploid tumours. Cell proliferation also appears prognostically important in this disease, with low cell proliferation tumours behaving less aggressively (VOLM et al. 1985a).

Similar findings are reported in endometrial carcinoma (IVERSEN and LAERUM 1985). Cytophotometric studies of DNA content also support the flow cytometric studies (ERDHART et al. 1984; MOBERGER et al. 1984).

The relationship to survival in carcinoma of the cervix is more complex, probably due to the effectiveness of surgery and radiotherapy. An early DNA cytophotometric study by NG and ATKIN (1973) suggested that diploid tumours behave more aggressively than DNA aneuploid tumours. This has not been confirmed by other workers, who have reported a higher incidence of lymph node involvement and of recurrence after radiotherapy or surgery at 2 years if the tumour had a DNA index of >1.5 (JAKOBSEN 1984) and an association between increased risk of recurrence and high S phase values, though these authors did not demonstrate a relationship between ploidy and survival (STRANG et al. 1987).

In a recent prospective series of 250 patients treated by radiotherapy (DYSON et al. 1984b, 1985, 1987) we have demonstrated a difference in the sites of tumour recurrence after treatment; DNA aneuploid tumours recurred more frequently at distant sites whilst diploid tumours tended to recur locally.

### *5.1.6 Lung Carcinoma*

Few studies have been reported on this very common tumour. Those that have suggest that DNA aneuploidy and high cell proliferation predict a poor prognosis (VOLM et al. 1985b; ZIMMERMANN et al. 1987). ZIMMERMANN et al. (1987) reported 90% survival at 20 months in diploid surgically resected lung carcinoma compared to 50% of patients with DNA aneuploid tumours ($P < 0.0005$). On multiple regression analysis ploidy was the most important predictor of survival ($P < 0.0001$), with nodal status ($P = 0.02$) and operation type ($P = 0.04$) of lesser importance. Further studies are clearly required on these tumours.

### *5.1.7 Bone and Soft Tissue Tumours*

Classical osteogenic and other high-grade sarcomas appear to have a high incidence of DNA aneuploidy whilst parosteal osteosarcomas and low-grade chondrosarcomas are usually diploid (KREICBERGS et al. 1984; MANKIN et al. 1985). DNA aneuploidy has been suggested to play a useful role in confirming the histological opinion of grade with the requirement that the pathologist critically reviews the histology if a tumour is found to be DNA aneuploid. Recent work on smooth muscle tumours of the gastrointestinal tract and soft tissue has shown an enhanced ability to predict the behavior of the former when the presence of DNA aneuploidy and mitosis counting are combined (COOPER et al. 1987). A substantial difference in the frequency of DNA aneuploidy was found between a series of soft tissue leiomyosarcomas (70%) and smooth muscle tumours of the gastrointestinal tract (20%) (QUIRKE and MRC Soft Tissue Tumour Panel, unpublished observations).

In a preliminary study on 21 cases of Ewing's sarcoma we failed to demonstrate a relationship between DNA aneuploidy and survival, but found to our surprise a very strong relationship between cell proliferation and survival, with diploid tumours above the median cell proliferation (%S + $G_2$ 24.7%) surviving significantly longer ($P < 0.003$) than those below this value (ANDREW et al. unpublished observation). If this is confirmed in a larger series then flow cytometric measurements may provide the first accurate grading method in this disease.

## 5.2 Premalignant Lesions

Less work has been performed on premalignant and benign tumours, but the literature is interesting as it appears to show variations in the incidence and type of DNA stemline between different pathways of neoplastic progression. Most work has been performed on gastrointestinal and genitourinary neoplasia; therefore these will be discussed at some length.

### *5.2.1 Gastrointestinal Tract*

#### 5.2.1.1 The Colorectum: The Adenoma-Carcinoma Sequence

As stated above, between 60% and 85% of colorectal carcinomas are DNA aneuploid, this work being supported by strong cytogenetic evidence that about half of carcinomas have chromosomal abnormalities in the peridiploid range and the other half major abnormalities of 56–91 chromosomes (REICHMANN et al. 1981). Chromosomal abnormalities have also been detected in adenomas of the colorectrum (ENTERLINE and ARVAN 1967; LUBS and KOTLER 1967) and in familial adenomatous polyposis (FAP) both in the adenomas themselves and in fibroblast and lymphocyte cultures (MARK et al. 1973; MITELMAN et al. 1974; GARDNER et al. 1982, 1985).

Several flow cytometric studies have now been performed on sporadic adenomas which show DNA aneuploidy rates of 6% (QUIRKE et al. 1986b), 9% (WEISS et al. 1985) and 13% (GOH and JASS 1986). VAN DEN INGH et al. (1985) reported a higher incidence of 27%, but only 15% of adenomas in this series were below 1 cm compared to a more representative 45% in our series. The size distributions are not available in the other studies for comparative purposes. The prevalence of DNA aneuploidy correlated with the size of the adenoma, with large adenomas having a higher prevalence of DNA aneuploidy (VAN DEN INGH et al. 1985; QUIRKE et al. 1986b; GOH and JASS 1986), with the type of adenoma (QUIRKE et al. 1986b) and with the degree of dysplasia in one study (GOH and JASS 1986) but not the others (VAN DEN INGH et al. 1985; QUIRKE et al. 1986b). It is not surprising that agreement on dysplasia is lacking with the reported problems of inter- and intraobserver variation in grading practices in adenomas (BROWN et al. 1985) and the marked variation in incidence of mild, moderate and severely dysplastic adenomas reported in different series from the same institution (KONISHI and MORSON 1982; GOH and JASS 1986).

The median degree of cell proliferation (%$S + G_2$) in 100 adenomas was 16.4%, (QUIRKE et al. 1988) significantly lower than that found in the carcinomas (%$S + G_2$ 24.7%) using the same cell cycle analysis program. There was no correlation between the %$S + G_2$ and the histological parameters.

In a series of 210 samples from 20 cases of FAP (QUIRKE et al. 1988), DNA aneuploidy was found at a much earlier stage than in sporadic adenomas. It was present within adenomas <5 mm in size and in mucosa bearing very small early "multicryptal lesions"; the prevalence remained similar at approximately 15% for adenomas up to 10 mm in size. The small number of adenomas found above this size made assessments of its incidence unreliable in this group. Similar DNA stemlines were seen in different adenomas derived from the same region in one case in which the whole colon was available for study. An abdominoperineal resection for three adenocarcinomas of the rectum 8 years earlier from this patient also revealed the same abnormal stemline, suggesting that a field change had occurred affecting the DNA content of these cells. Measurement of cell proliferation (%$S + G_2$) in the FAP adenomas revealed a significantly elevated level in the larger (>5 mm) adenomas, at 18.6%, as opposed to the smaller adenomas (<5 mm), at 15.4%.

DNA aneuploidy is absent from normal mucosa in children and young adults, but it can be demonstrated as a field change in the transition zone and resection margins of patients undergoing resection for adenocarcinoma (NORTH and QUIRKE, unpublished observations). The possible implications for metachronous disease are currently unknown.

### 5.2.1.2 The Colorectum: Ulcerative Colitis

Few studies have been performed in ulcerative colitis (UC) and the current value of the detection of DNA aneuploidy is open to question. HAMMARBERG et al. (1984) demonstrated DNA aneuploidy in pooled colonoscopic biopsies, confirming cytogenetic evidence of chromosomal aneuploidy described by XAVIER et al. (1974). They went on to suggest that its presence might provide a valuable predictor of the presence of malignancy. To fulfil the requirement of a screening test it should be a sensitive and specific test with a low false positive and negative rate. In a retrospective study of 297 samples by FOZARD et al. (1986) three groups of patients were investigated: short-term colitics (12), long-term colitics (>10 years) with no evidence of carcinoma (12) and long-term colitics (>10 years) complicated by carcinoma (14). DNA aneuploidy was found in 21%, 67%

and 67% of each group of patients. It did not occur prior to 4 years, but increased in frequency after this time. No difference was found between carcinoma- and non-carcinoma-bearing groups in the longstanding colitics. DNA aneuploidy appeared to arise many years before the development of carcinoma. A surprising feature was the appearance of similar abnormal DNA stemlines within the colon, suggesting the colorectrum had suffered a field change similar to that seen in the intensively studied FAP patients. Prospective studies are underway to confirm or refute our findings in this disease.

Simple flow cytometric estimates of cell proliferation in UC are complicated by the admixture of large populations of inflammatory cells theoretically rendering this technique unsuitable unless a marker of epithelial cell lineage is utilised. However, estimates of cell proliferation by Ki 67 labelling do appear to suggest a correlation between cell proliferation and disease activity (FRANKLIN et al. 1985) and further studies to elucidate its predictive value are required.

#### 5.2.1.3 Stomach

Few studies have investigated the dysplasia-carcinoma sequence in the stomach. MACARTNEY and CAMPLEJOHN (1986) in a small series of patients found cases of mild and moderate dysplasia to be diploid and a high incidence of DNA aneuploidy in severe dysplasia (five of seven cases). These findings must be judged in the light of work be TEODORI et al. (1984), who reported DNA aneuploidy in 9 of 20 (45%) gastric biopsies showing gastritis. The purported high incidence of DNA aneuploidy in gastritic mucosa is of great interest in the light of the incidence in ulcerative colitis, another chronic inflammatory high cell proliferation state (FOZARD et al. 1986).

### *5.2.2 Genitourinary System*

#### 5.2.2.1 Bladder

Difficulties arise in the assessment of premalignant conditions of the bladder as transitional cell carcinomas do not have an easily identifiable precursor lesion; however, early stage bladder tumours (pTa) are usually diploid (CHIN et al. 1985) and on progression (pT1) 30%–40% become DNA aneuploid (TRIBUKAIT et al. 1982; CHIN et al. 1985). Carcinoma in situ of the bladder appears very different in that 90%–100% of cases appear DNA aneuploid and as in the colorectum the same DNA index can be demonstrated in different areas of the bladder, suggesting a field change has occurred (GUSTAFSON et al. 1982a; MELAMED and KLEIN 1984). Patients with multiple DNA aneuploid stemlines in carcinoma in situ have been reported to have a high incidence of clinical recurrence (GUSTAFSON et al. 1982b).

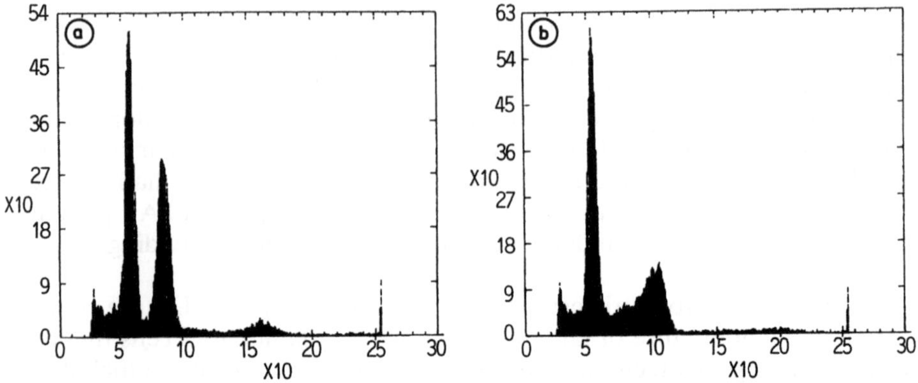


**Fig. 6. a** DNA histogram of a triploid partial hydatidiform molar pregnancy (DNA index 1.5). **b** DNA histogram of a diploid complete hydatidiform molar pregnancy. Note the very high level of cell proliferation. (Reproduced by kind permission of the Editor of the Journal of Clinical Pathology)

5.2.2.2 Cervix

Abnormalities of DNA in cervical intraepithelial neoplasia have been described (LINDEN et al. 1979; GOERTTLER and STOHR 1979; VALET et al. 1981; JAKOBSEN et al. 1983), but only one group has correlated flow cytometric findings with the histological grade, finding an increase in the degree of DNA aneuploidy from CIN I through to invasive carcinoma (JAKOBSEN et al. 1983; JAKOBSEN 1984).

5.2.2.3 Ovary

An investigation of ovarian tumours of borderline malignancy found a very low incidence of DNA aneuploidy (FRIEDLANDER et al. 1984c). Two of 44 tumours (5%) were DNA aneuploid and one of the two rapidly progressed to an aggressive ovarian carcinoma, whereas only 1 of 38 (3%) diploid tumours progressed.

5.2.2.4 Molar Neoplasia

A valuable role for flow cytometry has emerged in the study of molar neoplasia. Partial hydatidiform moles are cytogenetically triploid and easily detected by flow cytometry, with an abnormal stemline from the molar tissue with a DNA index around 1.5 (Fig. 6a; HEMMING et al. 1987; FISHER et al. 1987). Complete moles are diploid with a single $G_0/G_1$ peak (Fig. 6b; HEMMING et al. 1987; FISHER et al. 1987). When the percentage of cells above the $G_0/G_1$ peak of a diploid complete molar pregnancy is quantitated and compared to that found in first, second and third trimester pregnancies then complete moles demonstrate a level of cell proliferation far in excess of normal placental tissue (Fig. 7; HEMMING et al. 1987). Similarly, cases of persistent trophoblastic disease also demonstrate a high

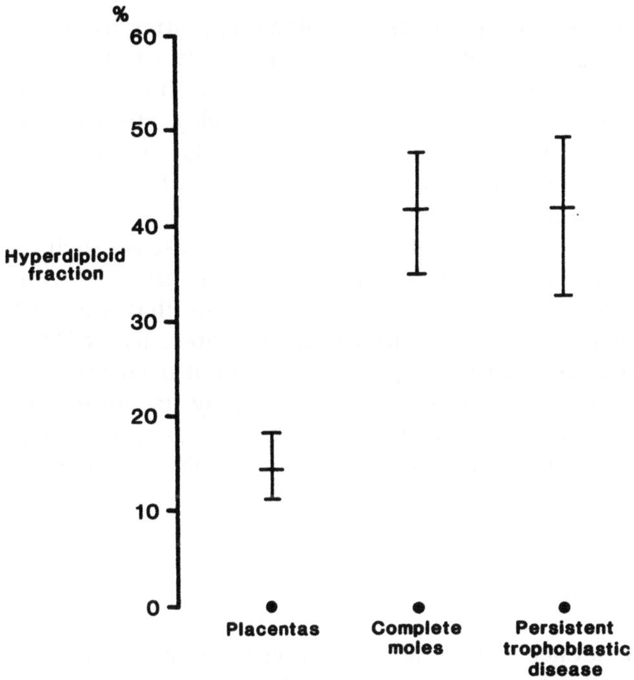


**Fig. 7.** Comparison of the hyperdiploid fraction (the number of cells above the $G_0/G_1$ peak) of normal placentas, complete hydatidiform molar pregnancies and cases of persistent trophoblastic disease. *Bars* represent median value, 25th and 75th centiles. (Reproduced by kind permission of the Editor of Placenta)

level of cell proliferation outside that of normal placental tissue but this falls within the range of complete moles and does not predict those likely to progress (HEMMING et al. 1988). Such measurements strongly suggest that a major cause of the premalignant potential of hydatidiform molar pregnancy is the high level of cell proliferation seen in this condition.

## 5.3 Benign Endocrine Tumours

Several studies have recently been published on benign endocrine tumours. These are of interest since in sites such as the pituitary gland they have a very low incidence of progression to a malignant state, whereas in others (e.g. the thyroid and adrenal) they may have a more aggressive behaviour. ANNIKO et al. (1984) found 23 of 47 (49%) pituitary tumours to be DNA aneuploid. A higher incidence was seen in those secreting growth hormone or prolactin than TSH or ACTH. A study of parathyroid adenomas (IRVIN and BAGWELL 1979) found 2 of 37 (5%) adenomas to be DNA aneuploid and a higher incidence of small populations of polyploid (DNA tetraploid) cells in these tumours when compared to normals. A study of 67 follicular adenomas and seven adenomatous goitres found 18 of 67 (27%) of the former and one of seven (14%) of the latter to be DNA aneuploid. Follow-up of these patients did not reveal any useful prognostic information from these measurements (JOENSUU et al. 1986). However, in the adrenal BOWLBY et al. (1986) found five of six (83%) adrenal carcinomas to be DNA aneuploid and none of 16 adenomas. A large tumour size correlated with aneuploidy.

The frequent finding of DNA aneuploidy in a benign disease such as pituitary adenoma suggests that its presence per se is not confined to conditions that are likely to progress to invasive malignancy and that site must always be considered when assessing its importance. The presence of DNA aneuploidy in an adenomatous goitre supports its potential occurrence in long-standing hyperplastic condidtions. Whether it may be found in hyperplastic conditions with no neoplastic potential is unknown, but studies to date on psoriatic epithelium have not revealed DNA aneuploidy (BAUER et al. 1980, 1981; GELFANT et al. 1983).

## 5.4 DNA Aneuploidy and Treatment

A substantial body of evidence is now accumulating to support the view that DNA aneuploid tumours are more susceptible to radiotherapy and chemotherapy.

In a series of over 250 patients with cervical carcinoma assessed using multiple biopsies, a significantly higher level of cell death was found during radiotherapy in DNA aneuploid tumours than in their diploid counterparts (DYSON et al. 1987). This was reflected in a more rapid clinical response and an increased rate of loss of tumour from the biopsies as noted by histopathological examination (DYSON et al. 1985). Further support for this position is provided by WYJKSTROM et al. (1984), who found an enhanced response to radiotherapy in DNA aneuploid bladder tumours. Enhanced susceptibility to chemotherapy has been reported in studies on neuroblastoma (LOOK et al. 1984; GANSLER et al. 1986), ALL (Third International Workshop on Chromosomes in Leukaemia 1981; LOOK et al. 1985; SMETS et al. 1987) and possibly AML (HIDDEMANN et al. 1986) although in the latter due to small numbers of patients the difference was not significant. Increased chemosensitivity has also been shown in vitro in DNA aneuploid malignant gliomas (SHAPIRO and SHAPIRO 1985). DNA aneuploid cells in this study were also found to be genetically unstable, and to have shorter doubling times than their "diploid" counterparts. Overall the above studies provide good evidence that the detection of DNA aneuploidy may be of value in the prediction of the response to treatment of certain malignancies.

# 6 The Biology of DNA Aneuploidy

## 6.1 Formation of DNA Aneuploidy

The mapping of the presence of DNA aneuploidy in a variety of malignant and premalignant conditions has given clues to the mode of formation of DNA aneuploid stemlines from which early hypotheses can be formulated. Carcinogenesis is a multistep pathway with the cell suffering sequential changes that enable it to escape from local growth control mechanisms. The occurrence of DNA aneuploidy is but one step on this road, albeit from a prognostic viewpoint an apparently important one in certain sites.

Minor changes from a diploid to a peridiploid cell can occur via a variety of chromosomal changes, e.g. partial or complete deletion, endoreduplication, unbalanced translocations, the production of double minutes or other abnormal chromosome structures such as rings or dicentrics, etc. The formation of a DNA aneuploid cell calls for much greater changes in karyotype. The most likely mechanism for this is a mitotic abnormality or possibly cell fusion leading to DNA tetraploidy; both of these aberrations have been described as occurring as the result of exposure to carcinogens or viruses (WOLMAN 1983). The natural occurrene of polyploidy in certain cells (e.g. liver, megakaryocytes, trophoblast, cardiac muscle etc.) also suggests the existence of an internal mechanism for the control of DNA content. This might be switched off by DNA damage following exposure to chemical carcinogens or viruses and provides a further method for the formation of tetraploid cells during carcinogenesis.

There is good experimental evidence and also indirect evidence from chromosomal studies that progression of cells from a diploid/peridiploid to a tetraploid/peritetraploid state occurs. This has been demonstrated in the Dunning rat prostatic carcinoma model and during the induction of neoplasia in the mouse salivary gland (COWELL and WIGLEY 1980; ISAACS and SANDBERG 1982) and is supported by karyotypic evidence of doubling of many chromosome classes in malignant tumours with a high chromosome number (e.g. SANDBERG 1980; SHAPIRO and SHAPIRO 1985). The generation of such gross abnormalities is likely to be due to the genomic abnormalities discussed above but a second possible mechanism is that of cell fusion (BARSKI et al. 1961; BARSKI and CORNEFERT 1962; HARRIS and KLEIN 1969), where a tumour cell fuses with either a second tumour cell or a host cell to give rise to a tretraploid cell. The generation of DNA aneuploid cells from tetraploid cells has been shown to occur in the animal model (COWELL and WIGLEY 1980; ISAACS and SANDBERG 1982; ISAACS et al. 1982) and in cell hybrids by the spontaneous loss of chromosomes from cells (KLEIN et al. 1971). Results from cell fusion experiments also demonstrate that chromosome loss in a non-random phenomenon from hybrids, that suppression of the malignant cell phenotype can occur with fusion of a malignant and a normal cell, and that reversion to the malignant state occurs with subsequent loss of chromosomes (HARRIS et al.

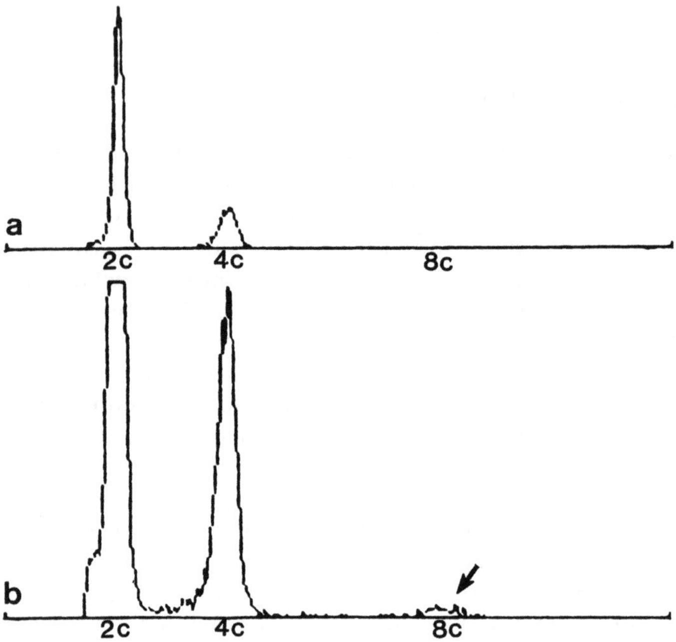


**Fig. 8.** **a** DNA histogram showing a polypoid population in the left ventricle of a post-mortem heart. **b** Scaled up histogram of **a** to show the octaploid population (*arrow*)

1969; HARRIS and KLEIN 1969; HARRIS 1970; KLEIN et al. 1971). The controlling mechanism for this non-random loss is unknown. Such experimental results provide evidence for the possible role of suppressor genes (antioncogenes) in the control of the malignant phenotype. The importance of an unbalanced genome is supported by the modal DNA index of 1.5 found in most solid DNA aneuploid tumours and the apparently less aggressive behaviour of certain DNA tetraploid tumours (RONSTROM et al. 1981; TRIBUKAIT et al. 1982; QUIRKE et al. 1987).

## 6.2 Polyploidy

Before considering the possible biological advantages of DNA aneuploid tumours, the concept of polyploidy must be discussed. Polyploidy is widely found throughout the plant and animal kingdoms, and its presence in these situations must confer a biological advantage. In humans, natural polyploidy is restricted to sites such as the liver, megakaryocytes, and trophoblast (BRODSKY and URYVAEVA 1976; QUIRKE, unpublished observations) but does occur in certain situations when the organism is stressed, such as in the myocardium with ventricular hypertrophy (see Fig. 8) (SANDRITTER and SCOMAZZONI 1964; EISENSTEIN and WIED 1970).

The biological advantages of polyploidy have been discussed in depth by others (BRODSKY and URYVAEVA 1977) and will only be briefly reviewed here. Cell functions can be divided into three levels: those essential for survival, those required for proliferation and reproduction, and lastly those needed for differentiation and function. The transcriptional and translational capabilities are finite in any given cell; therefore the first priority will be cell survival and the remaining transcriptional/translational activity will be utilised for either proliferation or differentiation. If the cell is stressed then differentiation and functional activities are the first to be lost, followed by proliferative capacity in order to preserve the viability of the cell. A polyploid cell has the great advantage not only of doubling its DNA content but also of doubling its transcriptional and translational activity, leading to a greater functional reserve (BRODSKY and URYVAEVA 1977). This has been demonstrated in the rat liver by following the level of amino acid incorporation and enzyme activity during changes of cell DNA content. Polyploidy therefore confers a positive biological advantage on a cell under certain environmental conditions.

## 6.3 Potential Biological Advantages of DNA Aneuploidy in Cells

The increased chromosome number in DNA aneuploid tumours may lead to a biological advantage through two mechanisms: those related to the genome and others related to cell function. Genomic advantages could be genetic instability with an increased tendency to give rise to mutations with a concomitant reduction in the occurrence of random lethal mutations due to the larger number of chromosomes. The threshold for lethal DNA damage may be higher in a DNA aneuploid cell and the ability to give rise to a higher number of new mutations would enable a tumour to adapt more rapidly to a change in its local environment or a change of environment following metastasis. At the same time as doubling its DNA and volume, the cell surface area only increases by a factor of 1.59 due to the geometrical properties of a sphere (BRODSKY and URYVAEVA 1977). It therefore has relatively less surface area to maintain as well as an increased functional reserve. Functional advantages could accrue from a greater transcriptional/translational reserve, especially if adverse microenvironmental conditions arise locally or in a metastatic desposit; or these could occur from changes in the cell surface. Such changes might take the form of a relative reduction in the cell surface receptors for growth factors that suppress the growth of the cell or an increased number of receptors for growth-promoting growth factors due to the enhanced synthetic capacity of the cell acting in conjunction with the relative reduction of the cell surface area. The potential growth advantage of tetraploid cells has been confirmed by FOURNIER and PARDEE (1975), who induced polyploidy with cytochalasin-B in fibroblasts and found an increased cell turnover in these polyploid cells. Any of the above changes might lead to enhanced cell proliferation or enhanced survival for the DNA aneuploid tumour cell and thus a biological advantage.

If the DNA aneuploid cell arises on the basis of cell fusion then incorporation of foreign epitopes into the tumour cell membrane may occur. Experimental evidence suggests that this incorporation could lead to the production of highly aggressive metastatic variants (for a review see DE BAETSELIER et al. 1984).

## 6.4 Evidence for Non-random DNA Abnormalities

The formation of related DNA stemlines in UC (FOZARD et al. 1986), FAP (QUIRKE 1987) and urinary bladder (GUSTAFSON et al. 1982a; MELAMED and KLEIN 1984) provides strong evidence that the development of these abnormalities may be non-random. The clustering of a variety of tumours at different modal DNA indices such as 1.4 and 2.0 for seminomas, 1.3 for teratomas (QUIRKE et al. 1986a), 1.5 for the majority of solid tumours, 1.3 and 2.0 for lymphomas (MORGAN et al. 1986) and 1.2 for AML (HIDDEMANN et al. 1986) is further evidence of preferred stemlines in these tumours. In UC and FAP (FOZARD et al. 1986; QUIRKE et al. 1988) the DNA indices of abnormal stemlines were related to an individual colorectum, making these both site- and person-dependent abnormalities. Evidence exists to support the non-random nature of chromosome changes in malignancy and has been reviewed by MITELMAN (1980). Investigation of Rous sarcoma virus-induced fibrosarcoma revealed the sequential development of a series of non-random abnormalities involving three chromosomes, numbers 7, 12 and 13 (MITELMAN 1972). A totally different chromosome pattern emerged, however, when morphologically identical fibrosarcomas were induced by carcinogens such as 7,12-dimethylbenz(α)anthracene in the same rat strain, leading to a trisomy of chromosome 2 (LEVAN and MITELMAN 1976). The same was found for 20-methylcholanthrene- and 3-4-benzpyrene-induced sarcomas and *N*-nitroso-*N*-butylurea-induced leukaemias with DNA damage also demonstrable on chromosome 2 (LEVAN 1974; LEVAN and LEVAN 1975). T-cell murine leukaemias induced by a variety of chemical agents have also been shown to develop a specific abnormality, trisomy 15 (DOFUKU et al. 1975; CHAN 1978). This work provides strong evidence for non-random DNA damage by carcinogens and viruses, and in the Rous sarcoma model for a pattern of sequential changes induced by one oncogenic virus. It is now known that the action of certain carcinogens may be highly specific, causing single base pair changes in oncogenes, leading to their activation (BALMAIN 1985).

The non-random DNA indices seen in man may represent chromosome loss due to specific carcinogen-induced abnormalities of mitosis such as are seen using chloramphenicol, streptonigrin or colchicine. Colchicine is of special interest as in high doses it is known to cause endoreduplication of chromosomes (BRODSKY and URYVAEVA 1977). The importance of chemical carcinogens in the induction of DNA aneuploidy is hinted at by a study of carcinogen-exposed acute non-lymphoblastic leukaemia patients and non-exposed patients (MITELMAN et al. 1978). Only 25% of patients with the disease in the non-exposed group demonstrated chromosomal aberrations whereas 80% of the exposed group did so.

Supportive evidence is provided by the findings that patients treated with chlorambucil develop cytogenetic abnormalities, that 75% of patients with leukaemia after previous chemotherapy treatment have abnormal chromosomes, especially numbers 5 and 7, and that there is a relationship between abnormalities of these two chromosomes and exposure to chemicals, solvents and other petroleum products (MITELMAN et al. 1981; GOLOMB et al. 1982; Fourth International Workshop on Chromosomes in Leukaemia 1984; ROWLEY 1984).

Substantial evidence exists, therefore, of non-random changes in individual chromosomes during carcinogenesis, with accruing evidence of larger non-random changes which are dependent on tumour type (e.g. lymphoma vs carcinoma, and seminoma vs teratoma) and possibly the mode of exposure or type of carcinogen. Further studies on the initiation of DNA aneuploidy in cells in different neoplasms may be rewarded with a better understanding of the causation and progression of human tumours.

## 7 DNA Content in Combination with Other Staining Methods

Two broad categories will be discussed, those tests aimed at detecting subpopulations of cycling cells and secondly staining techniques enabling the correlation of other cellular properties to DNA content.

### 7.1 Quantitation of Subpopulations of Cells in the Cell Cycle

Two techniques are available; the most frequently used relies upon bromodeoxyuridine whilst the second utilises a monoclonal antibody, Ki 67.

#### *7.1.1 Bromodeoxyuridine*

This technique can be performed either by in vitro incubation of cells in a solution of bromodeoxyuridine or by intravenous administration of the compound, though the latter method has not yet achieved wide application. The cells take up bromodeoxyuridine and incorporate it into their DNA during DNA synthesis. After a period of time, usually 30 min, the reaction is halted by withdrawal of the bromodeoxyuridine or fixation of the tissue. By varying the time of exposure to bromodeoxyuridine it is pos-

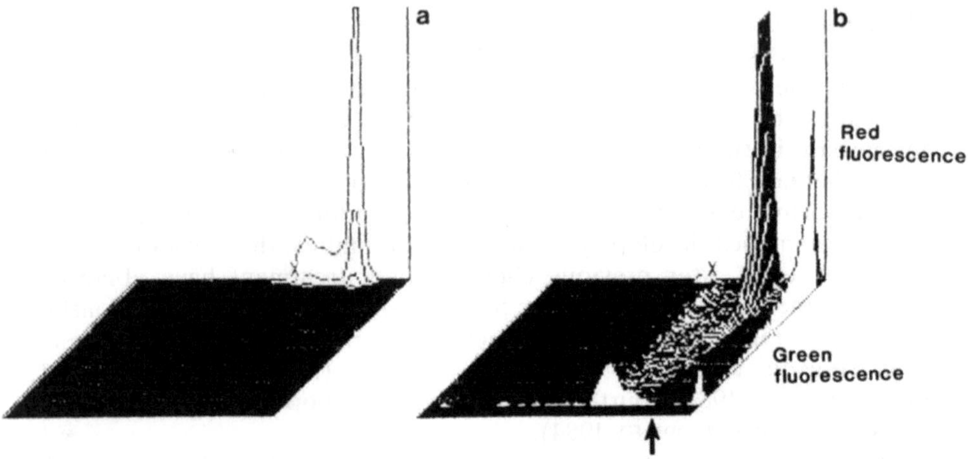


**Fig. 9 a, b.** Bromodeoxyuridine labelling. **a** Control DNA histogram of RAJI lymphoma cells showing a high cell proliferation. **b** Same cells identified by a fluorescein-labelled monoclonal antibody to bromodeoxyuridine. The labelled cells project from the baseline as a horseshoe enabling quantitation of the number of cells in S phase (*arrow*)

sible to conduct cell kinetic experiments and obtain estimates of cell cycle times. After removal from the bromodeoxyuridine the nuclear DNA is denatured by one of a variety of techniques to expose the incorporated bromodeoxyuridine. Incubation of the cells with a fluorescent-labelled monoclonal antibody to bromodeoxyuridine labels those cells in S phase during the incubation period, allowing their quantitation. Labelled cells traverse a semicircular path arising from the $G_0/G_1$ peak through S phase to $G_2/M$ (Fig. 9).

Bromodeoxyuridine survives formalin fixation and paraffin embedding and can be combined with the technique of HEDLEY et al. (1983) for the measurement of DNA content in paraffin-embedded tissue (DE FAZIO et al. 1987). Immunocytochemical labelling within tissue sections is also possible. Whilst bromodeoxyuridine has many advantages it must be remembered that it relies upon the same principles as thymidine uptake and is prone to identical disadvantages. Indeed, with bromodeoxyuridine labelling a population of cells in S phase that fails to label can be identified, the so-called $S_0$ cells. The importance of these cells is currently unknown. Many experimental papers have appeared utilising bromodeoxyuridine labelling but few clinical reports have yet emerged.

### *7.1.2 Ki67*

In 1983 GERDES et al. reported the production of a monoclonal antibody immunoreactive with a cell cycle-related nuclear antigen. Flow cytometric characterisation suggested that this antigen was present in $G_{1B}$, S, $G_2$ and mitosis (GERDES et al. 1984). Flow cytometric studies are possible using dual staining of DNA content and Ki67 but its superiority over standard cell proliferation measurements is currently unknown. Immunocytochemical studies have been performed on a wide range of tissues using this antibody, with early reports confirming the flow cytometric results of differences between the levels of cell proliferation of low- and high-grade non-Hodgkin's lymphomas (WEISS et al. 1987) and of a poor prognosis in patients with high cell proliferation tumours (HALL et al. 1988). Reports are emerging in other conditions such as breast (GERDES et al. 1986; BARNARD et al. 1987), lung (GATTER et al. 1986) and brain tumours (BURGER et al. 1986) but investigations are seriously hindered by the requirement for fresh tissue. Monoclonal antibodies to cell cycle-related antigens immunoreactive in paraffin-embedded tissue are required in order to determine rapidly the prognostic value of these techniques. A variety of other antibodies of potential use in the flow cytometric or immunohistochemical quantitation of cell proliferation react with proliferating cell nuclear antigen (PCNA) (MATHEWS et al. 1983; CHAN et al. 1983; SMETANA et al. 1983), $C_5F_{10}$ (LLOYD et al. 1985), an antibody that reacts with mitosing cells, and possibly p145, a nucleolar antigen (FREEMAN et al. 1986, 1987).

## 7.2 DNA Content and Quantitation of Non-Cell Cycle Related Cellular Properties

A wide range of physical and chemical properties of a cell can be quantitated together with DNA content. Physical properties such as cell size (Fig. 2) or granularity can be easily obtained from a Coulter volume accessory, forward angle-light scatter (FALS) or 90° light scatter. Estimates of the level of cell death can be derived by utilising the exclusion of a fluorescent stain such as ethidium bromide or propidium iodide by the cell membrane of viable cells. Dead and dying cells take up the stain and can be identified seperately from viable cells (DYSON et al. 1984a, b), thus allowing the monitoring of the effects of treatment (DYSON et al. 1985, 1987). The degree of hypoxia of cells might also be measurable using fluorescent nitroheterocycles (OLIVE and DURAND 1983).

Antibodies to cell surface antigens and receptors, or after fixation or cell membrane permeablisation, to internal constituents of the cell or nucleus, can be fluorescently labelled, enabling quantitation of their target antigens. Little work on clinical tumour material has yet been reported apart from cell surface antigen detection in lymphomas and leukaemias and combined DNA and cytokeratin studies (OUD et al. 1985).

WATSON's group (WATSON et al. 1985; ELIAS JONES et al. 1986) have reported the flow cytometric measurement of the c-*myc* oncoprotein in nuclei isolated from paraffin-embedded material. Whilst identifying an area of great future interest, their work is open to two major criticisms: (a) the unknown length of time between removal and fixation of their surgical specimens prior to measurement, an important factor in a protein with a half-life of 20 min, and (b) more importantly, the loss of an unknown amount of c-*myc* oncoprotein during its redistribution from the nucleus to the cytoplasm upon fixation and paraffin embedding (JONES et al. 1988c).

The combination of biochemical techniques with flow cytometric assessments of DNA content is likely to emerge as a powerful technique since it enables biochemical estimates to be performed on individual cells (see MELAMED et al. 1979; WATSON 1980). The study of enzyme reactions within cells is based upon the principle of enzymatic conversion of a non-fluorescent substrate to a fluorescent end-product, such measurements allowing quantitation of the activity of the target enzyme.

Flow cytometric estimation of oestrogen receptors has been approached in a different fashion by quantitating the receptor binding of fluorescent oestrogen probes (KUTE et al. 1983). Such techniques may lead to a combined estimate of DNA content, S phase fraction and oestrogen receptors in breast carcinoma. Measurement of intracellular pH (VISSER et al. 1979) and calcium concentration (VALET et al. 1985), membrane potentials, mitochondrial function (DARZYNKIEWICZ et al. 1982) and many other properties of the cell are currently possible.

## 8 Conclusion

Potential clinical applications of flow cytometry are currently emerging, but it should be stressed that retrospective studies must be confirmed by well-designed prospective studies prior to clinical application. Areas of great interest are the prediction of prognosis in common solid malignancies such as the lung, colorectum, prostate, bladder, ovary etc. where measurement of DNA content appear superior to subjective histological assessments. Flow cytometry has rejuvenated interest amongst pathologists in cell proliferation and it is time that energy was diverted away from the perpetual scramble after endless uncharacterised antigens in conditions such as non-Hodgkin's lymphoma towards refinement of the role of a proven prognostic indicator of cell kinetics.

Measurements of DNA content in premalignant disease have also identified the development of major aberrations of DNA content in the early stages of neoplasia in certain tumours and suggest that such abnormalities may be more widespread than previously thought. Differences have also been revealed between various pathways of neoplasia within the same organ such as the adenoma-carcinoma and dysplasia-carcinoma sequences in the colorectum and de novo transitional cell carcinomas and carcinoma in situ of the bladder.

This review has sought to concentrate on the measurement of DNA content, but a wide range of new techniques are now available for application in medical research. Flow cytometry is a versatile technique and has much to offer in many areas of medicine.

## References

Akerman M, Brandt L, Johnson A, Olsson H (1987) Mitotic activity in non Hodgkin's lymphoma. Relation to the Kiel classification and to prognosis. Br J Cancer 55:219–223

Anniko M, Tribukait B, Wersall J (1984) DNA ploidy and cell phase in human pituitary tumours. Cancer 53:1708–1713

Arends JW, Schutte B, Wiggers T, Verstijnen CPHJ, Blijham GH, Bosman FT (1987) Comparison of phenotypic and genotypic features in human primary large bowel carcinomas and lymph node metastases. Cancer Res 47:4342–4344

Armitage NC, Robins RA, Evans DF, Turner DR, Baldwin RW, Hardcastle JD (1985) The influence of tumour cell DNA abnormalities on survival in colorectal cancer. Br J Surg 72:828–830

Arndt-Jovin DJ, Tovin TM (1977) Analysis and sorting of living cells according to deoxyribonucleic acid content. J Histochem Cytochem 25:585–589

Atkin NB (1972) Modal deoxyribonucleic acid value and survival in carcinoma of the breast. Br Med J I:271–272

Atkin NB, Kay R (1979) Prognostic significance of modal DNA value and other factors in malignant tumours based on 1465 cases. Br J Cancer 40:210–221

Auer G, Eriksson E, Azavedo E, Caspersson T, Wallgren A (1984) Prognostic significance of nuclear DNA content in mammary adenocarcinoma in humans. Cancer Res 44:394–396

Avery OT, Macleod CM, McCarty M (1944) Studies on the chemical nature of the substance inducing transformation of pneumococcal types. Induction of transformation by a deoxyribonucleic acid fraction isolated from pneumococcus type III. J Exp Med 79:137–157

Badalament RA, Hermansen DK, Kimmell M et al. (1987) The sensitivity of bladder wash flow cytometry, bladder wash cytology and voided cytology in the detection of bladder carcinoma. Cancer 60:1423–1427

Baildam AD, Zaloudik J, Howell A et al. (1987) DNA analysis by flow cytometry, response to endocrine treatment and prognosis in advanced carcinoma of the breast. Br J Cancer 55:553–559

Baisch H, Gohde W, Linden WA (1975) Analysis of PCP-data to determine the fraction of cells in the various phases of the cell cycle. Radiat Environ Biophys 12:31–39

Baisch H, Beck H-P, Christensen IJ et al. (1982) A comparison of mathematical methods for the analysis of DNA histograms obtained by flow cytometry. Cell Tissue Kinet 15:235–249

Ballantyne KC, James PD, Robins RA, Baldwin RW, Hardcastle JD (1987) Flow cytometric analysis of the DNA content of gastric cancer. Br J Cancer 56:52–54

Balmain A (1985) Guest editorial – transforming ras oncogenes and multistage carcinogenesis. Br J Cancer 51:1–7

Banner BF, Tomas-De La Vega JE, Roseman DL, Coon JS (1985) Should flow cytometric DNA analysis precede definitive surgery for colon carcinoma? Ann Surg 202:740–744

Barlogie B, Raber MN, Schuman J et al. (1983) Flow cytometry in clinical cancer research. Cancer Res 43:3982–3997

Barlogie B, Stass S, Dixon D et al. (1987) DNA aneuploidy in adult acute leukaemia. Cancer Genet Cytogenet 28:213–228

Barnard MJ, Hall PA, Lemoine NR, Kadar N (1987) Proliferative index in breast carcinoma determined in situ by Ki67 immunostaining and its relationship to clinical and pathological variables. J Pathol 152:287–295

Barski G, Cornefert F (1962) Characteristics of 'hybrid' type clonal cell lines obtained from mixed cultures in vitro. J Natl Cancer Inst 28:801–821

Barski G, Sorieul S, Cornefert F (1961) 'Hybrid' type cells in combined cultures of two different mammalian cell strains. J Natl Cancer Inst 26:1269–1290

Bauer FW, Crombag NHCMN, de Grood RM, Boezeman JBM (1980) Quantification of deviations of the epidermal DNA distributions in psoriasis and the effect of therapy. In: Laerum OD, Lindmo T, Thorud E. (eds) Flow cytometry IV. European Press Ghent, pp 377–381

Bauer FW, Crombag NHCMN, de Grood RM, Boezeman JBM, de Grood RM (1981) Flow cytometry as a tool for the study of cell kinetics in skin. 2. Cell kinetic data in psoriasis. Br J Dermatol 104:271–276

Bauer KD, Merkel de Winter JN, Marder RJ, Hauck WW, Wallemark CB, Williams TJ, Variakojis D (1986) Prognostic implications of ploidy and proliferative activity in diffuse large cell lymphomas. Cancer Res 46:3173–3178

Bloch DP, Hew HCY (1960) Changes in nuclear histones during fertilization and early embryonic development in the pulmonate snail *Helix aspersa*. J Biophys Biochem Cytol 8:69–81

Blomjous CEM, Schipper NW, Baak JPA, Van Galen EM, De Voogt HJ, Meyer CJLM (1988) Retrospective study of prognostic importance of DNA flow cytometry of urinary bladder carcinoma. J Clin Pathol 41:21–25

Blumenfeld D, Braly PS, Ben-Ezra J, Klevecz RR (1987) Tumour DNA content as a prognostic feature in advanced epithelial ovarian carcinoma. Gynecol Oncol 27:389–398

Bocking A, Chatelain R, Auffermann W, Kruger GRF, Asmus B, Wohltmann D, Schuster C (1986a) DNA grading of malignant lymphomas. I. Prognostic significance, reproducibility, and comparisons with other classifications. Anticancer Res 6:1205–1216

Bocking A, Chatelain R, Auffermann W, Lohr G-W, Reif M, Rossner R, Becker H (1986b) DNA grading of malignant lymphomas. II. Correlation with clinical parameters. Anticancer Res 6:1217–1224

Bowlby LS, de Bault LE, Abraham SR (1986) Flow cytometric analysis of adrenal cortical tumour DNA. Relationship between cellular DNA and histopathological classification. Cancer 58:1499–1505

Braylan RC, Diamond LW, Powell ML, Harty-Golder B (1980) Percentage of cells in the S phase of the cell cycle in human lymphoma determined by flow cytometry. Correlation with labelling index and patient survival. Cytometry 1:171–174

Brodsky WY, Uryvaeva IV (1977) Cell polyploidy: its relation to tissue growth and function. Int Rev Cytol 50:275–332

Brown LJR, Smeeton NC, Dixon MF (1985) Assessment of dysplasia in colorectal adenomas: an observer variation and morphometric study. J Clin Pathol 38:174–179

Bunn PA, Krasnow S, Makuch RW, Schlam ML, Schecter GP (1982) Flow cytometric analysis of DNA content of bone marrow cells in patients with plasma cell myeloma: clinical implications. Blood 59:528–535

Burger PC, Shibata T, Kleihues P (1986) The use of the monoclonal antibody Ki 67 in the identification of proliferating cells: application to surgical neuropathology. Am J Surg Pathol 10:611–617

Caspersson TO (1950) Cell growth and cell function. Norton, New York

Caspersson TO (1979) Quantitative tumour cytochemistry. Cancer Res 39:2341–2355

Caspersson TO, Schultz J (1938) Nucleic acid metabolism of the chromosomes in relation to gene reproduction. Nature 142:294–295

Chan FP (1978) Chromosome studies in induced murine thymomas. Dissertation Abstract International, Section B, 38:3994–3995

Chan PK, Frakes R, Tan EM, Brattain MG, Smetana K, Busch H (1983) Indirect immunofluorescence studies of proliferating cell nuclear antigen in nucleoli of human tumour and normal tissues. Cancer Res 43:3770–3777

Chin JL, Huben RP, Nava E, Rustum YM, Greco JM, Poutes JE, Frankfurt OS (1985) Flow cytometric analysis of DNA content in human bladder tumours and irrigation fluids. Cancer 56:1677–1681

Coon JS, Deitsch AD, de Vere White RW et al. (1988) Inter-institutional variability in DNA flow cytometric analysis of tumours. The National Cancer Institute's flow cytometry research experience. Cancer 61:126–130

Cooper PN, Quirke P, Hardy GJ, Bird CC, Dixon MF (1987) Prediction of the behaviour of smooth muscle tumours of the gastrointestinal tract. J Pathol 152:217–218 A

Cornelisse CJ, van de Velde CJH, Caspers RJC, Moolenaar AJ, Hermans J (1987) DNA ploidy and survival in breast cancer patients. Cytometry 8:225–234

Cowell JK, Wigley CB (1980) Changes in DNA content during in vitro transformation of mouse salivary gland epithelium. J Natl Cancer Inst 64:1443–1448

Crissman HA, Tobey RA (1974) Cell cycle analysis in 20 minutes. Science 184:1297
Crissman HA, Steinkamp JA (1982) Rapid, one step staining procedures for analysis of cellular DNA and protein by single and dual laser flow cytometry. Cytometry 3:84–90
Crissman HA, Van Egmond J, Holdrinet RS, Pennings A, Haanen C (1981) Simplified method for DNA and protein staining of human hematopoietic cell samples. Cytometry 2:59–62
Darzynkiewicz Z, Traganos F, Staiano-Coico L, Kapuscinski J, Melamed MR (1982) Interactions of rhodamine 123 with living cells studied by flow cytometry. Cancer Res 42:799–806
de Baetselier P, Roos E, Brys L et al. (1984) Non-metastatic tumour cells acquire metastatic properties following somatic hybridization with normal cells. Cancer Metastasis Rev 3:5–24
de Fazio A, Leary JA, Hedley DW, Tattersall MHN (1987) Immunohistochemical detection of proliferating cells in vivo. J Histochem Cytochem 35:571–577
Diamond LW, Nathwani BN, Rappaport H (1982) Flow cytometry in the diagnosis and classification of malignant lymphoma and leukemia. Cancer 50:1122–1135
Dofuku R, Biedler JL, Spengler BA, Old LJ (1975) Trisomy of chromosome 15 in spontaneous leukaemia of AKR mice. Proc Natl Acad Sci 72:1515–1517
Dolbeare FA, Smith RE (1979) Flow cytoenzymology: rapid enzyme analysis of single cells. In: Melamed MR, Mullaney PF, Mendelsohn ML (eds) Flow cytometry and sorting. John Wiley, New York, pp 317–333
Dowle CS, Owainati A, Robins A, Burns K, Ellis IO, Elston CW, Blamey RW (1987) Prognostic significance of the DNA content of human breast cancer. Br J Surg 133–136
Dyson JED, Quirke P, Bird CC, McLaughlin JB, Surrey CR (1984a) Relationship between cell ploidy and glucocorticoid induced cell death in human lymphoid cell lines. Br J Cancer 49:731–738
Dyson JED, Joslin CAF, Quirke P, Bird CC (1984b) Flow cytofluorometric analysis of serial biopsies of tumours of the uterine cervix. Eur J Cancer Clin Oncol 20:1249–1259
Dyson JED, Joslin CAF, Quirke P, Rothwell RI, Bird CC (1985) Quantitation by flow cytofluorometry of response of tumours of the uterine cervix to radiotherapy. Br J Radiol 58:41–50
Dyson JED, Joslin CAF, Rothwell RI, Quirke P, Khoury GG, Bird CC (1987) Flow cytofluorometric evidence for the differential radioresponsiveness of aneuploid and diploid cervix tumours. Radiother Oncol 8:263–272
Edmin SO, Stenling R, Roos G (1987) Prognostic value of DNA content in colorectal carcinoma. A flow cytometric study with some methodological aspects. Cancer 60:1282–1287
Eisenstein R, Wied GL (1970) Myocardial DNA and protein in maturing and hypertrophied human hearts. Proc Soc Exp Biol Med 133:176–179
Elias-Jones J, Hendy-Ibbs P, Cox H, Evan GI, Watson JV (1986) Cervical brush biopsy specimens suitable for DNA and oncoprotein analysis using flow cytometry. J Clin Pathol 39:577–581
Enblad P, Glimelius B, Bengtsson A, Ponten J, Pahlman L (1985) DNA content in carcinoma of the rectum and rectosigmoid. Acta Path Microbiol Scand [A] 93:277–284
Enterline HT, Arvan DA (1967) Chromosome constitution of adenoma and adenocarcinoma of the colon. Cancer 20:1746–1759
Erdhart K, Auer G, Bjorkholm E et al. (1984) Prognostic significance of nuclear DNA content in serous ovarian tumours. Cancer Res 44:2198–2202
Farsund T, Hoestmark JG, Laerum OD (1984) Relation between flow cytometric DNA distribution and pathology in human bladder cancer. A report on 69 cases. Cancer 54:1771–1777
Feulgen R, Rossenbeck H (1924) Mikroskopisch-chemischer Nachweis einer Nucleinsäure von Typus der Tymonucleinsäure und auf die darauf beruhende elektive Färbung von Zellkernen in mikroskopischen Präparaten. Hoppe-Seyler's Z Physiol Chem 135:203–248
Finan PJ, Quirke P, Dixon MF, Dyson JED, Giles GR, Bird CC (1986) Is DNA aneuploidy a good prognostic indicator in patients with advanced colorectal cancer? Br J Cancer 54:327–330
Fisher RA, Lawler SD, Omerod MG, Imrie PR, Povey S (1987) Flow cytometry used to distinguish between complete and partial moles. Placenta 8:249–256

Fordham MVP, Burdge AH, Matthews J, Williams G, Cooke T (1986) Prostatic carcinoma cell DNA content measured by flow cytometry and its relation to clinical outcome. Br J Surg 73:400–403
Forsslund G, Cedermark B, Ohman U, Erhardt K, Zetterberg A, Auer G (1984). The significance of DNA distribution pattern in rectal carcinoma. A preliminary study. Dis Colon Rectum 27:579–584
Fournier RE, Pardee AB (1975) Cell cycle studies of mononuclear and cytochalasin-B-induced binucleate fibroblasts. Proc Natl Acad Sci USA 72: 869–873
Fozard JBJ, Quirke P, Dixon MF, Giles GR, Bird CC (1986) DNA aneuploidy in ulcerative colitis. Gut 27:1414–1418
Fourth International Workshop on Chromosomes in Leukaemia (1984) The correlation of karyotype and occupational exposure to potential mutagenic/carcinogenic agents in acute non-lymphocytic leukaemia. Cancer Genet Cytogenet 11:326–331
Frankfurt OS, Slocum HK, Rustum YM et al. (1984) Flow cytometric analysis of DNA aneuploidy in primary and metastatic human solid tumours. Cytometry 5:71–80
Franklin WA, McDonald GB, Stein HO, Gatter KC, Jewell DP, Clarke LC, Mason DY (1985) Immunohistological demonstration of abnormal crypt cell kinetics in ulcerative colitis. Human Pathol 16:1129–1132
Freeman JW, McRorie DK, Busch RK et al. (1986) Identification and partial purification of a nucleolar antigen with a molecular weight of 145 000 found in a broad range of human cancers. Cancer Res 46:3593–3598
Freeman J, Dowell BL, Ochs RL, Ross BE, Busch H (1987) Effect of differentiation on the expression of a nucleolar antigen with a molecular weight of 145 000 in HL-60 cells. Cancer Res 47:586–591
Friedlander ML, Taylor IW, Russell P, Musgrove EA, Hedley DW, Tattersall MHN (1983) Ploidy as a prognostic factor in ovarian carcinoma. Int J Gynaecol Pathol 2:55–63
Friedlander ML, Hedley DW, Taylor IW (1984 a) Clinical and biological significance of aneuploidy in human tumours. J Clin Pathol 37:961–974
Friedlander ML, Hedley DW, Taylor IW, Russell P, Coates AS, Tattersall MNH (1984 b) Influence of cellular DNA content on survival in advanced ovarian cancer. Cancer Res 44:397–400
Friedlander ML, Russell P, Taylor IW, Hedley DW, Tattersall MHN (1984 c) Flow cytometric analysis of cellular DNA content as an adjunct to the diagnosis of ovarian tumours of borderline malignancy. Pathology 16:301–306
Gansler T, Chatten J, Varello M, Bunin GR, Atkinson B (1986) Flow cytometric DNA analysis of neuroblastoma. Correlation with histology and clinical outcome. Cancer 58:2453–2458
Gardner EJ, Rogers SW, Woodward SR (1982) Numerical and structural chromosome aberration in cultured lymphocytes and cutaneous fibroblasts of patients with multiple adenomas of the colorectum. Cancer 49:1413–1419
Gardner EJ, Woodward SR, Preston Hughes J (1985) Evaluation of chromosomal diagnosis for hereditary adenomatosis of the colorectrum. Cancer Genet Cytogenet 15:321–334
Gatter KC, Dunnill MS, Gerdes J, Stein H, Mason DY (1986) New approach to assessing lung tumours in man. J Clin Pathol 39:590–593
Geisinger KR, Kute TE, Marshall RB, Homesley HD, Morgan TM (1986) Analysis of the female sex steroid hormone receptors in adenocarcinoma of the endometrium. Am J Clin Pathol 85:536–541
Gelfant S, Drewinko B, Darzynkiewicz Z, Magdelena E (1983) Cycling and non-cycling human germinative epidermal cells – the continuous $^3$H-thymidine labelling method in vivo and flow cytometric studies. In: Wright NA, Camplejohn RS (eds) Psoriasis: Cell proliferation. Churchill Livingstone, Edingburgh, pp 209–217
Gerdes J, Schwab Lemke H, Stein H (1983) Production of a monoclonal antibody reactive with a human nuclear antigen associated with cell proliferation. Int J Cancer 31:13–20
Gerdes J, Lemke H, Baisch H, Wacker H-H, Schwab U, Stein H (1984) Cell cycle analysis of a cell proliferation-associated human nuclear antigen defined by the monoclonal antibody Ki-67. J Immunol 133:1710–1715
Gerdes J, Pileri S, Bartels H, Stein H (1986) Proliferation marker Ki-67: correlation with histological tumour grading and prognosis. Verh Dtsch Ges Pathol 70:152–158

Goerttler K, Stohr M (1979) Quantitative cytology of the positive region in flow sorted vaginal smears. J Histochem Cytochem 27:567–572

Goh HS, Jass JR (1986) DNA content and the adenoma-carcinoma sequence in the colorectum. J Clin Pathol 39:387–392

Goh HS, Jass JR, Atkin WS, Cuzick J, Northover JMA (1987) Value of flow cytometric determination of ploidy as a guide to prognosis in operable rectal cancer: a multivariate analysis. Int J Colorectal Dis 2:17–21

Golomb HM, Alimena G, Rowley JD, Vardiman JW, Testa JR, Sovik C (1982) Correlation of occupation and karyotype in adults with non-lymphocytic leukaemia. Blood 60:404–411

Griffin NR, Howard MR, Quirke P, O'Brien CJ, Childs JA, Bird CC (1988) Prognostic indicators in centroblastic lymphoma. J Clin Pathol 41:866–870

Gustafson H, Tribukait B, Esposti PL (1982 a) The prognostic value of DNA analysis in primary carcinomas in situ of the urinary bladder. Scand J Nephrol 16:141–146

Gustafson H, Tribukait B, Esposti PL (1982 b) DNA pattern, histological grade and multiplicity related to recurrence rate in superficial bladder tumours. Scand J Urol Nephrol 16: 135–139

Gustafson H, Tribukait B, Esposti PL (1982 c) DNA profile and tumour progression in patients with superficial bladder tumours. Urol Res 10:13–18

Hall PA, Richards MA, Gregory WM, d'Ardenne AJ, Lister TA, Stansfeld AG (1988) The prognostic value of Ki67 staining in non-Hodgkin's lymphoma. J Pathol 154:223–235

Hamada S, Fujita S (1983) DAPI staining improved for quantitative flow cytofluorometry. Histochemistry 79:219–226

Hamilton VT, Habbersett MC, Herman CJ (1980) Flow microfluorometric analysis of cellular DNA: Critical comparison of mithramycin and propidium iodide. J Histochem Cytochem 28:1125–1128

Hammarberg C, Slezak, P, Tribukait B (1984) Early detection of malignancy in ulcerative colitis: a flow-cytometric DNA study. Cancer 53:291–295

Hammarberg C, Tribukait B, Ohman U (1986) Early effects of preoperative irradiation upon the cell cycle composition in rectal adenocarcinomas. A flow-cvtometric DNA investigation. Acta Radiol Oncol 35:46–50

Harris H (1970) Cell fusion. The Dunham lectures. Clarendon Press, Oxford

Harris H, Klein G (1969) Malignancy of somatic cell hybrids. Nature 224:1315–1316

Harris H, Miller OJ, Klein G, Worst P, Tachibana T (1969) Suppression of malignancy by cell fusion. Nature 223:363–368

Hattori T, Sugihara H, Fukuda M, Hamada S, Fujita S (1986) DNA ploidy patterns of minute carcinomas in the stomach. Jpn J Cancer Res 77:276–281

Hawkes SP, Bartholomew JC (1977) Quantitative determination of transformed cells in a fixed population by simultaneous fluorescence analysis of cell surface and DNA in individual cells. Proc Natl Acad Sci USA 74:1626–1630

Hedley DW, Friedlander ML, Taylor IW, Rugg CA, Musgrove EA (1983) Method for analysis of cellular DNA content of paraffin-embedded pathological material using flow cytometry. J Histochem Cytochem 31:1333–1335

Hedley DW, Rugg CA, Ng ABP, Taylor IW (1984) Influence of cellular DNA content on disease-free survival of stage II breast cancer patients. Cancer Res 44:5395–5398

Hedley DW, Friedlander ML, Taylor IW (1985) Application of DNA flow cytometry to paraffin-embedded archival material for the study of aneuploidy and its clinical significance. Cytometry 6:327–333

Hedley DW, Rugg CA, Gelber R (1987) Association of DNA index and S-phase fraction with prognosis of nodes positive early breast cancer. Cancer Res 47:4729–4735

Hemming JD, Quirke P, Womack C, Wells M, Elston CW, Bird CC (1987) Diagnosis of molar pregnancy and persistant trophoblastic disease by flow cytometry. J Clin Pathol 40:615–620

Hemming JD, Quirke P, Womack C, Wells M, Elston CW, Bird CC (1988) Flow cytometry in persistent trophoblastic disease. Placenta 9:615–621

Hery M, Gioanni J, Lalanne C-M, Namer M, Courdi A (1987) The DNA labelling index: a prognostic factor in node negative breast cancer. Breast Cancer Res Treat 9:207–211

Hiddemann W, Schumann J, Andreeff M et al. (1984) Convention on nomenclature for DNA cytometry. Cytometry 5:445–446
Hiddemann W, Wormann B, Gohde W, Buchner T (1986) DNA aneuploidies in adult patients with acute myeloid leukaemia. Incidence and relation to patient characteristics and morphological subtypes. Cancer 57:2146–2152
Inokuchi K, Kodama Y, Sasaki O, Kamegawa T, Okamura T (1983) Differentiation of growth patterns of early gastric cancer determined by cytophotometric DNA analysis. Cancer 51:1138–1141
Irvin GL, Bagwell CB (1979) Identification of histologically undetectable parathyroid hyperplasia by flow cytometry. Am J Surg 138:567–571
Isaacs JT, Sandberg AA (1982) Abnormal changes associated with progression of the Dunning R-3327 rat prostatic adenocarcinoma system. Cancer Res 42:4131–4132
Isaacs JT, Wake N, Coffey DS, Sandberg AA (1982) Genetic instability coupled to clonal selection as a mechanism for tumour progression in the Dunning R-3327 rat prostatic adenocarcinoma system. Cancer Res 42:2353–2361
Iversen OE, Laerum OD (1985) Ploidy disturbances in endometrial and ovarian carcinomas. A review. Anal Quant Cytol 7:327–336
Jakobsen A (1984) Ploidy level and short-time prognosis of early cervix cancer. Radiother Oncol 1:271–275
Jakobsen A, Kristensen PB, Poulsen HK (1983) Flow cytometric classification of biopsy specimens from cervical intraepithelial neoplasia. Cytometry 4:9–14
Joensuu H, Klemi P, Eerola E (1986) DNA aneuploidy in follicular adenomas of the thyroid gland. Am J Pathol 124:373–376
Jones DJ, Moore M, Schofield PF (1988a) Refining the prognostic significance of DNA ploidy states in colorectal cancer: a prospective flow cytometric study. Int J Cancer 41:206–210
Jones DJ, Moore M, Schofield P (1988b) Prognostic significance of DNA ploidy in colorectal cancer: a prospective flow cytometric study. Br J Surg 75:28–33
Jones DJ, Ghosh AK, Moore M, Schofield PF (1988c) A critical appraisal of the immunohistochemical detection of the *c-myc* oncogene product in colorectal cancer. Br J Cancer 56:779–783
Kallioniemi O-P, Blanco G, Alavaikko M, Hietanen T, Mattilla J, Lauslahti K, Koivula T (1987) Tumour DNA ploidy as an independent prognostic factor in breast cancer. Br J Cancer 56:637–642
Kamegawa T, Okamura T, Sugimachi K, Inokuchi K (1986) Preoperative detection of a highly malignant type of early gastric cancer by cytophotometric DNA analysis. Jpn J Surg 16:169–174
Klein FA, Herr HA, Sogani PC, Whitmore WF, Melamed MR (1982) Detection and follow-up of carcinoma of the urinary bladder by flow cytometry. Cancer 50:389–395
Klein G, Bregola U, Weiner F (1971) The analysis of malignancy by cell fusion. I. Hybrids between tumour cells and derivatives. J Cell Sci 8: 659–672
Kokal W, Sheibani K, Terz J, Harada JR (1986) Tumour DNA content in the prognosis of colorectal carcinoma. JAMA 255:3123–3127
Konishi F, Morson BC (1982) Pathology of colorectal adenomas: a colonoscopic survey. J Clin Pathol 35:830–841
Korenaga D, Okamura T, Sugimachi K, Inokuchi K (1985) Prognostic study of intramucosal carcinoma of the stomach with DNA aneuploidy. Jpn J Surg 15:443–448
Kreicbergs A, Silvfersward C, Tribukait B (1984) Flow analysis of primary bone tumours. Relationship between cellular DNA content and histopathologic classification. Cancer 53:129–136
Krishan A (1975) Rapid flow cytofluorometric analysis of mammalian cell cycle by propidium iodide staining. J Cell Biol 66:188–193
Kute TE, Linville C, Barrows G (1983) Cytofluorometric analysis for estrogen receptors using fluorescent estrogen probes. Cytometry 4:132–140
Kute TE, Muss HB, Hopkins M, Marshall R, Case D, Kammire L (1985) Relationship of flow cytometry results to clinical and steroid receptor status in human breast cancer. Breast Cancer Res Treat 6:113–121

Lalande ME, Miller RG (1979) Fluorescence flow analysis of lymphocyte activation using Hoechst 33342 dye. J Histochem Cytochem 27:394–397

Latt SA, Stetten G (1976) Spectral studies on 33258 Hoechst and related benzimidazole dyes useful for fluorescent detection of deoxyribonucleic acid synthesis. J Histochem Cytochem 24:24–33

Lauren P (1965) The two histological main types of gastric carcinoma: diffuse and so-called intestinal-type carcinoma. An attempt at a histo-clinical classification. Acta Pathol Microbiol Scand 64:31–49

Lehtinen T, Lehtinen M, Aine R et al. (1987) Nuclear DNA content of non endemic Burkitt's lymphoma. J Clin Pathol 40:1201–1205

Levan G (1974) The detailed chromosome constitution of a benzpyrene-induced rat sarcoma. A tentative model for G-band analysis in solid tumours. Hereditas 79:273–289

Levan G, Levan A (1975) Specific chromosome changes in malignancy: studies in rat sarcomas induced by two polycyclic hydrocarbons. Hereditas 79:161–198

Levan G, Mitelman F (1976) G banding in Rous rat sarcomas during serial transfer: significant chromosome aberrations and incidence of stromal mitoses. Hereditas 84:11–14

Linden WA, Ochlich K, Baisch H et al. (1979) Flow cytometric prescreening of cervical smears. J Histochem Cytochem 27:529–535

Lloyd RV, Wilson BS, Varani J, Gaur PK, Moline S, Makari JG (1985) Immunocytochemical characterisation of a monoclonal antibody that recognises mitosing cells. Am J Pathol 121: 275–283

Look AT, Hayes FA, Nitschke R, McWilliams NB, Green AA (1984) Cellular DNA content as a predictor of response to chemotherapy in infants with unresectable neuroblastoma. N Engl J Med 311:231–235

Look AT, Robertson PK, Williams DL et al. (1985) Prognostic importance of blast cell DNA content in childhood acute lymphoblastic leukaemia. Blood 65:1079–1086

Lubs HA, Kotler S (1967) The prognostic significance of chromosome abnormalities in colon tumours. Ann Intern Med 67:328–336

Ludberg S, Carstensen J, Rundquist I (1987) DNA flow cytometry and histopathological grading of paraffin embedded prostate biopsy specimens in a survival study. Cancer Res 47:1973–1977

Macartney JC, Camplejohn RS (1986) DNA flow cytometry of histological material from dysplastic lesions of human gastric mucosa. J Pathol 150: 113–118

Macartney JC, Camplejohn RS, Powell G (1986 a) DNA flow cytometry of histological material from human gastric cancer. J Pathol 148:273–277

Macartney JC, Camplejohn RS, Alder J, Stone MG, Powell G (1986 b) Prognostic importance of DNA flow cytometry in non-Hodgkin's lymphomas. J Clin Pathol 39:542–546

Mankin HJ, Connor JF, Schiller AL, Perlmutter N, Alho A, McGuire M (1985) Grading of bone tumours by analysis of nuclear DNA content using flow cytometry. J Bone Joint Surg [Am] 404–413.

Mark J, Mitelman F, Dencker H, Norryd C, Tranberg KG (1973) The specificity of the chromosomal abnormalities in human colonic polyps. A cytogenetic study of multiple polyps in a case of Gardner's syndrome. Acta Pathol Microbiol Scand 81:85–90

Mathews MB, Bernstein RM, Franza BR, Garrels JL (1983) Identification of the proliferating cell nuclear antigen and cyclin. Nature 309:374–376

McGuire WL, Dressler LG (1985) Emerging impact of flow cytometry in predicting recurrence and survival in breast cancer patients. J Natl Cancer Inst 75:405–410

McGuire WL, Meyer JS, Barlogie B, Kute TE (1985) Impact of flow cytometry on predicting recurrence and survival in breast cancer patients. Breast Cancer Res Treat 5:117–128

Meisingset KK, Steen HB (1981) Intracellular binding of fluorescein in lymphocytes. Cytometry 1:272–278

Melamed MR, Klein FA (1984) Flow cytometry of urinary bladder irrigation specimens. Human Pathol 15:302–305

Melamed MR, Mullaney PF, Mendelsohn MR (1979) Flow cytometry and sorting. Wiley, New York

Melamed MR, Enker WE, Banner P, Janov AJ, Kessler G, Darzynkiewics Z (1986) Flow cytometry of colorectal carcinoma with three year follow up. Dis Colon Rectum 29:184–186
Meyer JS, Prioleau PG (1981) S-phase fractions of colorectal carcinomas related to pathologic and clinical features. Cancer 48:1221–1228
Meyer JS, Friedman E, McCrate MM, Bauer WC (1983) Prediction of early course of breast carcinoma by thymidine labelling. Cancer 51:1879–1886
Meyer JS, McDivitt RW, Stone KR, Prey MU, Bauer WC (1984) Practical breast carcinoma cell kinetics: review and update. Breast Cancer Res Treat 4:79–88
Miller RG, Lalande ME, Keystone EC (1984) Hoechst 33342 dye uptake as a probe of membrane permeability in mammalian cells. In: Eisert WG, Mendelsohn ML (eds) Biological dosimetry. Springer, Berlin Heidelberg New York, pp 229–234
Mitelman F (1972) Predetermined sequential chromosome changes in serial transplantation of Rous rat sarcomas. Acta Pathol Microbiol Scand [A] 80:313–328
Mitelman F (1980) Cytogenetics of experimental neoplasms and non random chromosome correlation in man. Clin Haematol 9:195–219
Mitelman F (1984) Restricted number of chromosomal regions implicated in aetiology of human cancer and leukaemia. Nature 310:325–327
Mitelman F, Mark PG, Nilsson PG, Dencker H, Norryd C, Tranberg KG (1974) Chromosome banding pattern in human colonic polyps. Hereditas 78:63–68
Mitelman F, Brandt L, Nilsson PG (1978) Relation among occupational exposure to potential mutagenic/carcinogenic agents, clinical findings, and bone marrow chromosomes in acute non lymphocytic leukaemia. Blood 52:1229–1237
Mitelman F, Nilsson PG, Brandt L et al. (1981) Chromosome pattern, occupation and clinical features in patients with acute non-lymphocytic leukaemia. Cancer Genet Cytogenet 4:197–214
Moberger B, Auer G, Forsslund G, Moberger B (1984) The prognostic significance of DNA measurements in endometrial carcinoma. Cytometry 5:430–436
Moran RE, Black MM, Alpert L, Straus MJ (1984) Correlation of cell cycle kinetics, hormone receptors histopathology, and nodal states in human breast cancer. Cancer 54:1586–1590
Morgan DR, Williamson JMS, Quirke P et al. (1986) DNA content and prognosis of non-Hodgkin's lymphoma. Br J Cancer 54:643–649
Morgan K, Quirke P, O'Brien CJ, Bird CC (1988) Hodgkin's disease: a flow cytometric study. J Clin Pathol 41:365–369
Muller W, Gautier F (1975) Interactions of heteroaromatic compounds with nucleic acids. Eur J Biochem 54:385–394
Ng A, Atkin NA (1973) Histological cell type and DNA value in the prognosis of squamous cell cancer of the uterine cervix. Br J Cancer 28:322–331
Nicolini C, Belmont A, Parodi S, Lessin S, Abraham S (1979) Mass action and acridine organe staining: static and flow cytofluorometry. J Histochem Cytochem 27:102–113
Nowell PC, Hungerford DA (1960) Chromosome studies on normal leukaemic human leukocytes. J Natl Cancer Inst 25:85–93
O'Brien CJ, Quirke P, Holgate CS et al. (1988a) Flow cytometric parameters in high grade NHL and their relationship to immunological phenotype, remission induction and survival. J Pathol 154:56 A
O'Brien CJ, Quirke P, Holgate CS et al. (1988b) Neither histological subtype nor immunological phenotype are associated with remission induction or survival in high grade NHL. J Pathol 154::38 A
Okamura T, Korenaga D, Haraguchi M et al. (1987) Growth mode and DNA ploidy in mucosal carcinomas of the stomach. Cancer 59:1154–1160
Olive PL, Durand RE (1983) Fluorescent nitroheterocycles for identifying hypoxic cells. Cancer Res 43:3276–3280
Orihuela E, Varadachay S, Herr HW, Melamed M, Whitmore WF (1987) The practical use of tumour marker determination in bladder washing specimens. Assessing the urothelium of patients with superficial bladder cancer. Cancer 60:1009–1016
Oud PS, Henelerik JBJ, Beck HLM, Veldhuizen JAM, Vooijs GP, Herman CJ, Ramaekers

FCS (1985) Flow cytometric analysis and sorting of human endometrial cells after immunocytochemical Labelling for cytokeratin using a monoclonal antibody. Cytometry 6:159–164

Owainati AAR, Robins RA, Hinton C et al. (1987) Tumour aneuploidy, prognostic parameters and survival in primary breast cancer. Br J Cancer 55:449–454

Owden RR, Curtis SK (1983) Supravital experiments with pyronin Y, a fluorochrome of mitochondria and nucleic acids. Histochemistry 77:535–542

Pears AGE (ed) (1968) Histochemistry: theoretical and applied, 3rd edn, vol 1. Little Brown, Boston, pp 254–257

Petersen SE, Friedrich U (1986) A comparison between flow cytometric ploidy investigation and chromosome analysis of 32 human colorectal tumours. Cytometry 7:307–312

Petersen SE, Lorentzen M, Bichel P (1980) A mosaic subpopulation structure of human colorectal carcinomas demonstrated by flow cytometry. In: Flow cytometry IV. Universitetsforlaget, Oslo, pp 412–416

Petrova AS, Subrichina GN, Tschistjakova OV, Lukina TA, Weiss H, Wildner G (1980) Flow cytofluorometry, cytomorphology and histology in gastric adenocarcinoma. Oncology 37:318–324

Poccia DL, Palerity BA, Campisi J, Lyman H (1979) Fluorescence staining of living cells with fluorescamine. Protoplasma 98:91–113

Pollack A, Prudhomme DL, Greenstein DB, Irvin GL, Claflin AJ, Block NL (1982) Flow cytometric analysis of RNA content in different cell populations using pyronin Y and methyl green. Cytometry 3:28–35

Pollack A, Moulis H, Block NL, Irvin GL III (1984) Quantification of cell kinetic responses using simultaneous flow cytometric measurements of DNA and nuclear protein. Cytometry 5:473–481

Quirke P (1987) The role of flow cytometry in the assessment of the pathobiology of colorectal neoplasia. PHD Thesis, University of Leeds

Quirke P, Dyson JED (1986) Flow cytometry: methodology and application in pathology. J Pathol 149:79–87

Quirke P, Dyson JED, Dixon MF, Bird CC, Joslin CAF (1985) Heterogeneity of colorectal adenocarcinomas evaluated by flow cytometry and histopathology. Br J Cancer 51:99–106

Quirke P, Dyson JED, Sutton J, Anderson CK, Joslin CAF, Bird CC (1986a) Assessment of germ cell tumours of the testes by flow cytometry and histopathology. In: Jones WG, Ward AM, Anderson CK (eds) Germ cell tumours. Pergamon Press, Oxford, pp 45–54

Quirke P, Fozard JBJ, Dixon MF, Dyson JED, Giles GR, Bird CC (1986b) DNA aneuploidy in colorectal adenomatous polyps. Br J Cancer 53:477–481

Quirke P, Durdey P, Dixon MF, Dyson JED, Williams NS, Bird CC (1987) The prognostic significance of DNA aneuploidy and cell proliferation in rectal adenocarcinomas. J Pathol 151:285–292

Quirke P, Dixon MF, Day DW, Fozard JBJ, Talbot IC, Bird CC (1988) DNA aneuploidy and cell proliferation in familial adenomatous polyposis. Gut 29:603–607

Raber MN, Barlogie B, Luna M (1984) Flow-cytometric analysis of DNA content in post mortem tissue. Cancer 53:1705–1707

Reichmann A, Martin P, Levin B (1981) Chromosomal banding patterns in human large bowel. Int J Cancer 28:431–440

Robbins BA, de la Vega D, Ogata K, Tau EM, Nakamura RM (1987) Immunohistochemical detection of proliferating cell nuclear antigens in solid human malignancies. Arch Pathol Lab Med 111:841–845

Rodenburg CJ, Ploem-Zaaijer JJ, Cornelisse CJ et al. (1987) Use of DNA image cytometry in addition to flow cytometry for the study of patients with advanced ovarian cancer. Cancer Res 47:3938–3941

Rognum TO, Thorud E, Elgjo K, Brandtzaeg P, Orjasaeter H, Nygaard K, Clausen OPF (1980) DNA flow cytometry (FCM) in carcinomas of the large bowel compared with the two functional cell markers secretory component (SC) and carcinoembryonic antigen (CEA), the histological tumour grade and the clinical stage. A preliminary communication. In: Laerum OD, Lindmo T, Thorud E (eds) Flow cytometry IV. Universitetsforlaget, Oslo, pp 417–423

Rognum TO, Thorud E, Elgjo K, Brandtzaeg P, Orjasaeter H, Nygaard K (1982) Large bowel carcinomas with different ploidy related to secretory component, IgA and CEA in epithelium and plasma. Br J Cancer 45:921–934
Rognum TO, Thorud E, Lund E (1987) Survival of large bowel carcinoma patients with different DNA ploidy. Br J Cancer 56:622–636
Röhmer R-M, Ellwart J (1981) Cell cycle analysis by combining the 5-bromodeoxyuridine/33258 Hoechst technique with DNA specific ethidium bromide staining. Cytometry 2:31–34
Ronstrom L, Tribukait B, Esposti PL (1981) DNA pattern and cytological findings in fine needle aspirates of untreated prostatic tumors. A flow cytofluorometric study. Prostate 2:79–88
Roos G, Dige U, Lenner P, Lindh J, Johansson H (1985) Prognostic significance of DNA-analysis by flow cytometry in non-Hodgkin's lymphoma. Hematol Oncol 3:233–242
Roti-Roti JL, Higashikubo R, Blair OC, Uygur N (1982) Cell cycle position and nuclear protein content. Cytometry 3:91–96
Rowley JD (1984) Biological implications of consistent chromosome rearrangements in leukaemia and lymphoma. Cancer Res 44:3159–3168
Ryan DH, Fallon MA, Horan PK (1988) Critical review: flow cytometry in the clinical laboratory. Clin Chim Acta 171:125–174
Sandberg AA (1980) The chromosomes in human cancer and leukaemia. Elsevier, New York
Sandberg AA, Turc-Carel C (1987) The cytogenetics of solid tumours in relation to diagnosis. Classification and pathology. Cancer 59:387–395
Sandritter W, Scomazzoni G (1964) DNA content (Feulgen photometry) and dry weight (interference microscopy) of normal and hypertrophic heart muscle fibres. Nature 202:100–101
Schutte B, Reynders MMJ, Bosman FT, Blijham GH (1985) Flow cytometric determination of DNA ploidy level in nuclei isolated from paraffin-embedded tissue. Cytometry 6:26–30
Schutte B, Reynders MMJ, Wiggers T, Arends JW, Volovics L, Bosman F, Blijham GH (1987) Retrospective analysis of the prognostic significance of DNA content and proliferative activity in large bowel carcinoma. Cancer Res 47:5494–5496
Scott NA, Rainwater LM, Wieand HS, Weiland LH, Pemberton JH, Beart RW, Lieber MM (1987) The relative prognostic value of flow cytometric DNA analysis and conventional clinicopathologic criteria in patients with operable rectal carcinoma. Dis Colon Rectum 30:513–520
Shackney SE, Levine AM, Fisher RI et al. (1984) The biology of tumor growth in the non-Hodgkin's lymphomas. A dual parameter flow cytometry study of 220 cases. J Clin Invest 73:1201–1214
Shapiro HM (1981) Flow cytometric estimation of DNA and RNA content in intact cells stained with Hoechst 33342 and pyronin Y. Cytometry 2:143–150
Shapiro HM (1985) Practical flow cytometry. Alan R, Liss, New York
Shapiro JR, Shapiro WR (1985) The subpopulations and isolated cell types of freshly resected high grade human gliomas: their influence on the tumour's evolution in vivo and behavior and therapy in vitro. Cancer Metastasis Rev 4:107–124
Silvestrini R, Daidone MG, Gasparini G (1985) Cell kinetics as a prognostic marker in node-negative breast cancer. Cancer 56:1982–1987
Silvestrini R, Daidone MG, Valagussa P, Salvadori B, Rovini D, Bonadonna G (1987) Cell kinetics as a prognostic marker in locally advanced breast cancer. Cancer Treat Rep 71:375–379
Smetana K, Gyorkeyf G, Chan P-K, Tan E, Busch H (1983) Proliferating cell nuclear antigens (PCNA) and human malignant tumour nucleolar antigens (HMTNA) in nucleoli of human haematological malignancies. Blut 46:133–141
Smets LA, Homan-Blok J, Hart A, de Vaan G, Behrendt H, Hahlen K, de Waal FJ (1987) Prognostic implication of hyperdiploidy as based on DNA flow cytometric measurement in childhood acute lymphocytic leukaemia – a multicenter study. Leukaemia 1:163–166
Stephenson RA, James BC, Gay H, Fair WR, Whitmore WF, Melamed MR (1987) Flow cytometry of prostate cancer: relationship of DNA content to survival. Cancer Res 47:2504–2509

Stone KR, Craig RB, Palmer JO, Rivkin SE, McDivitt RW (1985) Short-term cryopreservation of human breast carcinoma cells for flow cytometry. Cytometry 6:357–361

Strang P, Lindgren A, Stendahl U (1987) Blood group antigens in relation to DNA content S-phase and heterogeneity and their prognostic values in cervical carcinoma. Anticancer Res 7:125–128

Streffer C, van Beuningen D, Gross E, Schabronath J, Eigler F-W, Rebmann A (1986) Predictive assays for the therapy of rectum carcinoma. Radiother Oncol 5:303–310

Takamatsu T, Nakanishi K, Fukuda M, Fujita S (1983) Cytofluorometric nuclear DNA determinations in infant, adolescent adult and aging human hearts. Histochemistry 77:485–494

Taylor IW (1980) A rapid single step staining technique for DNA analysis by flow microfluorimetry. J Histochem Cytochem 28:1021–1024

Taylor IW, Milthorpe BK (1980) An evaluation of DNA fluorochromes, staining techniques, and analysis for flow cytometry. J Histochem Cytochem 28:1224–1228

Teodori L, Capurso L, Cordelli E, de Vita R, Koch M, Tarquini M, Pallone F, Mauro F (1984) Cytometrically determined relative DNA content as an indicator of neoplasia in gastric lesions. Cytometry 5:63–70

Tetu B, Katz RL, Kalter SP, von Eschenbach AC, Barlogie B (1987) Acridine orange flow cytometry of urinary bladder washings in the detection of transitional cell carcinoma of the bladder. The influence of prior local therapy. Cancer 58:2494–2500

Third International Workshop on Chromosomes in Leukaemia (1981) Chromosomal abnormalities and their clinical significance in acute lymphoblastic leukaemia. Cancer Res 43:868–873

Thomas GDH, Dixon MF, Smeeton NC, Williams NS (1983) Observer variation in the histological grading of rectal carcinoma. J Clin Pathol 36:385–391

Tijo JH, Levan A (1956) The chromosome number in man. Hereditas 42:1–6

Traganos F, Darzynkiewicz Z, Sharpless T, Melamed MR (1977) Simultaneous staining of ribonucleic and deoxyribonucleic acids in unfixed cells using acridine organe in a flow cytofluorometric system. Histochem Cytochem 25:46–56

Tribukait B, Gustafson H, Esposti PL (1982) The clinical significance of ploidy and proliferation in the clinical and biological evaluation of bladder tumours: a study of 100 untreated cases. Br J Urol 54:130–135

Tribukait B, Hammarberg C, Rubio C (1983 a) Ploidy and proliferation patterns in colorectal adenocarcinomas related to Dukes' classification and to histopathological differentiation. A flow-cytometric DNA study. Acta Pathol Microbiol Scand [A] 91:89–95

Tribukait B, Ronstrom L, Esposti PL (1983 b) Quantitative and qualitative aspects of DNA measurements related to the cytologic grade in prostate carcinoma. Anal Quant Cytol 5:107–111

Tubiana M, Koscienly S (1988) Cell kinetics, growth rate and the natural history of breast cancer. The Heuson Memorial Lecture. Eur J Clin Oncol 24:9–14

Tubiana M, Pejovic MH, Chavandra N, Contesso G, Malaise EP (1984) The long-term prognostic significance of the thymidine labelling index in breast cancer. Int J Cancer 33:441–445

Valet G, Ormerod MG, Warnecke HH, Benker G, Rubenstroth-Bauer G (1981) Sensitive three parameter flow cytometric detection of abnormal cells in human cervical cancers: a pilot study. J Cancer Res Clin Oncol 102:177–184

Valet G, Raffael A, Russmann L (1985) Determination of intracellular calcium in vital cells by flow cytometry. Naturwissenschaften 72:600–601

Van den Ingh HF, Griffioen G, Cornelisse CJ (1985) Flow cytometric detection of aneuploidy in colorectal adenomas. Cancer Res 45:3392–3397

Vernon TO, Glazer AN, Stryer L (1982) Fluorescent phycobiliprotein conjugates for analysis of cells and molecules. J Cell Biol 93:981–986

Visser JWM, Jongeling AAM, Tanke HJ (1979) Intracellular pH determination by fluorescence measurements. J Histochem Cytochem 27:32–35

Volm M, Bruggemann A, Gunther M, Kleine W, Pfleiderer A, Vogt-Schaden M (1985 a) Prognostic relevance of ploidy, proliferation and resistance – predictive tests in ovarian carcinoma. Cancer Res 45:5180–5185

Volm M, Mattern J, Sonka J, Vogt-Schaden M, Wayss K (1985b) DNA distribution in non-small-cell lung carcinomas and its relationship to clinical behavior. Cytometry 6:348–356
Walle AJ, Niedemayer W (1985) Aneuploidy as a marker of minimal residual disease in leukaemia. Cancer Detection and Prevention 8:303–315
Wallen CA, Higashikubo R, Dethlefsen LA (1982) Comparison of two flow cytometric assays for cellular DNA-acridine orange and propidium iodide. Cytometry 3:155–160
Watson JD, Crick FHC (1953) A structure for deoxyribose nucleic acid. Nature 171:737–738
Watson JV (1980) Enzyme kinetic studies in cell populations using fluorogenic substrates and flow cytometric techniques. Cytometry 1:143–151
Watson JV, Sikora KE, Evan GI (1985) A simultaneous flow cytometric assay for c-*myc* oncoprotein and cellular DNA in nuclei from paraffin embedded material. J Immunol Methods 83:179–192
Weiss H, Wildner GP, Jacobasch K-H, Heinz U, Schaelicke W (1985) Characterisation of human adenomatous polyps of the colorectal bowel by means of DNA distribution patterns. Oncology 42:33–41
Weiss LM, Strickler JG, Medeiros LJ, Gerdes J, Stein H, Warnke RA (1987) Proliferative rates of non-Hodgkin's lymphomas as assessed by Ki67. Human Pathol 18:1155–1159
Wied GL (1966) An introduction to quantitative cytochemistry. Academic Press, London
Wied GL, Bahr GF (1970) Introduction to quantitative cytochemistry II. Academic Press, London
Wirsching RP, Lamerz R, Wiebecke B, Demmel N, Liewald F, Valet G (1987) Flow cytometric evaluation of colorectal carcinoma as completion of conventional tumour examination. J Exp Clin Cancer Res 6:117–128
Wolley RC, Hery F, Koss LG (1982a) Caution on the use of lymphocytes as standards in the flow cytometric analysis of cultured cells. Cytometry 2:370–373
Wolley RC, Schreiber K, Koss LG, Karas M, Sherman A (1982b) DNA distribution in human colon carcinomas and its relationship to clinical behavior. J Natl Cancer Inst 69:15–22
Wolman SR (1983) Karyotypic progression in human tumours. Cancer Metastasis Rev 2:257–295
Wyatt JI, Quirke P, Ward DC, Clayden AD, Dixon MF, Johnston D, Bird CC (1989) Comparison of histopathological and flow cytometric parameters in prediction of prognosis in gastric cancer. J Pathol 158:195–202
Wyjkstrom H, Gustafson H, Tribukait B (1984) Deoxyribonucleic acid analysis in the evaluation of transitional cell carcinoma before cystectomy. J Urol 132:894–898
Xavier RG, Prolla JC, Bemvenuti GA, Kirsner JB (1974) Tissue cytogenetic studies in chronic ulcerative colitis and carcinoma of the colon. Cancer 34: 684–695
Young GAR, Hedley DW, Rugg CA, Iland HJ (1987) The prognostic significance of proliferative activity in poor histology non-Hodgkin's lymphoma: a flow cytometry study using archival material. Eur J Cancer Clin Oncol 23:1497–1504
Zimmerman PV, Bint MH, Hawson GAT, Parsons PG (1987) Ploidy as a prognostic determinant in surgically treated lung caner. Lancet II:530–533

# Karyotypic Analysis of Solid Tumors

P. DAL CIN and A. A. SANDBERG

## 1 Introduction

Much progress has been made in and much published on the chromosome changes and their significance in cancer and leukemia during the last decade. This progress has included not only the cytogenetic definition and classification of various leukemias (ROWLEY 1984; SANDBERG 1986 a), but also the cytogenetic "dissection" of some solid tumor entities (SANDBERG and TURC-CAREL 1987; SANDBERG et al. 1988) and the application of the karyotypic findings to molecular approaches (CROCE 1986; DUESBERG 1987). The latter have uncovered rearrangements of genes (including oncogenes) in some conditions, increased or abnormal gene expression in others, and the need for more knowledge regarding the exact genes affected in most human neoplastic states.

The classification of the various acute leukemias has, for a number of years, been based on the so-called FAB systems (BENNET et al. 1976, 1981, 1985), which have relied heavily on the characteristics of cellular morphology and immunology. The cytogenetic demonstration that FAB categories thought to be relatively homogeneous contained subsets characterized by specific chromosome changes ultimately led to workshops (First MIC Cooperative Study Group 1986; Second MIC Cooperative Study Group 1988; Third MIC Cooperative Study Group 1988) in which these changes were related to the various clinical, immunologic, and cytologic aspects of each leukemic subset. In these workshops members of the FAB group, immunologists, histochemists, clinicians, and cytogeneticists critically evaluated the information available in each area in an attempt to define as accurately and reliably as was possible the nature of the conditions associated with specific chromosome changes. Hopefully, these results will find usefulness in the diagnostic and clinical aspects of various hematologic neoplastic disorders. It is certain that at least some of the decisions reached at these workshops will have to be modified in the future as cogent cytogenetic and particularly molecular data are brought to bear on each subset; in addition there is a strong likelihood that new subsets will be defined cytogenetically and otherwise.

Though progress in solid tumor cytogenetics has not been as rapid or clinically meaningful as that in the leukemias, advances in methodology have made possible the detailed examination of the karyotypes of a number of tumour entities (sarcomas, testicular and kidney cancers, neuroblastoma), with most of the adenocarcinomas still to be elucidated cytogenetically. These advances have led to the description of a number of specific chromosome changes in solid tumors, as well as to two workshops (1986, 1987) dealing with the karyotypic aspects of such tumors. Future workshops will undoubtedly combine progress at the cytogenetic and at the molecular and other levels. As in the case of the leukemias, the combination of cytogenetic and molecular results will most likely lead to the description of subtypes within existing tumor entities and the utilization of the findings in the diagnosis, classification and prognosis of these tumors, and the development of therapeutic approaches.

The present chapter will deal with chromosomal changes in solid tumors and their meaning, application, and relationship with specific histologic entities and subentities.

## 2 Primary (Specific) and Secondary Chromosome Changes

The identification of specific cytogenetic events in solid tumors has lagged behind that in the leukemias for several reasons, including the frequent encounter of multiple and complex chromosomal changes, varying in nature from tumor to tumor and which may reflect differences in the responsible causative agents, as well as in the karyotypic evolution of secondary chromosome changes in the malignant cells. These complex karyotypic changes may confuse the picture as they render difficult the identification of the primary chromosome changes associated with specific tumor types. The establishment of the primary karyotypic event, particularly when present as a single change, may provide cogent information (e.g., finding and understanding the genes involved) in the genesis and maintenance of the neoplastic state.

Cytogeneticists and molecular biologists will ultimately have to comprehend the nature, meaning, and mechanisms responsible for the additional (secondary) chromosome changes seen in leukemia and cancer. In the latter, these additional changes are often so complex and numerous that they tend to mask the primary karyotypic change. Except in some leukemias [i(17)q in chromic myelogenous leukemia, +8 in acute leukemia, −Y or −X in acute myelogenous leukemia with t(8;21)], the additional chromosome changes have not been found to follow a nonrandom pattern and invariably appear to be associated with the biologic progression of the disease. In other words, a leukemia or tumor is at its lowest level of aggressiveness when it is associated with the primary karyotypic change only; the leukemia or tumor shows more aggressive behavior with the acquisition of additional chromosome changes (Nowell 1986). Since the latter appear to be responsible for (or at least associated with) progression of the disease state, cytogeneticists and molecular biologists will have to address the nature, meaning, and consequences of the additional chromosome changes. This is a challenge that will have to be met, particularly in the case of, for example, adenocarcinomas of the breast, lung, colon, and prostate, if progress is to be made in the control and, ultimately, the cure of these diseases. The genetic and hence molecular and metabolic disturbances caused by the primary karyotypic changes may turn out to be relatively simple when compared to those associated with the secondary or additional chromosome changes. These disturbances reflect the complexity and large number of such changes often seen in neoplasia, particularly in solid tumors.

With the above background it must be kept in mind that significant proportions (and in the case of some neoplastic conditions, the majority) of solid tumors have not had a primary karyotypic change established. Furthermore, the wide array of additional cytogenetic changes which have so far defied classification or organization must be kept in mind when evaluating the role of the chromosome changes in various conditions.

When a leukemia or tumor is associated with a sole karyotypic change (translocation, loss or gain of a chromosome, insertion, deletion, etc.), it can be assumed that this change represents the primary or specific chromosome change characterizing the particular leukemia or tumor. When a group of leukemias or tumors of similar origin and pathology contains the same primary karyotypic change, it can be safely assumed that they constitute a subset or subtype within the leukemia or tumor type. Generally, and particularly in solid tumors (adenocarcinomas), a large number of tumors may have to be studied in order to ascertain whether a unique karyotypic change occurs with sufficient frequency in a number of tumors to qualify as the primary chromosome change.

Though the primary chromosome change may play a key role in the process of leukemogenesis or carcinogenesis, and in some cases the evidence is more than suggestive (e.g., the *bcr-abl* abnormal gene in chronic myelogenous leukemia), there is the possibility that the primary karypotypic change reflects events at the molecular level which may have preceded the chromosome change and, in fact, be the cause of it. Thus, however essential the karyotypic change is in the development of the neoplastic state, there may exist an underlying molecular event necessary for the genesis of the specific chromosome change.

## 3 Methodologic Aspects of Cytogenetic Analysis of Solid Tumors

### 3.1 The Role of the Cytogeneticist, Surgeon, and Pathologist

As in the case of leukemias, to establish the chromosome changes in a tumor, the actual tumor cells must be examined. Generally, 0.5 g or more of tumor tissue is required for cytogenetic analysis. The full cooperation of the surgeon and pathologist is a sine qua non for success in the cytogenetics of solid tumors. That of the surgeon is necessary in order to make sure that viable and representative specimens are obtained for chromosomal analysis. That of the pathologist is crucial in order to postpone fixation of the specimen until part of the tumor for the cytogenetic examination is secured and to make certain that appropriate tissue is sent in a proper medium to the cytogenetic laboratory with, ultimately, a detailed description of the histology of the tumor. The surgeon is also responsible for the details regarding the clinical aspects of each case, the findings at operation, and follow-up. Rapport must be established between the cytogeneticist, surgeon, and pathologist, not only regarding the importance of establishing the karyotypic changes in various tumors, but also for a fuller understanding of the problems associated with these examinations (e.g., failure to obtain metaphases for analysis, length of time required for a meaningful cytogenetic evaluation of a tumor, the cytogenetic complexity of some tumors). Communication at all levels and about results is a key to success in tumor cytogenetics.

### 3.2 Problems Specific to Solid Tumors

Despite the fact that the chromosomes of human solid tumors were first observed more than 100 years ago (see SANDBERG 1980 for historical background) and the remarkable advances in cytogenetic techniques, these tumors are still difficult to examine in detail cytogenetically, compared with leukemic cells. The major obstacle in cytogenetic analyses of solid tumors has been the lack of a reliable technique that can not only disaggregate the cells of the tumor tissue, but also yield an adequate number of mitotic divisions in primary short-term cultures.

A significant percentage of solid tumors have a very low mitotic index, necessitating culture of the cells, which may not be successful in all cases or may lead to overgrowth by normal (diploid) cells. Additionally, the dominant cells following prolonged culture do not necessarily represent the major clone in the tumor in vivo and generally tend to have a large number of chromosomes with an array of numerical and morphologic chromosomal anomalies.

The morphology of the chromosomes in solid tumors is often indistinct and less than optimal for detailed analysis as compared to the cell morphology encountered in leukemias.

Solid tumors are not infrequently infected (particularly those of the gastrointestinal tract, lung, and cervix), so that upon culture, even for a brief period of time, the infecting organisms destroy the tumor cells or make cytogenetic examination impossible. In addition, for optimal chromosome results, viable tumor tissue is necessary; however, some tumors are necrotic and may not yield sufficient metaphases for analysis.

The presence of cytogenetically normal cells in tumors is not unexpected since occasionally stromal and supporting elements in carcinomas and sarcomas appear to be of normal tissue origin. Another possibility is infiltration of the tumors by normal leukocytes, not a rare finding in some cancers. Although it may be contended that such cytogenetically normal cells are neoplastic, it is our view that if they are, the changes are beyond the resolution of cytogenetic techniques presently used.

Recent advances in cell culture and cytogenetic techniques applied to tumor cells have largely obviated some of the above-mentioned difficulties (KUSYK et al. 1979; WAKE et al. 1981; GIBAS LM et al. 1984; LIMON et al. 1986). These advances include a number of steps established in our labora-

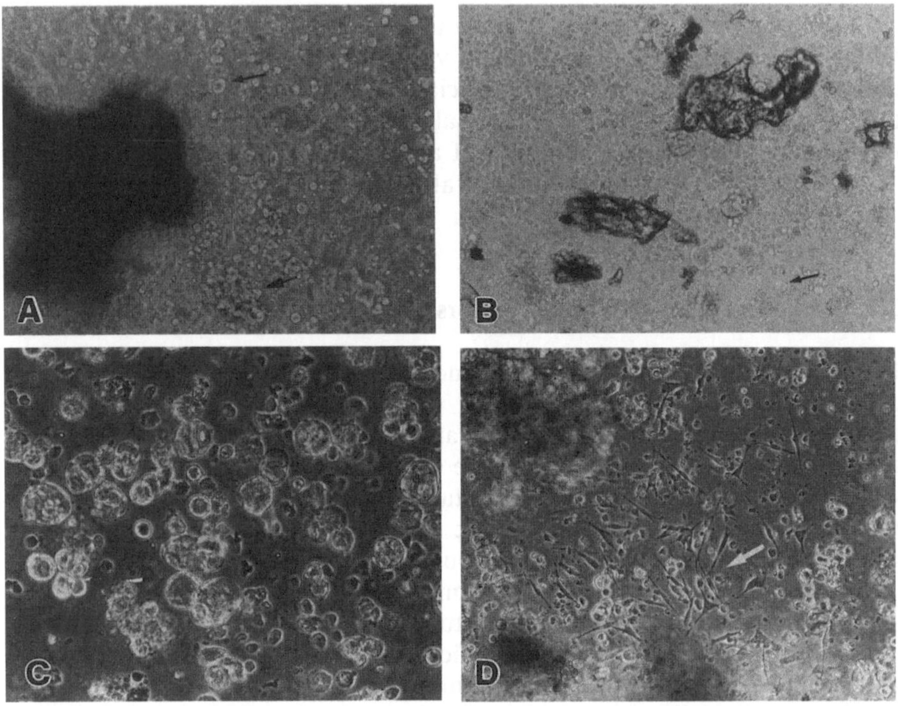

**Fig. 1 A–D.** Disaggregation of tumor cells by collagenase. **A** Osteosarcoma tumor cells immediately after mincing. *Slender arrow* points to single cell, *other arrow* to a cluster of cells. **B** The same tumor cells after 16 h of collagenase treatment. *Arrow* points to a single cell. **C** Adenocarcinoma of colon tumor cells after 16 h of collagenase treatment. **D** The same tumor cells after 48 h of collagenase treatment. *Arrow* points to a group of cells attached to the bottom of the flask. Single and clustered cells are also visible

tory, the most crucial of which consist of: chronic exposure of the cells to collagenase through the incubation or culture period; the use in situ of a modified hypotonic solution for the swelling of cells; the use of methotrexate for clearer banding of chromosomes; and, most importantly, frequent examination of the incubated or cultured cells for their mitotic activity in order to establish the peak of such activity.

### *3.2.1 Enzymatic Disaggregation*

Collagenase isolated from *Clostridium histolyticum* is a noncytotoxic hydrolytic enzyme which achieves maximum efficiency in isotonic media in a pH range from 6.5 to 7.8 (LASFARGUES 1973). Unlike trypsin, collagenase does not alter the cell surface. There are two approaches to collagenase disaggregation of solid tumors: one is a short-time enzymatic treatment with high concentration of collagenase (1000 U/ml) (WAKE et al. 1981; GIBAS LM et al. 1984) and the second utilizes low concentrations of col-

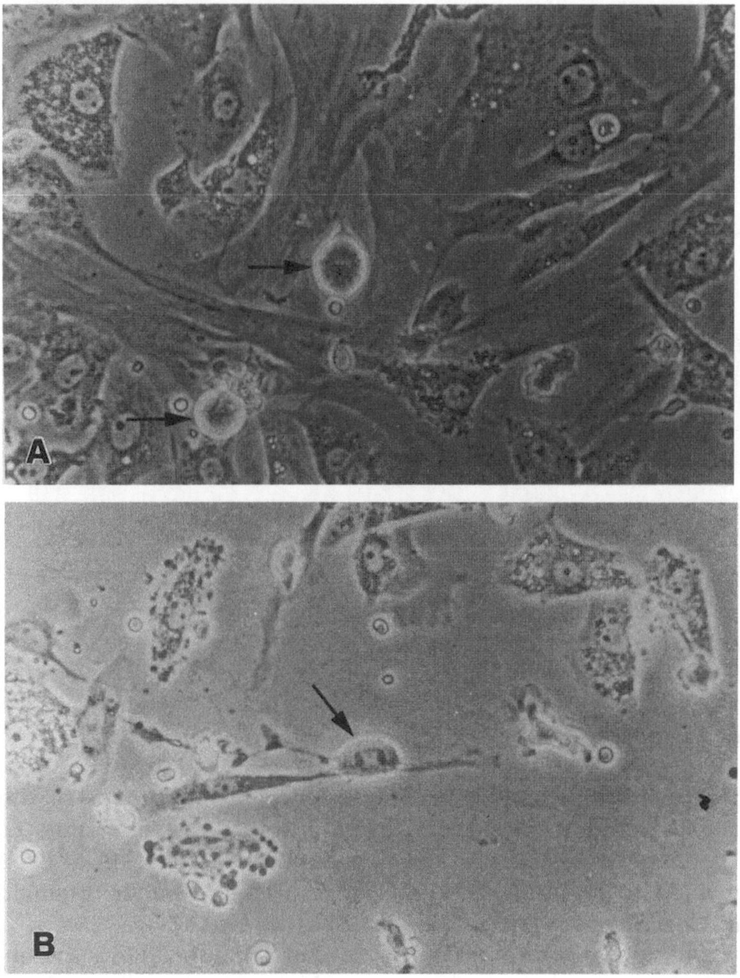


**Fig. 2 A, B.** Primary tissue culture of an adrenocortical carcinoma. **A** *Arrows* point to cells at metaphase. **B** *Arrow* points to a cell in anaphase

lagenase (200 U/ml) for 16–24 h (LIMON et al. 1986). Resistance of some tumors to collagenase treatment may be the result of too brief collagenase treatment. With prolonged collagenase treatment, almost all types of tumor can be disaggregated in 24 h. In some tumors (basal cell carcinoma, bladder cancer) this treatment can be longer, but this does not affect the growth of the tumor cells in culture. Connective tissue cells are almost completely disassociated by collagenase and remain in suspension, but may be separated by allowing the epithelial cancer to sediment for 5–10 min. Thus, this approach offers a possibility for release of some of the connective tissue cells and eliminating growth of normal cells in culture.

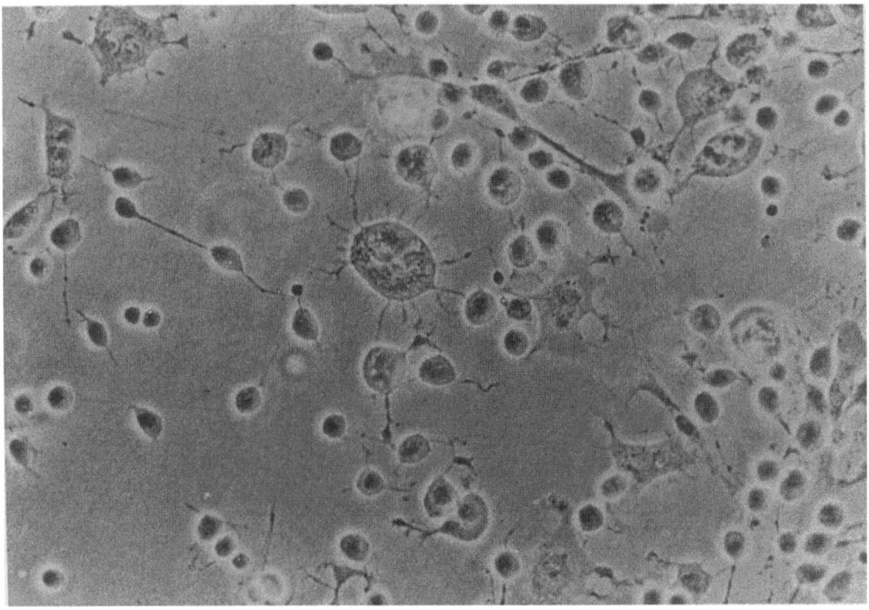

**Fig. 3.** Hypotonic treatment of primary culture after 15 min. Note round shape of cells

### *3.2.2 Microscopic Observations*

Frequent and detailed observations under an inverted microscope of the tumor disaggregation and/or culture are essential for successful results. In the former it is possible to ascertain the degree of disaggregation of each type of tumor and, in the latter, the time when a large number of single cells and small clusters are observed floating (Fig. 1).

A frequent and detailed observation of the culture flasks allows the observer to choose the optimal time of harvest, which often is individualized for each flask and should be the height of the mitotic activity (Fig. 2).

### *3.2.3 Hypotonic Solution In Situ*

The introduction of a new hypotonic solution used in situ for human solid tumor cytogenetic analysis by Gibas LM et al. (1984) was an important improvement in the methodology, for it led to more optimal spreading of chromosomes. This hypotonic solution works excellently on the chromosomes of tumors, not only resulting in the spread of the chromosomes with minimal overlapping, but also in improving their morphology. It should be reiterated that it is necessary to use different exposure times to the hypotonic solution for different types of tumor. Careful observation of changes in the shape of cells in situ under the microscope (Fig. 3) allows the observer the opportunity to ascertain the optimum time of hypotonic treatment. Generally, we use 60 min of hypotonic treatment for epithelial tumors and 30 min for sarcomas. We have found the same phenomenon as KUSYK et al. (1979), i.e., for good spreading of metaphases it is necessary to use a volume of hypotonic solution at least twice that of the medium in the flask.

### *3.2.4 The Use of Methotrexate*

Generally, cell synchronization techniques yield a higher number of cells in various early stages of division (late prophase, prometaphase, and early metaphases) than do other techniques. Originally described for phytohemagglutinin-stimulated peripheral blood cells (YUNIS 1976), they also have been used for the cytogenetic examination of malignant cells (HAGEMEIJER et al. 1979; YUNIS 1981; GIBAS LM et al. 1984). These synchronization techniques not only yield elongated chromosomes but also a greatly improved quality of metaphase spreads.

These refinements in cytogenetic techniques have led to higher rates of success in establishing the karyotypic changes in solid tumors. The result is that not only are more and more tumors being characterized cytogenetically, but also, as in the case of the leukemias, subtypes have been shown to exist within previously assumed entities (SANDBERG and TURC-CAREL 1987). Thus, it is our contention that all cancers (or any other neoplastic entity) will be shown cytogenetically to consist of a number of well-defined subtypes, each being of different or unique causation or histogenesis and inviting individual approaches to their classification and therapy.

## 4 Chromosome Changes in Solid Tumors and Their Relation to Pathology

The number of solid tumors characterized by consistent specific chromosome markers is still small (Table 1), as is the number of cases of chromosomally investigated tumors of each subtype. However, these few examples have already proven the usefulness of the cytogenetic findings in the diagnosis and classification of tumors.

### 4.1 Chromosome Markers and Subtypes of Soft Tissue Sarcomas

Soft tissue (mesenchymal) neoplasms are rare entities which represent less than 1% of all human neoplasms; they constitute a relatively heterogeneous group of tumors with regard to origin, pathology, and clinical course, and not infrequently defy a clear-cut histologic designation (ENZINGER and WEISS 1983).

**Table 1.** Specific (primary) chromosome changes in tumors

| Tumors | Chromosome changes |
|---|---|
| *Benign* | |
| Meningioma | −22 or 22 q− |
| Mixed tumors of salivary glands | t (3; 8) (p 21; q 12) |
| | t (9; 12) (p 13−22; q 13−15) |
| Lipoma | t with 12 q 14 |
| Colonic adenoma | 12 q− and/or +7 |
| | 12 q− and/or +8 |
| Leiomyoma | t (12; 14) (q 14−15; q 22−24) |
| Cortical adenoma of kidney | +7, +7, +17, −Y |
| *Carcinomas* | |
| Bladder | i (5 p) |
| | +7 |
| | −9/9 q− |
| | 11 p− |
| Prostate | del(10) (q 24) |
| Lung (SCLC) | del(3) (p 14 p 23) |
| Colon | 12 q−[a] |
| | +7[a] |
| | +8[a] |
| | +12[a] |
| | 17 (q 11)[a] |
| | 17 p−[a] |
| | −18[a] |
| Kidney | del(3) (p 11→p 21) |
| Uterus | 1 q−[a] |
| Ovary | 6 q−[a] |
| | t (6; 14) (q 21; q 24) |
| Endometrium | Trisomy 1 q |
| | +10 |
| *Sarcomas* | |
| Liposarcoma (myxoid) | t (12; 16) (q 13; p 11) |
| Synovial sarcoma | t (X; 18) (p 11.2; q 11.2) |
| Rhabdomyosarcoma (alveolar) | t (2; 13) (q 37; q 14) |
| Extraskeletal myxoid chondrosarcoma | t (9; 22) (q 31; q 12.2) |
| *Embryonal and other* | |
| Testicular (germ cell tumors) | i (12 p) |
| Retinoblastoma | del (13) (q 14)[b] |
| | i(6p) |
| Wilms' tumor | del(11) (p 13)[b] |
| Neuroblastoma | del(1) (p 32 p 36) |
| Malignant melanoma | del (6) (q 11 q 27)[a] |
| | i (6 p)[a] |
| | del (1) (p 11 p 22)[a] |
| | t (1; 19) (q 12; q 13) |
| Mesothelioma | del (3) (p 13→p 23) |
| Ewings's sarcoma and peripheral neuroepithelioma | t (11; 22) (q 24; q 12) |

[a] Not yet proved to be primary chromosome changes
[b] Associated with a constitutional chromosome change

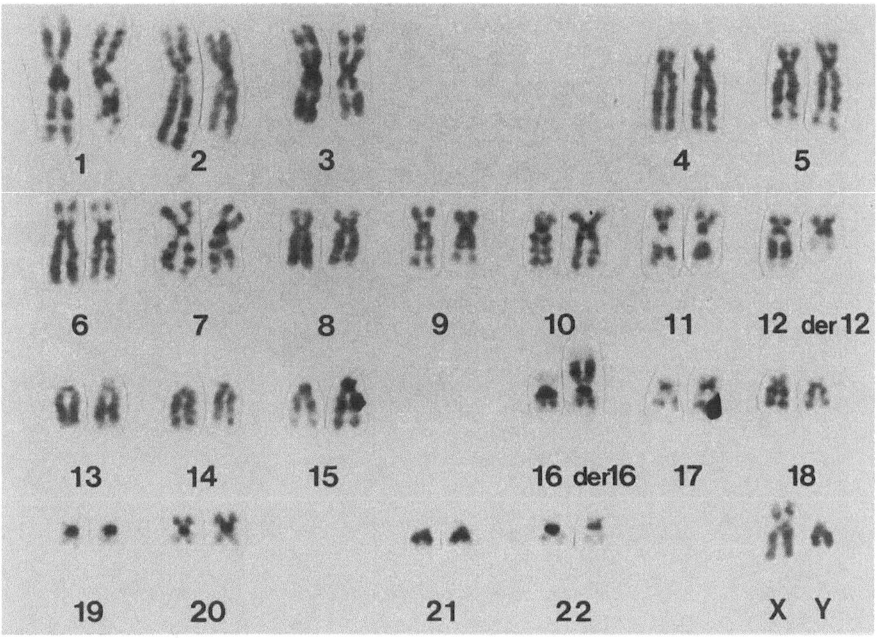


**Fig. 4.** Full karyotype showing the t(12; 16) (q 13; p 11) as the sole karyotypic change in a myxoid liposarcoma

An estimated 10%–20% of soft tissue tumors cannot be accurately classified by light microscopy though it is possible that the use of immunohistochemical techniques may reduce this number (DU BOULAY 1985).

### *4.1.1 Myxoid Liposarcomas, Malignant Fibrous Histiocytomas, and Other Liposarcomas*

Myxoid tumors of soft tissue are a small but broad group that may cause various problems in histologic interpretation (ALLEN 1921, MACKENZIE 1981). In particular, the distinction between myxoid liposarcomas (LPSs) and the myxoid variants of malignant fibrous histiocytomas (MFHs) presents a problem, especially since both subtypes are considered the most frequent of soft tissue sarcomas. Cytogenetics may be a powerful tool in the differential diagnosis. A translocation t(12; 16) (q 13; p 11) (Fig. 4) has been found to be specifically restricted to myxoid LPSs (TURC-CAREL et al. 1986c; SMITH et al. 1987; MERTENS et al. 1987) whereas rings, dicentrics, and telomeric associations have been described in MFH, pleomorphic and myxoid subtypes (MANDAHL et al. 1988). Rings, long markers, and telomeric associations also have been found to be a constant characteristic in a distinct subtype of LPS, well-differentiated LPS (KARAKOUSIS

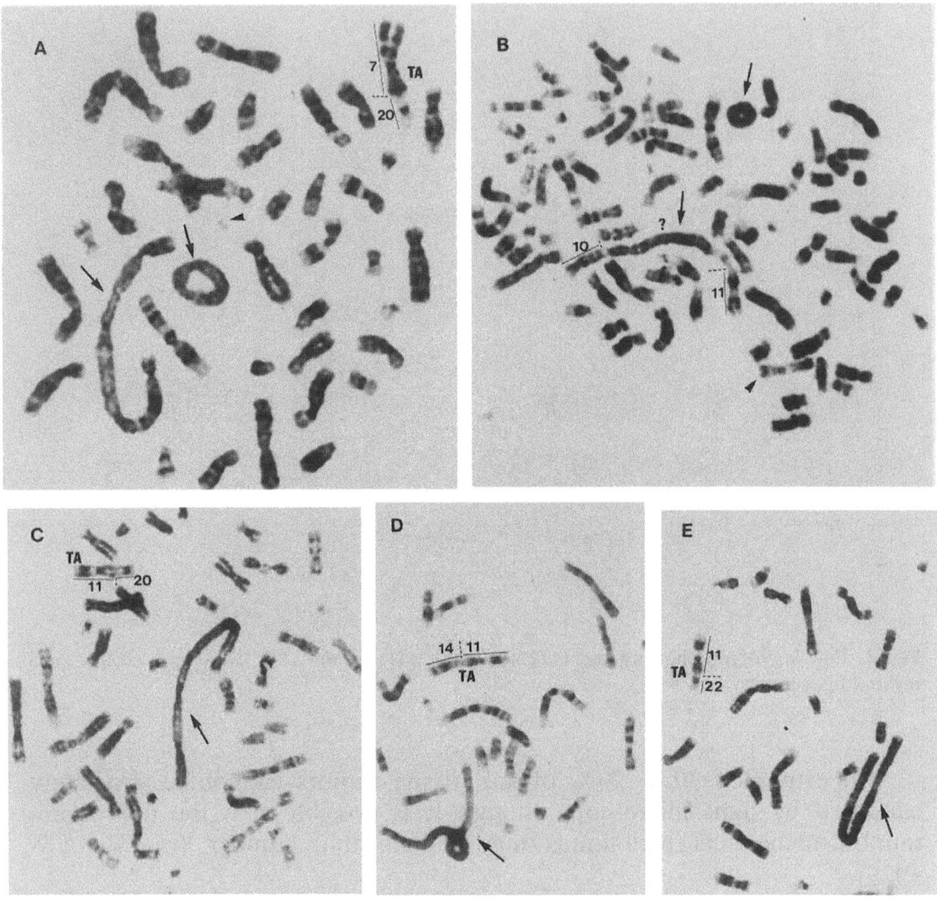


**Fig. 5 A–E.** Ring chromosomes, giant markers, and telomeric associations (*TA*) observed in five distinct cases of well-differentiated LPS

et al. 1987) (Fig. 5). These cytogenetic findings could be very useful to the pathologist, particularly when faced with one of the most frustrating situations in pathology, the lack of any specific evidence of cellular differentiation in some tumors (ROSS et al. 1986). For the clinician, the prognostic implications with regard to survival or response to treatment may be associated with the primary chromosome change, which probably plays a key role in formulating the basic biology of the tumor, as well as the additional chromosome changes, which are associated with progression (invasion, metastases, nature of response to therapy) of the tumor (KARAKOUSIS et al. 1987).

### *4.1.2 Ring(s) in Liposarcomas and Lipomas*

Within the adipose tissue tumors, subcutaneous lipomas are benign proliferations and LPSs (myxoid, round cell, pleomorphic) (ENZINGER and WEISS 1983) are malignant, according to well-established clinical and histologic criteria.

In contrast to this clear-cut behavior, some lipomatous neoplasms in the subcutis or within muscles are clinically benign, whereas similar proliferations in the retroperitoneum have clinically an invasive behavior, LPS well-differentiated, although they display histologic patterns similar to those of subcutaneous and intramuscular lipomas. In fact, some authors have named the subcutaneous and intramuscular categories as "atypical LPS" and "atypical intramuscular LPS," respectively (EVANS et al. 1979); the last entity has been called "well-differentiated lipoma-like LPS" by others (AZUMI et al. 1987). However, the term used to designate the nonretroperitoneal tumors constitutes a matter of controversy among histopathologists in classifying mature fat cell neoplasms.

Cytogenetic findings can be used by the pathologist as guides for the correct diagnosis of such tumors. Often other criteria have not been of help.

Though the number of lipomas chromosomally investigated so far is relatively small, four subgroups may be characterized cytogenetically: (a) those with a normal diploid chromosome set; (b) those with translocations involving 12q13–q14; (c) those with translocations not involving 12q13–q14; and (d) others with ring chromosome(s). All of these lipomas are considered as true benign lipomas (TURC-CAREL et al. 1986b; HEIM et al. 1986, 1987; MANDAHL et al. 1987; DAL CIN et al. 1988a).

From what was stated above, several conclusions can be inferred:

1. Consistent chromosome changes in tissue proliferations are no longer a criterion of malignancy. Similar conclusions have been arrived at by those studying meningiomas (ZANG 1982), mixed tumors of salivary glands (MARK et al. 1983), and leiomyomas (HEIM et al. 1988; GIBAS et al. 1988; TURC-CAREL et al. 1988; VANNI and LECCA 1988).
2. Lipoma and myxoid LPSs, both tumors of adipose tissue but one benign and the other malignant, share a consistent chromosomal translocation involving very close, if not identical, regions on chromosome 12, i.e., band 12q13 in myxoid LPSs and region 12q13 or q14 in typical lipomas. This observation, though it may favor a possible common basis in at least one of the steps responsible for the neoplastic process in adipose tissue, will have to be confirmed or refuted by appropriate molecular studies.
3. Ring chromosomes characterize a subgroup of atypical intramuscular lipomas (HEIM et al. 1987) and well-differentiated LPSs (KARAKOUSIS et al. 1987). Moreover, the latter tumors are all retroperitoneal, a location that EVANS et al. (1979) have used as a criterion to differentiate

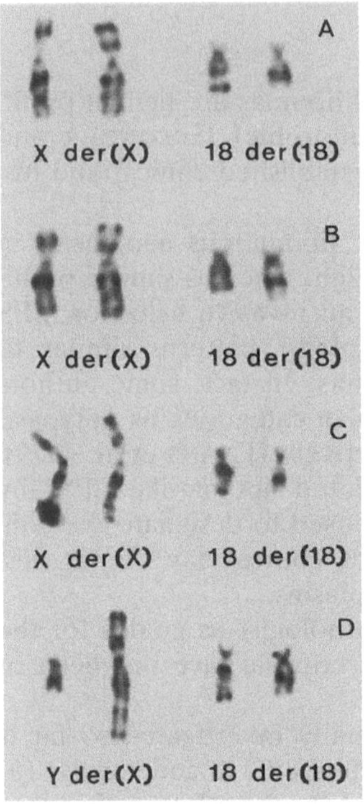


**Fig. 6 A–D.** Partial karyotypes showing the nonrandom translocation (X; 18) (p 11.2; q 11.2) in four synovial sarcomas

the malignant entity from the atypical intramuscular lipoma. A long-term follow-up of these patients is necessary to establish whether these atypical intramuscular lipomas with ring(s) chromosomes have a malignant behavior and whether retroperitoneal well-differentiated LPSs evolve from a "silent" atypical lipoma in the retroperitoneal area.

### *4.1.3 Synovial Sarcomas*

In many instances, synovial sarcomas are so poorly differentiated that they do not show specific features sufficient to suggest their true origin, and hence they may be confused with other poorly differentiated sarcomas. Histochemical stains have been the primary aid in the histologic diagnosis of these poorly differentiated neoplasms. The presence of keratin proteins in the cytoplasm of the tumor cells is a characteristic property of synovial sarcoma and aids in the identification of monophasic synovial sarcoma (CORSON et al. 1984). However, tissue fixation and processing can profoundly affect the results of the immunohistochemical staining procedures. Hence, in some cases the cytogenetic analysis of the tumor cells of these poorly differentiated tumors can be of great help in clarifying the histogenesis.

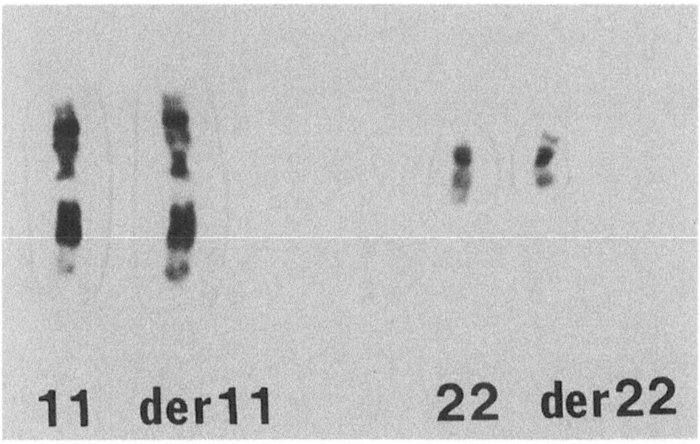


**Fig. 7.** Partial karyotype showing t(11;22) (q24;q12) in Ewing's sarcoma

A specific translocation, t(X;18) (p11.2;q11.2), characterizes synovial sarcomas irrespective of their histologic patterns, i.e., biphasic, monophasic, or poorly differentiated (Fig. 6) (TURC-CAREL et al. 1987; SMITH et al. 1987; GRIFFIN and EMANUEL 1987; WANG-WUU et al. 1987; UEDA et al. 1988). This translocation has been found in tumor cells from primary tumors as well in metastases, particularly in those to the lung.

## 4.2 Small Round Cell Tumors

There is a group of tumors in children genetically referred to as primitive, undifferentiated or small, round blue cell tumors, e.g., Ewing's sarcoma, rhabdomyosarcoma, neuroblastoma, and lymphoma of the bone, in which the diagnosis is often difficult (TRICHE and ASKIN 1983; TRICHE et al. 1986). The specific chromosome findings can be used to differentiate these tumors (WHANG-PENG et al. 1986). Since 1983 (AURIAS et al. 1983; TURC-CAREL et al. 1983), a total of 70 cases of Ewing's sarcoma have been chromosomally investigated (see SANDBERG et al. 1988) and the translocation t(11;22) (q24;q12) (Fig. 7) has been seen in 90%.

Such a chromosome abnormality has not been described in the other tumors of the small round cell group and, at the present time, is the only reliable criterion for the diagnosis of Ewing's sarcoma.

A consistent chromosome change, t(2;13) (q37;q14), characterizes the alveolar histologic subtype of rhabdomyosarcoma (SEIDEL et al. 1982; TURC-CAREL et al. 1986a; LAI et al. 1987; LIZARD-NACOL et al. 1987; WANG-WUU et al. 1988).

Structural chromosome abnormalities involving 1q and/or 1p and 3p14–12, as well as numerical changes such as trisomies 2 and 3, have been reported in embryonal and unclassified rhabdomyosarcomas (POTLURI and GILBERT 1985; TRENT et al. 1985; DOUGLASS et al. 1985; WANG-WUU et al. 1988).

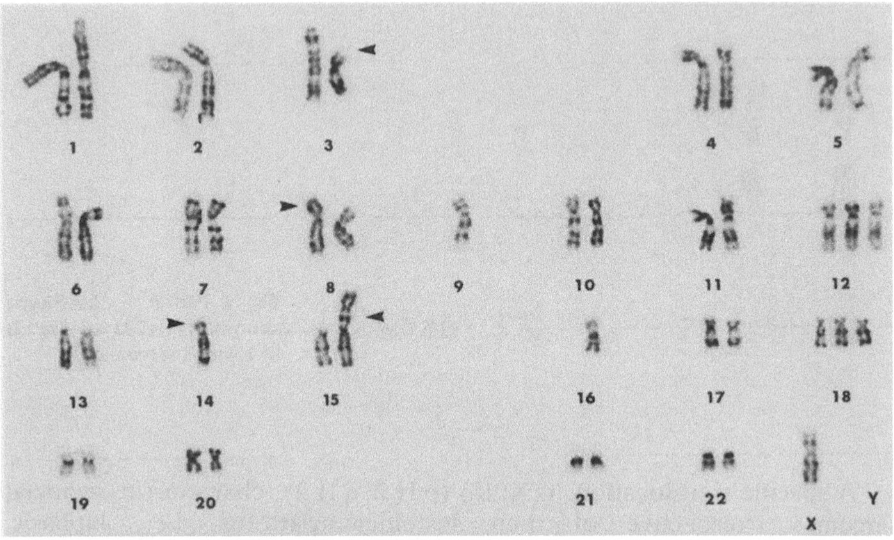


**Fig. 8.** Full karyotype from a small cell carcinoma. Note the del(3) (p14), a primary chromosome change characteristic of this kind of tumor. Other additional anomalies are also present: −9, +12, −14, −16, +18, 8p+, 14p+, 15p+

Partial monosomy for the short arm of chromosome 1 is the most consistent structural abnormality in human neuroblastoma (BRODEUR 1982). The region of the short arm of chromosome 1 deleted is variable but the portion distal to band 1p32 is consistently lost.

### 4.3 Renal Adenoma Versus Adenocarcinoma

Differentiation of renal adenoma (benign) from an adenocarcinoma (malignant) may be very difficult because no gross anatomic, histologic, histochemical, immunologic, or ultrastructural features are known which can discriminate between the two entities (LEDER et al. 1979; OLSEN 1984). They virtually overlap one another without any sharp, definable boundary. In the past, tumor size was the main criterion of malignancy. It was observed that tumors less than 3 cm in diameter usually did not metastasize. Thus, tumors that were larger than 3 cm in diameter were considered malignant (BELL 1950). However, this criterion has been strongly criticized by some authors (BENNINGTON and BECKWITH 1975); many patients with tumors larger than 3 cm have no metastases at the time of nephrectomy and vice versa.

It is a generally supported and accepted view that renal adenomas are small or structurally early forms of carcinoma which have not yet produced metastases (BENNINGTON and BECKWITH 1975; BANNAYAN and LOMM 1980; PASSEGE et al. 1987). This view has no practical consequence if this

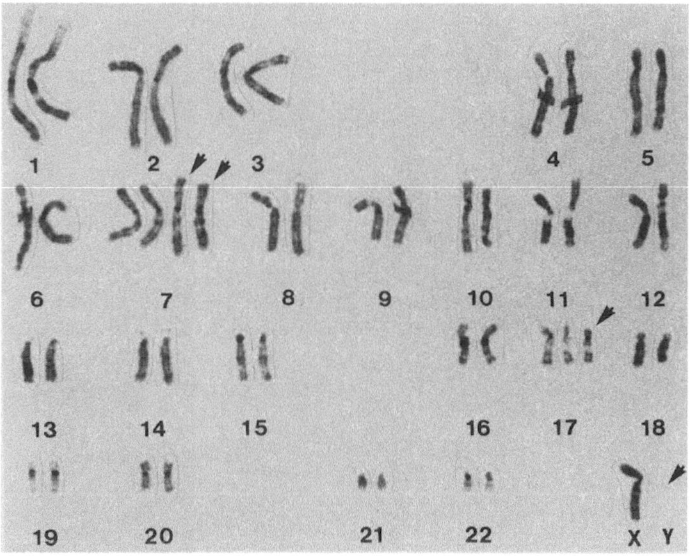


**Fig. 9.** Full karyotype from a renal cortical adenoma showing characteristic changes for such a tumor: 48, X, −Y, +7, +7, +17

tumor is diagnosed at autopsy. However, if it is found at surgery incidentally or otherwise, it has been suggested that it is better to diagnose it as a potential adenocarcinoma and follow up the patient carefully (BENNINGTON and BECKWITH 1975).

Cytogenetic analysis can play an important role in the differential diagnosis of these two tumors whose histologic features are so similar (FITE 1945).

Consistent rearrangements involving the short arm of chromosome 3 (3p11−p21) have been found in renal cell carcinoma of the hereditary (COHEN et al. 1979; PATHAK et al. 1982; WANG and PERKINS 1984; YOSHIDA et al. 1986) as well as of sporadic forms (YOSHIDA et al. 1986; BERGER et al. 1986; KOVACS et al. 1987; SZÜCS et al. 1987; CARROLL et al. 1987) (Fig. 8). The involvement of the long arm of chromosome 5, the gain of one chromosome 7, and the loss of one sex chromosome were also observed in renal cell carcinoma; however, their role in renal carcinoma development merits further study.

Recently, a combination of chromosome abnormalities (+7, +7, +17, −Y) was described by us to be associated with a renal tumor histologically defined as renal adenoma (DAL CIN et al. 1988b) (Fig. 9). It is interesting that the above combination of chromosome changes can be found among previously reported cytogenetic findings in relatively large series of renal cell carcinomas (DE JONG et al. 1988; MILES et al. 1988) without their significance being realized.

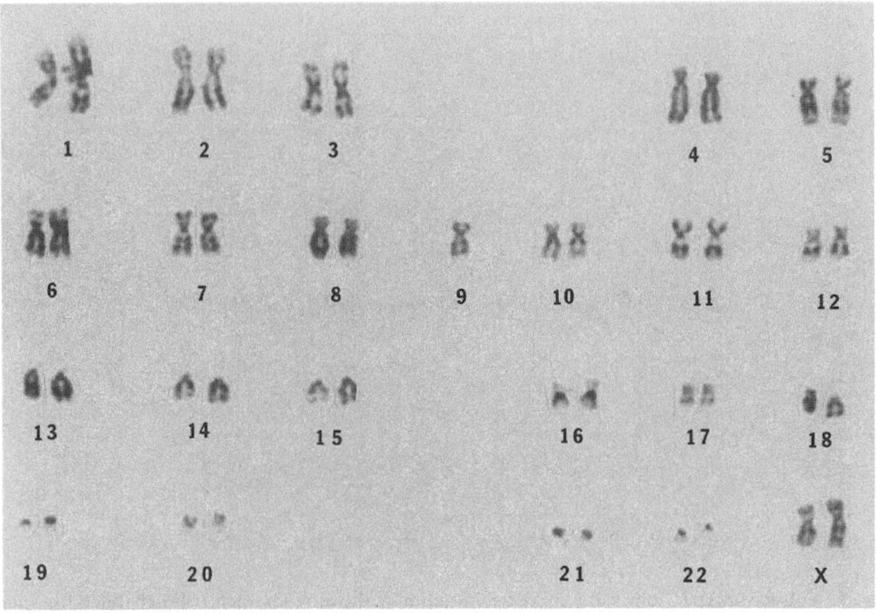


**Fig. 10.** Full karyotype showing monosomy 9 (−9) as the sole abnormality in one of the cytogenetically defined subentities of bladder cancer

## 5 Chromosome Changes Defining Subtypes Within a Tumor Category, Same Lineage, and Histogenesis

### 5.1 Tumor Subtypes

Though transitional cell carcinoma (TCC) may appear to be histologically a rather uniform entity, specific karyotypic changes point to subentities within this category.

An isochromosome for the short arm of chromosome 5 or 5q-, monosomy 9 or 9q-, deletion of the short arm of chromosome 11, and trisomy 7 have been observed in association with bladder TCC (GIBAS et al. 1984a, 1986, 1987; ATKIN and BAKER 1985b) (Fig. 10). That each of these chromosome markers may characterize subsets within the bladder with distinct biologic behavior (invasiveness, metastatic spread) is supported by the higher aggressiveness of those bladder tumors with morphologic chromosome changes than those with trisomy 7 or monosomy 9 (SANDBERG 1986b).

It is possible that other entities similar to TCC may ultimately be "dissected" cytogenetically and that a more detailed correlation may be found between the cytogenetic findings and such parameters as stage and grade of the cancer, response to therapy, survival, invasiveness, metastatic spread, and anatomic location of the tumor.

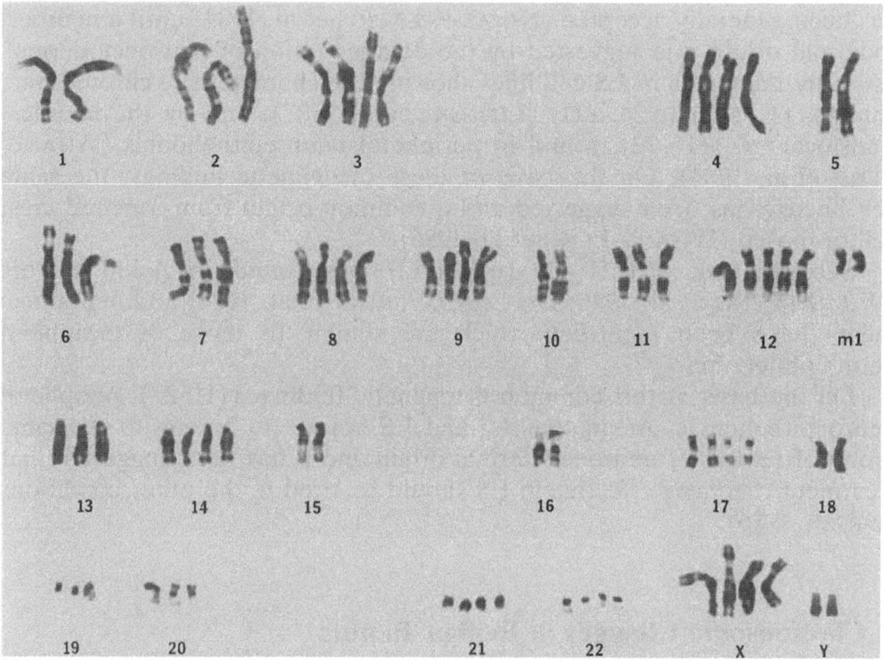


**Fig. 11.** Full karyotype showing i(12p), characteristic of germ cell tumors of the testes. Most of the groups had extra chromosomes, including three extra X chromosomes which account for the positive sec chromatin pattern seen in most of these tumors

## 5.2 Same Lineage

A specific chromosome rearrangement may occur in an undifferentiated stem (primitive) cell which subsequently proceeds along a different pathway of differentiation and/or malignant transformation. Almost all germ cell tumors of the testis, such as seminomas, teratomas, embryonal cells, and choriocarcinomas, exhibit at least one copy of an isochromosome for the short arm of chromosome 12, i(12p) (Fig. 11) (ATKIN and BAKER 1982, 1983, 1985a; GIBAS et al. 1984b; DELOZIER-BLANCHET et al. 1985). The presence of this characteristic chromosome marker indicates a similar origin of these tumors.

It is interesting that the same chromosome abnormality, i(12p), has been recently detected in ovarian dysgerminomas; this is not surprising since these ovarian tumors are identical histologically to seminomas of the testis (JENKYN and MCCARTNEY 1987; ATKIN and BAKER 1987).

## 5.3 Histogenesis of Ewing's Sarcoma

The origin of Ewing's sarcoma (ES) has been controversial since its initial description (EWING 1921). An origin from a primitive mesenchymal cell has been generally accepted (NAVAS-PALACIOS et al. 1984) until a neuroectodermal origin was suggested by the demonstration of neuroectodermal-associated antigens in ES cell lines showing the characteristic chromosome change, t(11;22) (q24;q11) (LIPINSKI et al. 1987), and by the identical translocation, t(11;22), found in peripheral neuroepitheliomas (WHANG-PENG et al. 1984). On the basis of these cytogenetic findings, the same cell lineage has been suggested and a common origin from a neural crest cell proposed (WHANG-PENG et al. 1986).

Subsequently, a t(11;22) (q24;q12) was found in Askin tumors (WHANG-PENG et al. 1986; DE CHADAVERIAN et al. 1984) and neural elements have been identified which are similar to those of peripheral neuroepithelioma.

On the basis of this common cytogenetic finding, t(11;22), peripheral neuroepitheliomas, Askin tumors, and ES appear to belong to the same group of tumors of neuroectodermal origin and it has been suggested that treatment regimens effective in ES should be tried in the other conditions (ISRAEL 1986).

## 6 Chromosome Changes in Benign Tumors

The chromosomal studies in human malignant neoplasms have shown that most, if not all, malignant cells exhibit chromosome abnormalities, often occurring in a nonrandom fashion (SANDBERG 1985). In contrast, the chromosomal status of so-called benign neoplasms has been far less defined, due essentially to the paucity of the studies reported to date. Though it had been generally accepted that benign tumors have a normal chromosome constitution (SANDBERG 1980), nonrandom chromosome abnormalities have been described in a number of benign tumors, i.e., in meningiomas, mixed tumors of the salivary glands, lipomas, leiomyomas, adenomas of the colon, and cortical adenomas of the kidney. Such data highlight the fact that the concept of an abnormal chromosome pattern being related to a malignant or premalignant state versus a normal diploid karyotype related to a benign state had to be qualified. On the other hand, cytogenetic data on benign tumors may give clues to the biological basis for the distinctive behavior of benign versus malignant neoplasms.

### 6.1 Meningioma

Meningiomas are considered to be histologically benign tumors and have been the most extensively studied, cytogenetically, of the human benign solid tumors.

The loss of all or part of chromosome 22 is a consistent chromosome finding in more than half of meningiomas cytogenetically investigated

(MARK 1977; ZANG 1982; YAMADA et al. 1980; ZANKL and ZANG 1980; KATSUYAMA et al. 1986; AL SAADI et al. 1987; CASALONE et al. 1987; REY et al. 1988). In addition to monosomy 22, which would be considered as the primary karyotypic event, nonrandom loss of chromosomes 8, 14, and, perhaps, a sex chromosome, and structural rearrangements involving 1 p and 11 p, are characteristic secondary features of meningiomas.

More studies on correlations of karyotypes, histologic characteristics, and biologic behavior are needed to determine the parameters which maintain the bulk of the meningiomas as benign tumors in spite of the significant chromosomal abnormalities.

### 6.2 Mixed Salivary Gland Tumors

Pleomorphic salivary gland adenomas are the most common benign tumors of the human parotid and other salivary glands. More than 100 such tumors have been investigated cytogenetically (MARK and DAHLENFORS 1986; MARK et al. 1982; BULLERDIEK et al. 1987 a, b). A normal karyotype has been reported in 55%–60% of these cases. The remaining tumors can be divided into: (a) those with chromosome 8 involvement, usually translocation of the segment distal to 8 q 12; (b) those with translocations involving the 12 q 13–15 region; (c) those with translocations and/or deletions affecting a distal segment of either the short or the long arm of chromosome 3; and (d) those with normal cells and one (or more) variant cell(s).

### 6.3 Lipoma

As described above (Sect. 4.1.2), at least four kinds of karyotype may be encountered in lipoma, a benign proliferation of adipose tissue: (a) a normal karyotype; (b) involvement of band 12 q 13–14; (c) abnormalities not involving 12 q 13–14; and (d) presence of ring(s) (TURC-CAREL et al. 1986 b; HEIM et al. 1986, 1987; MANDAHL et al. 1987; DAL CIN et al. 1988 a) (Fig. 12).

Such abnormalities were not recorded in other types of benign adipose tumor previously reported by SANDBERG et al. (1986), i.e., a lipoblastoma with a t(7;8) and del(4), and a fibrolipoma with a del(6) and inv(13). Malignant transformation in lipoma is very exceptional, however; it is also interesting that the same region, 12 q 13–14, is involved in one of the malignant proliferations of adipose tissue, i.e., myxoid LPS.

### 6.4 Leiomyoma

Recently, benign tumors of the smooth muscle of the uterus, leiomyomas, have been analyzed cytogenetically. A consistent involvement of regions

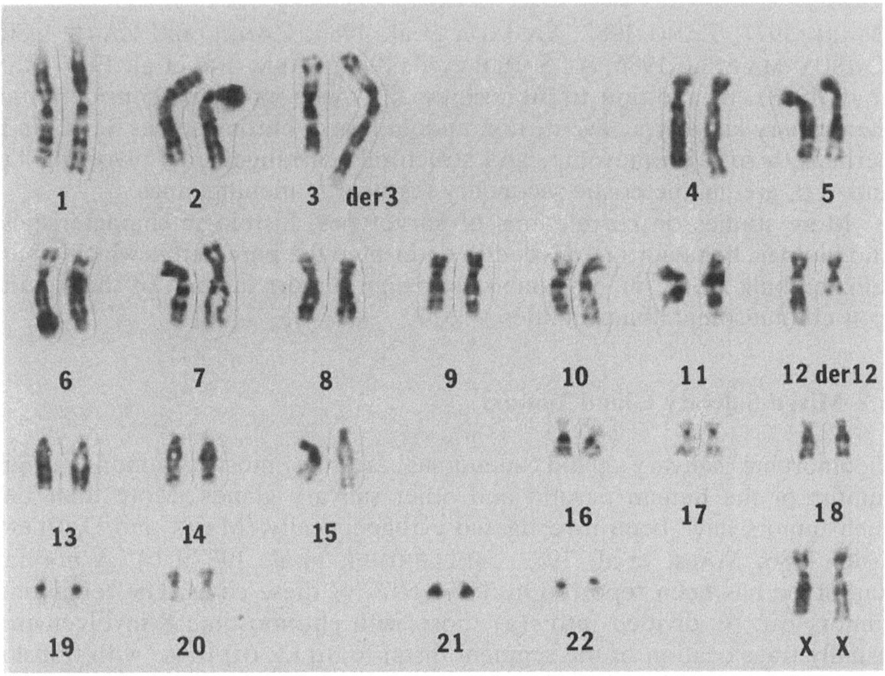


**Fig. 12.** Karyotype showing t(3; 12) (q27; q14), the most frequent chromosome change found in lipoma

12q14–q15 and 14q22–q24 has been reported (HEIM et al. 1988; GIBAS et al. 1988; TURC-CAREL et al. 1988; VANNI and LECCA 1988).

A balanced translocation, t(12; 14) (q14–q15; q22–q24), seems to be the most frequent chromosome change found in, at least, a subgroup of leiomyomas (Fig. 13), indicative that region 12q14–15 may show a preferential involvement (VANNI and LECCA 1988).

## 6.5 Adenomas of Colon

The possible meaning of chromosome changes in colon adenomas needs to be further investigated, not only for the paucity of such tumors cytogenetically analyzed (MARK et al. 1973; REICHMANN et al. 1982, 1985), but also because identical changes (+7, +8, 12q–) have been described in colonic adenocarcinomas (BECHER et al. 1983; OCHI et al. 1983; FERTI-PASSANTONOPOULOU et al. 1986). Therefore, it was tempting to correlate the presence of these chromosome changes (+7, +8, and/or +12q–) in adenomas with a possible premalignant status, though the relation of these karyotypic changes to cancer development remain unclear.

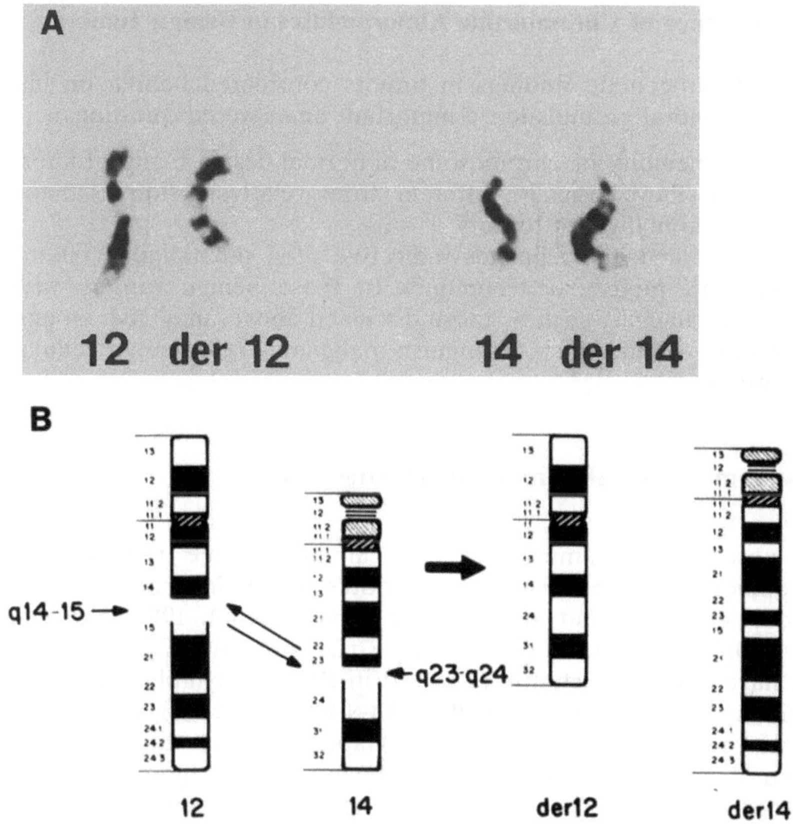


**Fig. 13 A, B.** Partial karyotype showing t(12; 14) (q 14–15; q 22–24), a rather characteristic karyotypic change in leiomyoma of the uterus

## 6.6 Cortical Adenoma of Kidney

As previously described in Sect. 4.3, renal cortical adenoma and adenocarcinoma overlap one another without any sharp, definable boundary (LEDER et al. 1979; OLSEN 1984).

A tetrasomy 7 (+7, +7), trisomy 17 (+17), and loss of one sex chromosome (−Y) has been seen by us as a recurrent combination of chromosome changes found only in renal cortical adenoma (DAL CIN et al. 1988 b).

The generally accepted view regards renal adenoma as an early form of carcinoma (BENNINGTON and BECKWITH 1975). Often these tumors are found and diagnosed only at autopsy. However, they can also be found at surgery incidentally and it has been suggested that they should be treated as potential adenocarcinoma with careful follow-up of the patient. At best cytogenetically, this is not true because the two entities have distinct chromosome changes.

### 6.7 The Significance of Chromosome Abnormalities in Benign Tumors

The abnormal cytogenetic findings in tumors considered benign on histologic and behavioral grounds leave important unanswered questions:

1. What is the meaning of chromosome abnormalities in benign tumors and how are they basically different from relatively simple abnormalities seen in malignant tumors?
2. Are these chromosome changes indicative of a premalignant state? Application of molecular techniques to these benign tumors with chromosome changes, such as those discussed above, may give an answer as to the events which distinguish malignant from benign cellular proliferations.

## 7 Rearrangement Breakpoints and Oncogenes

There are now several examples of reciprocal chromosome translocations in hematopoietic diseases, such as Burkitt's lymphoma and chronic myelogenous leukemia, that alter specific genes (CROCE 1986; NOWELL and CROCE 1986). Translocations involving 8q24 with either 2p13, 22q11, or 14q32 have been consistently observed in Burkitt's lymphoma (ZECH et al. 1976; VAN DEN BERGHE et al. 1976; LENOIR et al. 1982). It has been demonstrated that these translocations juxtapose the *myc* oncogene, located at 8q24, with the immunoglobulin genes, $\varkappa$ and $\lambda$ light chains, on chromosomes 2 and 22 respectively, or the heavy chain locus on chromosome 14 (CROCE et al. 1979; CROCE and NOWELL 1985). The effect of these rearrangements can lead to transcriptional deregulation of the expression of the c-*myc* gene (CROCE and NOWELL 1985).

In chronic myelogenous leukemia a reciprocal translocation between chromosomes 9 and 22, at bands q34 and q11 respectively, characterizes the disease (ROWLEY 1983). In this translocation c-*abl* is regularly translocated from its normal site on chromosome 9 to a very restricted region of chromosome 22, called "breakpoint cluster region" (bcr) (GROFFEN et al. 1984). This new and novel chimeric gene, *bcr/abl*, produces a novel mRNA, larger than the normal transcript, that appears to code for an abnormal protein of high molecular weight having tyrosine kinase activity greatly increased over that of the normal protein (SHTIVELMAN et al. 1985; HEISTERKAMP et al. 1985; GALE and CANAANI 1984; KONOPKA et al. 1984; KLOETZER et al. 1985).

In the case of most solid tumors, cytogenetic analysis has not yet provided the molecular biologist with the information necessary to identify the critical genes involved in the malignant transformation. However, there is little doubt that some of the nonrandom chromosome changes found in some of the solid tumors described above will provide the basis for a rational molecular genetic approach.

In adipose tissue tumors, both benign (lipoma) and malignant (myxoid LPS) lesions share the same chromosome region in their consistent rearrangements (12q13–q14). It appears to be difficult to investigate. The proximity of the breakpoints in these two entities precludes discrimination of the break in lipoma from that in myxoid LPS using cytogenetics only. On the other hand, if these breakpoints fall within the same band, what are the molecular events underlying the benign versus the malignant proliferation?

## 8 Fragile Sites and Cancer Breakpoints

Many fragile sites have been identified in human chromosomes, some of which may also be specific cancer chromosome breakpoints. The idea naturally arose that these phenomena might be connected and a fragile site hypothesis of cancer has been postulated (SUTHERLAND and HECHT 1985; LEBEAU 1986; YUNIS 1983; LEBEAU and ROWLEY 1984; HECHT and SUTHERLAND 1984; HECHT and GLOVER 1984; DEBRAEKELEER et al. 1985). However, there are now some experimental data suggesting that no such relationship exists, whereas other evidence supports an association. In a special issue of *Cancer Genetics and Cytogenetics* (1988), the findings on both sides have been well summarized and the question is still open.

## 9 Concluding Remarks

Establishing the primary (specific) karyotypic events in the leukemias has significantly expanded the classification and diagnostic parameters of these diseases. Not only have these primary chromosome changes been proven to be independent diagnostic parameters, but they have also more clearly defined the biology of these diseases at the clinical and molecular levels. Such progress has not been made in most solid tumors, particularly the adenocarcinomas and other epithelial tumors. The chromosome changes in these tumors are usually complex and numerous, tending to mask the primary change. Thus, epithelial tumors with a sole cytogenetic change are very rare, though the information they yield is crucial since it can be assumed that such a primary change characterizes at least a subtype within a known entity. Falling short of that, the best approach is to examine a sufficiently large group of tumors which hopefully will yield reliable information on recurrent chromosomal changes, particularly the primary ones, which may characterize subtypes within the tumors investigated. In some tumors this has not been achieved reliably, e.g., cancer of the breast, colon, and lung. Part of the problem may reside in the lack of detailed pathology supplied with most of the chromosome studies on adenocarcinomas and related tumors. Often, recently establishing refinements in pathologic approaches to tumor histology are lacking, for it is possible that the pathology of tumors which also includes cytochemical or cytoimmunologic aspects may define subtypes within a tumor entity and thus make correlation with the chromosome finding more readily accomplished and meaningful. This combination of approaches may tell us much more about tumor types and subtypes than is presently known.

In dealing with tumors in which the cytogenetic findings have been established, particularly the sarcomas and benign tumors, the pathologist should avail himself of these findings, thus expanding the diagnostic acumen, particularly on confusing or complex tumors.

A symbiotic relationship between cytogeneticists and pathologists will expedite the application of chromosome findings to achieve a more accurate diagnosis of solid tumors, such as has been done in some sarcomas and in differentiating benign from malignant tumors. Though it is realized that such a situation has yet to be established for most solid tumors, the application of effort in the design of appropriate studies will ultimately yield fruitful results similar to those already established in the leukemias.

Recent progress in the cytogenetics of soft tissue tumors and in some of the carcinomas already discussed above has not only yielded exciting and promising karyotypic information on specific tumor entities, but also paves the way for elucidation of similar cytogenetic parameters in the more common cancers. Often, success was attained through cooperative effort between the surgeon, pathologist, and cytogeneticist. This has led to the use of the chromosome findings as an aid in the diagnosis of complicated cases. Ultimately, often meaningful and practical parameters, e.g., prognosis, chance of recurrence, invasion, or metastases, and the planning of appropriate therapeutic approaches, may be related to the cytogenetic findings seen in tumor subtypes.

Some forms of cancer, e.g., that of the breast and lung, may require special national or international cooperative studies with the participation of pathologists, oncologists, molecular biologists, and cytogeneticists culminating in workshops akin to those which have dealt with the leukemias. Objective scrutiny and critique of the karyotypes, of the detailed and appropriate pathology, of the clinical data, and of other factors in a cooperative and friendly atmosphere will go a long way to deciphering the meaning of the complex cytogenetic findings characterizing most adenocarcinomas, and hopefully lead to the application of these findings in a useful way, as has been accomplished in the leukemias and sarcomas. This may not be an easy task, but with each participant pushing to the limit the realm of his or her expertise, including innovative approaches and new markers in the various disciplines, the task can be accomplished. For until such information is available on various tumors, we will be ignorant regarding important facets of the biology and possible etiology of these lesions.

*Acknowledgements.* This work was supported by grants from the National Institutes of Health CA-41183 and CA-14555. We wish to thank Ms. Jan Vaughan for clerical aspects.

## References


Allen PW (1921) Myxoid tumors of soft tissues. In: Pathology annual, part I. Appleton Century Crofts, New York, p 133

Al Saadi A, Latimer F, Madercic M, Robbins T (1987) Cytogenetic studies of human brain tumors and their clinical significance. II. Meningioma. Cancer Genet Cytogenet 26:127–141

Atkin NB, Baker MC (1982) Specific chromosome change, i(12p), in testicular tumours? Lancet II:1349

Atkin NB, Baker MC (1983) i(12p) Specific chromosomal marker in seminoma and malignant teratoma of the testis? Cancer Genet Cytogenet 10:199–204

Atkin NB, Baker MC (1985a) Chromosome analysis of three seminomas. Cancer Genet Cytogenet 17:315–323

Atkin NB, Baker MC (1985b) Cytogenetic study of ten carcinomas of the bladder: involvement of chromosomes 1 and 11. Cancer Genet Cytogenet 15:253–268

Atkin NB, Baker MC (1987) Abnormal chromosomes including small metacentrics in 14 ovarian cancers. Cancer Genet Cytogenet 26:355–361

Aurias A, Rimbaut C, Buffe D, Dubousset J, Mazabraud A (1983) Translocation of chromosome 22 in Ewing sarcoma. N Engl J Med 309:469–497

Azumi N, Curtis J, Kempson RL, Hendrickson MR (1987) Atypical and malignant neoplasms showing lipomatous differentiation. A study of 111 cases. Am J Surg Pathol 11:161–183

Bannayan GA, Lomm DL (1980) Renal cell tumors. Pathol Annu 15 (2):271

Becher R, Gibas Z, Sandberg AA (1983) Involvement of chromosomes 7 and 12 in large bowel cancer: trisomy 7 and 12q–. Cancer Genet Cytogenet 9:329–332

Bell ET (1950) Renal diseases, 2nd edn. Lea & Fibiger, Philadelphia

Bennett JM, Catovsky D, Daniel MT, Flandrin G, Galton DAG, Gralnik HR, Sultan C (1976) Proposals for the classification of the acute leukemias. Br J Haematol 33:451–458

Bennett JM, Catovsky D, Daniel MT, Flandrin G, Galton DAG, Gralnik HR, Sultan C (1981) Morphological classification of acute lymphoblastic leukemia: Concordance among observers and clinical correlations. Br J Haematol 47:553–561

Bennett JM, Catovsky D, Daniel MT, Flandrin G, Galton DAG, Gralnik HR, Sultan C (1985) Proposed revised criteria for the classification of acute myeloid leukemia: a report of the French-American-British Cooperative Group. Ann Intern Med 103:620–629

Bennington JL, Beckwith JB (1975) Tumors of the kidney, renal pelvis and ureter. In: Atlas of tumor pathology, 2nd series, fascicle 12. Armed Forces Inst Pathol, Washington, DC

Berger CS, Sandberg AA, Todd IAD, Pennington RD, Haddad FS, Hecht BK, Hecht F (1986) Chromosomes in kidney, ureter, and bladder cancer. Cancer Genet Cytogenet 23:1–24

Brodeur GM (1982) Genetics and cytogenetics of human neuroblastoma. In: Pochedly C (ed) Neuroblastoma: clinical and biological manifestations. Elsevier Biomedical, New York, pp 183–194

Bullerdiek J, Bartnitzke S, Weinberg M, Chilla R, Haubrich J, Schloot W (1987a) Rearrangements of chromosome region 12q13→q15 in pleomorphic adenomas of the human salivary gland (PSA). Cancer Cell Genet 45:187–190

Bullerdiek J, Böschen C, Bartnitzke S (1987b) Aberrations of chromosome #8 in mixed salivary gland tumours. Cancer Genet Cytogenet 24:205–212

Cancer Genetics and Cytogenetics (1988) 31:1–160. Fragile site issue.

Carroll PR, Murty VVS, Reuter V, Jhanwar S, Faiz WR, Whitmore WF, Chaganti RSK (1987) Abnormalities at chromosome region 3p12–14 characterize clear cell renal carcinoma. Cancer Genet Cytogenet 26:253–259

Casalone R, Granata P, Simi P et al. (1987) Recessive cancer genes in meningiomas? An analysis of 31 cases. Cancer Genet Cytogenet 27:145–159

Cohen AJ, Li FP, Berg S, Marchetto DJ, Jacobs SC, Brown RS (1979) Hereditary renal-cell carcinoma associated with a chromosomal translocation. N Engl J Med 301:592–595

Corson JM, Weiss LM, Banks-Schlegel SP, Pinkus GS (1984) Keratin versus epithelial membrane antigen in tumor diagnosis: an immunohistochemical comparison of five monoclonal antibodies. Hum Pathol 15:615–621

Croce CM (1986) Chromosome translocations and human cancer. Cancer Res 46:6019–6023
Croce CM, Nowell PC (1985) Molecular basis of human B cell neoplasia. Blood 65:1–7
Croce CM, Shander M, Martinis J, Cicurel L, D'Ancona GG, Dolby TW, Koprowski H (1979) Chromosomal location of the human immunoglobulin heavy chain genes. Proc Natl Acad Sci USA 76:3416–3419
Dal Cin P, Rao U, Turc-Carel C, Sandberg AA (1988a) Translocation (7;21) in a lipoma. Cancer Genet Cytogenet 30:17–22
Dal Cin P, Huben R, Li FP, Prout GR, Sandberg AA (1988b) New cytogenetic subtype of renal tumor. Cancer Genet Cytogenet 32:313
DeBraekeleer M, Smith B, Lin CC (1985) Fragile sites and structural rearrangements in cancer. Hum Genet 69:112–116
De Chadaverian JP, Vekemans M, Seemayer TA (1984) Reciprocal translocation in small-cell sarcomas. N Engl J Med 311:1702–1703
de Jong B, Oosterhuis JW, Idenburg VJS, Costeds SMMJ, Dam A, Mensink HJA (1988) Cytogenetics of 12 cases of renal adenocarcinomas. Cancer Genet Cytogenet 30:53–61
DeLozier-Blanchet CD, Engel E, Walt H (1985) Isochromosome 12p in malignant testicular tumors. Cancer Genet Cytogenet 15:375–376
Douglass EC, Green AA, Hayes FA, Etcubanas E, Horowitz M, Williams J (1985) Chromosome 1 abnormalities: a common feature of pediatric solid tumors. J Natl Cancer Inst 75:51–54
Du Boulay CEH (1985) Immunohistochemistry of soft tissue tumors: A review. J Pathol 146:74–94
Duesberg PH (1987) Retroviruses as carcinogens and pathogens: expectations and reality. Cancer Res 47:1199–1220
Enzinger FM, Weiss SW (1983) Soft tissue tumors. C.V. Mosby, St. Louis
Evans HL, Soule EH, Winkelmann RK (1979) Atypical lipoma, atypical intramuscular lipoma and well differentiated retroperitoneal liposarcoma. Cancer 43:574–584
Ewing J (1921) Diffuse endothelioma of bone. Proc NY Pathol Soc 21:17
Ferti-Passantonopoulou A, Panani A, Avgerinos A, Raptis S (1986) Cytogenetic findings in a large bowel adenocarcinoma. Cancer Genet Cytogenet 21:361–364
First International Workshop on Chromosomes in Solid Tumors (1986) Cancer Genet Cytogenet 19:1–197
First MIC Cooperative Study Group (1986) Morphologic, immunologic and cytogenetic (MIC) working classification of acute lymphoblastic leukemias. Cancer Genet Cytogenet 23:189–197
Fite GL (1945) Classification of tumors of the kidney. Arch Pathol 39:37–41
Gale RP, Canaani E (1984) An 8-kilobase *abl* RNA transcript in chronic myelogenous leukemia. Proc Natl Acad Sci USA 81:5648–5652
Gibas LM, Gibas Z, Sandberg AA (1984) Technical aspects of cytogenetic analysis of human solid tumors. Karyogram 10:25–27
Gibas Z, Prout GR Jr, Connolly JG, Pontes JE, Sandberg AA (1984a) Nonrandom chromosomal changes in transitional cell carcinoma of the bladder. Cancer Res 44:1257–1264
Gibas Z, Prout GR Jr, Sandberg AA (1984b) Malignant teratoma of the testis with an isochromosome no. 12, i(12p), as the sole structural cytogenetic abnormality. J Urol 131:762–763
Gibas Z, Prout GR, Pontes JE, Connolly JG, Sandberg AA (1986) A possible specific chromosome change in transitional cell carcinoma of the bladder. Cancer Genet Cytogenet 19:229–238
Gibas Z, Griffin CA, Emanuel BS (1987) Trisomy and i(5p) in a transitional cell carcinoma of the ureter. Cancer Genet Cytogenet 25:369–370
Gibas Z, Griffin CA, Emanuel BS (1988) Clonal chromosome rearrangements in a uterine myoma. Cancer Genet Cytogenet 32:19–24
Griffin CA, Emanuel BS (1987) Translocation (X;18) in a synovial sarcoma. Cancer Genet Cytogenet 26:181–183
Groffen J, Stephenson JR, Heistercamp N, DeKlein A, Barton CR, Grosveld G (1984)

Philadelphia chromosomal breakpoints are clustered within a limited region, bcr, on chromosome 22. Cell 36:93–99

Hagemeijer A, Smit EME, Bootsma D (1979) Improved identification of chromosomes of leukemic cells in methotrexate-treated cultures. Cytogenet Cell Genet 23:208–212

Hecht F, Glover TW (1984) Cancer chromosome breakpoints and common fragile sites induced by aphidicolin. Cancer Genet Cytogenet 13:185–188

Hecht F, Sutherland GR (1984) Fragile sites and cancer breakpoints. Cancer Genet Cytogenet 12:179–181

Heim S, Mandahl N, Kristoffersson U, Mitelman F, Rööser B, Rydholm A, Willén H (1986) Reciprocal translocation t(3;12)(q27;q13) in lipoma. Cancer Genet Cytogenet 23:301–304

Heim S, Mandahl N, Kristoffersson U, Mitelman F, Rööser B, Rydholm A, Willén H (1987) Marker ring chromosome. A new cytogenetic abnormality characterizing lipogenic tumors? Cancer Genet Cytogenet 24:319–326

Heim S, Nilbert M, Vanni R et al. (1988) A specific translocation, t(12;14)(q14–15;q23–24), characterizes a subgroup of uterine leiomyomas. Cancer Genet Cytogenet 32:13–17

Heisterkamp N, Stam K, Groffen J, de Klein A, Grosveld G (1985) Structural organization of the *bcr* gene and its role in the PH translocation. Nature 315:758–761

Israel MA (1986) The evolution of clinical molecular genetics. Neuroblastoma as a model tumor. Am J Pediatr Hematol Oncol 8:163–172

Jenkyn DJ, McCartney AJ (1987) A chromosome study of three ovarian tumors. Cancer Genet Cytogenet 26:327–337

Karakousis CP, Dal Cin P, Turc-Carel C, Limon J, Sandberg AA (1987) Chromosome changes in soft tissue sarcomas: a new diagnostic parameter. Arch Surg 122:1257–1260

Katsuyama J, Papenhausen PR, Herz F, Gazivoda P, Hirano A, Koss LG (1986) Chromosome abnormalities in meningiomas. Cancer Genet Cytogenet 22:63–68

Kloetzer W, Kurzrock R, Smith L, Talpaz M, Spiller M, Gutterman J, Arlinghaus R (1985) The human cellular *abl* gene product in the chronic myelogenous leukemia cell line K562 has an associated tyrosine protein kinase activity. Virology 140:230–238

Konopka JB, Watanabe SM, Witte ON (1984) An alteration of the c-*abl* protein in K562 leukemia cells unmasks associated tyrosine kinase activity. Cell 37:1035–1042

Kovacs G, Szücs S, De Riese W, Baumgartel H (1987) Specific chromosome aberration in human renal cell carcinoma. Int J Cancer 40:171–178

Kusyk C, Edwards CL, Arrighi FE, Romsdahl MM (1979) Improved method for cytogenetic studies of solid tumors. J Natl Cancer Inst 63:1199–1203

Lai J, Savary J, Deminatti M, Demaille M, Baranzelli M (1987) Translocation (2;13)(q37;q14) in rhabdomyosarcoma: a new case. Cancer Genet Cytogenet 25:371–372

Lasfargues EY (1973) Human mammary tumors. In: Kruse PF, Patterson MK (eds) Tissue culture. Academic Press, New York, pp 45–50

LeBeau MM (1986) Chromosomal fragile sites and cancer-specific rearrangements. Blood 67:849–858

LeBeau MM, Rowley JD (1984) Heritable fragile sites in cancer. Nature 308:607–608

Leder L-D, Richter HJ, Stambolis C (1979) Pathology of renal and adrenal neoplasm. In: Löhr E (ed) Renal and adrenal tumors. Pathology, radiology, ultrasonography, therapy, immunology. Springer, Berlin Heidelberg New York

Lenoir GM, Preud'Homme JL, Bernheim A, Berger R (1982) Correlation between immunoglobulin light chain expression and variant translocation in Burkitt's lymphoma. Nature 298:474–476

Limon J, Dal Cin P, Sandberg AA (1986) Application of long-term collagenase disaggregation for the cytogenetic analysis of human solid tumors. Cancer Genet Cytogenet 23:305–313

Lipinski M, Braham K, Philip I et al. (1987) Neuroectoderm-associated antigens on Ewing's sarcoma cell lines. Cancer Res 47:183–187

Lizard-Nacol S, Mugneret F, Volk C, Turc-Carel C, Favrot M, Philip T (1987) Translocation (2;13)(q37;q14) in alveolar rhabdomyosarcoma: a new case. Cancer Genet Cytogenet 25:373–374

Mackenzie DH (1981) The myxoid tumors of somatic soft tissues. Am J Surg Pathol 5:443–458

Mandahl N, Heim S, Johansson B et al. (1987) Lipomas have characteristic structural chromosomal rearrangements of 12q13–14. Int J Cancer 39:685–688
Mandahl N, Heim S, Arheden K, Rydholm A, Willén H, Mitelman F (1988) Rings, dicentrics, and telomeric association in histiocytomas. Cancer Genet Cytogenet 30:23–33
Mark J (1977) Chromosomal abnormalities and their specificity in human neoplasia: an assessment of recent observations by banding techniques. Adv Cancer Res 24:165–222
Mark J, Dahlenfors R (1986) Cytogenetical observations in 100 human benign pleomorphic adenomas: specificity of the chromosomal aberrations and their relationship to sites of localized oncogenes. Anticancer Res 6:299–308
Mark J, Mitelman F, Dencker H, Norryd C, Tranberg KG (1973) The specificity of the chromosomal abnormalities in human colonic polyps. Acta Pathol Microbiol Scand [A] 81:85–90
Mark J, Dahlenfors R, Ekedahl C, Stenman G (1982) Chromosomal patterns in a benign human neoplasm, the mixed salivary gland tumor. Hereditas 96:141–148
Mark J, Dahlenfors R, Ekedahl C (1983) Cytogenetics of the human mixed salivary gland tumor. Hereditas 99:115–129
Mertens F, Johansson B, Mandahl N, Heim S, Rydholm A, Willén H, Mitelman F (1987) Clonal chromosome abnormalities in two cases of liposarcoma. Cancer Genet Cytogenet 28:137–144
Miles J, Michalski K, Koube M, Weaver DJ (1988) Genomic defects in nonfamilial renal cell carcinoma: a possible specific chromosome change. Cancer Genet Cytogenet 34:135–142
Navas-Palacios JJ, Aparicio-Duque R, Valdes MD (1984) On the histogenesis of Ewing's sarcoma. An ultrastructural, immunohistochemical, and cytochemical study. Cancer 53:1882–1901
Nowell PC (1986) Mechanisms of tumor progression. Cancer Res 46:2203–2207
Nowell PC, Croce CM (1986) Chromosomes, genes, and cancer. Am J Pathol 125:8–15
Ochi H, Takeuchi J, Holyoke D, Sandberg AA (1983) Possible specific chromosome changes in large bowel cancer. Cancer Genet Cytogenet 10:121–122
Olsen S (1984) Tumors of the kidney and urinary tract. Color atlas and textbook. W.B. Saunders, Philadelphia
Passege E, Morone C, Pezzoli V, Ghislette M (1987) Massive renal adenomatosis: a case exhibiting low grade malignancy. Hum Pathol 18:859–863
Pathak S, Strong LC, Ferrell RE, Trinidade E (1982) Familial renal cell carcinoma with a 3;11 chromosome translocation limited to tumor cells. Science 217:939–941
Potluri VR, Gilbert F (1985) A cytogenetic study of embryonal rhabdomyosarcoma. Cancer Genet Cytogenet 14:169–173
Reichmann A, Martin P, Levin B (1982) Karyotypic findings in a colonic villous adenoma. Cancer Genet Cytogenet 7:51–57
Reichmann A, Martin P, Levin B (1985) Chromosomal banding patterns in human large bowel adenomas. Hum Genet 70:28–31
Rey JA, Bello JM, de Campos JM, Kusak E, Moreno S (1988) Chromosomal involvement secondary to −22 in human meningiomas. Cancer Genet Cytogenet 33:275–290
Ross JC, Hendrickson MR, Azumi N, Kempson RL (1986) The problem of the poorly differentiated sarcoma. In: Fer MF, Greco FA, Oldham RK (eds) Poorly differentiated neoplasms and tumors of unknown origin. Grune & Stratton, New York, pp 217–269
Rowley JD (1983) Identification of the constant chromosome regions involved in human hematologic malignant disease. Science 216:749–751
Rowley JD (1984) Biological implications of consistent chromosome rearrangements in leukemia and lymphoma. Cancer Res 44:3159–3168
Sandberg AA (1980) The chromosomes in human cancer and leukemia. Elsevier North-Holland, New York
Sandberg AA (1985) Application of cytogenetics in neoplastic diseases. In: CRC Critical Reviews in Clinical Laboratory Sciences, vol. 22. CRC Press, Ohio
Sandberg AA (1986a) The chromosomes in human leukemia. Semin Hematol 23:201–217
Sandberg AA (1986b) Chromosome changes in bladder cancer: clinical and other correlations. Cancer Genet Cytogenet 19:163–175

Sandberg AA, Turc-Carel C (1987) The cytogenetics of solid tumors. Relation to diagnosis, classification and pathology. Cancer 59:387–395
Sandberg AA, Gibas Z, Saren E, Li FP, Limon J, Tebbi CK (1986) Chromosome abnormalities in two benign adipose tumors. Cancer Genet Cytogenet 22:55–61
Sandberg AA, Turc-Carel C, Gemmill RM (1988) Chromosomes in solid tumors and beyond. Cancer Res 48:1049–1059
Second International Workshop on Chromosomes in Solid Tumors (1987) Cancer Genet Cytogenet 28:1–54
Second MIC Cooperative Study Group (1988) Morphologic, immunologic and cytogenetic (MIC) working classification of acute myeloid leukemias. Cancer Genet Cytogenet 30:1–15
Seidel T, Mark J, Hagmar B, Angervall L (1982) Alveolar rhabdomyosarcoma: a cytogenetic and correlated cytological and histological study. Acta Pathol Microbiol Immunol Scand 90:345–354
Shtivelman E, Lifshitz B, Gale RP, Canaani E (1985) Fused transcript of *abl* and *bcr* genes in chronic myelogenous leukaemia. Nature 315:550–554
Smith S, Reeves BR, Wong L (1987) Translocation t(12;16) in a case of myxoid liposarcoma. Cancer Genet Cytogenet 26:185–186
Sutherland GR, Hecht F (1985) Fragile sites on human chromosomes. Oxford University Press, New York
Szücs S, Müller-Brechlin R, De Riese W, Kovacs G (1987) Deletion 3p: the only chromosome loss in a primary renal cell carcinoma. Cancer Genet Cytogenet 26:369–373
Third MIC Cooperative Study Group (1988) Recommendations for a morphologic, immunologic and cytogenetic (MIC) working classification of the primary and therapy-related myelodysplastic disorders. Cancer Genet Cytogenet 32:1–10 (1988)
Trent J, Casper J, Meltzer P, Thompson F, Fogh J (1985) Nonrandom chromosome alterations in rhabdomyosarcoma. Cancer Genet Cytogenet 16:189–197
Triche TJ, Askin FB (1983) Neuroblastoma and the differential diagnosis of small-round-blue-cell tumors. Hum Pathol 14:569–595
Triche TJ, Askin FB, Kissane JM (1986) Neuroblastoma, Ewing's sarcoma, and the differential diagnosis of small-, round-, blue-cell tumors. In: Finegold M (ed) Pathology of neoplasia in children and adolescents, vol. 18. W.B. Saunders, Philadelphia, pp 145–195
Turc-Carel C, Philip I, Berger MP, Philip T, Lenoir GM (1983) Translocation (11;22) (q24;q12) in Ewing sarcoma cell lines. N Engl J Med 309:497–498
Turc-Carel C, Lizard-Nacol S, Justrabo E, Favrot M, Philip T, Tabone E (1986a) Consistent chromosomal translocation in alveolar rhabdomyosarcoma. Cancer Genet Cytogenet 19:361–362
Turc-Carel C, Dal Cin P, Rao U, Karakousis C, Sandberg AA (1986b) Cytogenetic studies of adipose tissue tumors. I. A benign lipoma with reciprocal translocation t(3;12) (q28;q14). Cancer Genet Cyotgenet 23:283–289
Turc-Carel C, Limon J, Dal Cin P, Rao U, Karakousis C, Sandberg AA (1986c) Cytogenetic studies of adipose tissue tumors. II. Recurrent reciprocal translocation t(12;16) (q13;p11) in myxoid liposarcomas. Cancer Genet Cytogenet 23:291–299
Turc-Carel C, Dal Cin P, Limon J et al. (1987) Involvement of chromosome X in primary cytogenetic change in human neoplasia: nonrandom translocation in synovial sarcoma. Proc Natl Acad Sci USA 84:1981–1985
Turc-Carel C, Dal Cin P, Boghosian L, Terk-Zakarian J, Sandberg AA (1988) Consistent breakpoints in region 14q22–q24 in uterine leiomyoma. Cancer Genet Cytogenet 32:25–31
Ueda T, Aozasa K, Tsujimoto M, Uchida A, Taniwaki M, Abe T (1988) Translocation X;18 and insertion 15;11 in a case of synovial sarcoma. Cancer Genet Cytogenet 30:183–185
van den Berghe H, Parloir C, Gosseye S, Eglebienne V, Cornu G, Sokal G (1976) Variant translocations in Burkitt lymphoma. Cancer Genet Cytogenet 1:9–14
Vanni R, Lecca U (1988) Involvement of the long arm of chromosomes 12 in chromosome rearrangements of uterine leiomyoma. Cancer Genet Cytogenet 32:33–34
Wake N, Slocum HK, Rustum YM, Matsui SI, Sandberg AA (1981) Chromosomes and causation of human cancer and leukemia. XLIV. A method for chromosome analysis of solid tumors. Cancer Genet Cytogenet 3:1–10

Wang N, Perkins KL (1984) Involvement of band 3p14 in t(3;8) hereditary renal carcinoma. Cancer Genet Cytogenet 11:469–481

Wang-Wuu S, Soukup SW, Lange BJ (1987) Another synovial sarcoma with t(X;18). Cancer Genet Cytogenet 26:179–181

Wang-Wuu S, Soukup S, Ballard E, Gotwals B, Lampkin B (1988) Chromosomal analysis of sixteen human rhabdomyosarcomas. Cancer Res 48:983–987

Whang-Peng J, Triche TJ, Knutsen T, Miser J, Douglass EC, Israel MA (1984) Chromosome translocation in peripheral neuroepithelioma. N Engl J Med 311:584–585

Whang-Peng J, Triche TJ, Knutsen T, Miser J, Kao-Shan S, Tsai S, Israel MA (1986) Cytogenetic characterization of selected small round cell tumors of childhood. Cancer Genet Cytogenet 21:185–208

Yamada K, Kondo T, Yoshioka M, Oami H (1980) Cytogenetic studies in twenty human brain tumors: association of no. 22 chromosome abnormalities with tumors of the brain. Cancer Genet Cytogenet 2:293–307

Yoshida MA, Ohyashiki K, Ochi H et al. (1986) Cytogenetic studies of tumor tissue from patients with nonfamilial renal cell carcinoma. Cancer Res 46:2139–2147

Yunis JJ (1976) High resolution human chromosomes. Science 191:1268–1270

Yunis JJ (1981) New chromosome techniques in the study of human neoplasia. Hum Pathol 12:540–549

Yunis JJ (1983) The chromosomal basis of human neoplasia. Science 221:227–236

Zang KD (1982) Cytological and cytogenetical studies on human meningioma. Cancer Genet Cytogenet 6:249–274

Zankl H, Zang KD (1980) Correlations between clinical and cytological data in 1980 human malignant meningiomas. Cancer Genet Cytogenet 1:351–356

Zech L, Haglund V, Nilson N, Klein G (1976) Characteristic chromosomal abnormalities in biopsies and lymphoid cell lines from patients with Burkitt and non-Burkitt lymphomas. Int J Cancer 17:47–56

# Techniques and Applications of In Situ Hybridisation

G. TERENGHI and R. A. FALLON

## 1 Introduction

During the last two decades, morphological investigations have changed considerably with the development of immunocytochemistry. However, the information gained by the demonstration of cellular antigens is limited as this technique does not give any insight into other intracellular metabolic processes. The introduction of in situ hybridisation (ISH) using nucleic acid probes provides further morphological understanding of gene expression in different physiological and pathological conditions.

The term ISH indicates the technique used to localise nucleic acid sequences in whole cells, tissue sections or chromosome preparations. The concept of this technique is intrinsically simple; it is based on the use of labelled nucleic acid probes which are able to associate with complementary RNA or DNA target sequences, to form a hybrid molecule. ISH was first applied by GALL and PARDUE (1969) to identify ribosomal intracellular DNA in cytological preparations by using a radiolabelled probe. Subsequently the technique was applied to a variety of experimental studies (see COGHLAN et al. 1985 for a review), but for years its widespread application has been limited by several factors, not least a lack of consensus concerning the methodology and, sometimes, the difficulty in reproducing accurately the results of different workers. However, ISH is capable of giving unique information which is not available from other techniques. With the advent of more sophisticated methods for preparing specific probes and recent reports of simpler and more cohesive methodologies, considerable interest has been focussed on ISH. Furthermore, the sensitivity of ISH has been considerably improved, with a limit of detection down to single copy gene sequences (BHATT et al. 1988; LAWRENCE et al. 1988).

In this chapter we give an updated review of ISH by examining separately the various aspects of methodology. Although the final choice of procedure details will obviously depend on specific applications, general guidelines are applicable to most types of investigation. As ISH requires the integration of methodologies from different disciplines (i.e. molecular biology, histology, histochemistry, virology etc.), it is essential to simplify the procedure since an easy method is the most likely to give consistently good results.

## 2 Tissue Preparation

The most important requirements for ISH are nucleic acid retention and morphology preservation. These points could be achieved through appropriate fixation, although the final hybridisation signal might be dependent on a variety of factors such as the type of sample, the rate of target degradation, the choice of fixative and the method of processing.

## 2.1 Target Preservation

The stability of the RNA or DNA target is related to external factors such as the presence of endogenous nucleases, handling and, most important of all, the time delay before fixation. The half-life of mRNA and DNA may vary considerably. Different nucleic acid sequences show different degradation curves, and from the studies carried out on this subject it seems likely that several factors may contribute to their stability (BELASCO et al. 1986; SHAW and KAMEN 1986; STANSSENS et al. 1986; BRAWERMAN 1987; PONTECORVI et al. 1988). Endogenous nucleases certainly play a major role in the degradation process, and it is important to inactivate them during processing of the sample and to avoid external nuclease contamination during ISH procedures. In particular RNase, a ubiquitous and very heat-stable enzyme, can degrade target mRNA and RNA probes at any stage of the ISH procedure. Hence all equipment used for ISH should be RNase-free, this being achieved in different ways: gloves should be worn throughout all manipulations, glassware and other heat-resistant equipment is to be baked at 250° C for a minimum of 4 h (MANIATIS et al. 1982; CUMMING and FALLON 1989), plastic disposable equipment and solutions should be autoclaved before use, and RNase inhibitors (i.e. human placental ribonuclease inhibitor, RNasin) should be added during probe preparation and ISH. It has also been noted that manipulation of unfixed samples could promote premature nucleic acid degradation, as sample dehydration and loss of cellular integrity through routine dissection and cutting may somehow accelerate the degradation processes, possibly by increasing the release of lysosomal contents and of endonucleases (JOHNSON et al. 1986; STOPA et al. 1989). Similar degradation events might also take place in necrotic areas of tumour samples.

It is recommended that fixation delay should be always kept to a minimum. Unfortunately there is no definite rule to define an acceptable time limit, as all the points examined above may contribute to the final result. However, it is probably acceptable to carry out the fixation within 10–20 min from tissue collection, as longer delays result in a decrease of hybridisation signal (HÖFLER et al. 1986 a). When dealing with experimental animals the tissue can be fixed immediately by immersion or perfusion. For clinical samples, arrangements should be made to collect the tissues from the ward or operating theatre in a closed container placed on water-ice in order to minimise tissue dehydration and premature nucleic acid degradation.

The use of post-mortem material is particularly important in neuropathology where there is a lack of suitable animal models for many diseases. Particular attention has to be paid to the stability of mRNA in these samples, as several non-controllable pre-mortem events (stress, pain, drug administration) might adversely affect the rate of nucleic acid degradation. The suitability of post-mortem tissue for ISH has been investigated by measuring the RNA yield and its integrity. There appears to be little variation of the yield of RNA with long post-mortem delay (up to 50 h), and of the stability of mRNA as measured by densitometry on Northern blots (JOHNSON et al. 1986; TAYLOR et al. 1986). Interestingly, the levels of RNA yield from human brain are generally lower than those from animal models, suggesting that pre- and/or post-mortem events might contribute to the RNA stability (JOHNSON et al. 1986; TAYLOR et al. 1986). Successful ISH for peptide mRNA has been carried out on human brain collected up to 4 h post-mortem and immediately fixed (TERENGHI et al. 1987; STOPA et al. 1989). However, for a correct interpretation of the hybridisation signal it is advisable to establish experimentally a post-mortem degradation curve for each type of mRNA, as there is evidence to show that different molecules behave differently.

## 2.2 Fixation

The collection and fixation of tissue represents an important and sometimes overlooked preparatory step for ISH. The aim is to preserve an intact tissue morphology throughout the hybridisation procedure, whilst the target nucleic acid must be retained in a form accessible to the probe.

In general fixatives for ISH can be grouped in two classes, cross-linking (e.g. formaldehyde glutaraldehyde) and precipitating (e.g. ethanol, methanol, acetone). Many authors prefer precipitating reagents, as it is thought that the cross-linking might prevent easy access of the probe to the target. However, precipitating reagents often do not preserve the tissue morphology efficiently, with a consequent possibility of target loss by diffusion (LAWRENCE and SINGER 1985; SINGER et al. 1986; MOENCH et al. 1985). Conversely, the use of cross-linking fixatives may require the use of permeabilisation steps before the hybridisation procedure in order to increase probe penetration. Several studies have been carried out on the subject (HAASE et al. 1984; MCALLISTER and ROCK 1985; MOENCH et al. 1985; HÖFLER et al. 1986a; SINGER et al. 1986; GUITTENY et al. 1988), but using different systems and showing conflicting results on the suitability of one fixative over the other. More recently paraformaldehyde has been used as fixative for hybridisation of peptide mRNA (HÖFLER et al. 1986a; HAMID et al. 1987; TERENGHI et al. 1987; LARSSON et al. 1988; STEEL et al. 1988). An added advantage of this fixative is the ability to perform immunocytochemical and hybridisation studies on the same tissue preparation (BRAHIC et al. 1984; HÖFLER et al. 1986b, 1987; SHIVERS et al. 1986a).

We have tested a panel of fixatives, previously shown to be suitable for ISH, on a range of tissue and cell preparations in order to determine whether one specific reagent could be recommended as suitable for different ISH applications. Table 1 summarises the type of fixatives tested and the corresponding fixation times. The experiment included both precipitating and cross-linking fixatives, as well as reagents which combine the two in different proportions. ISH was carried out on human and experimental

**Table 1.** Panel of reagents and fixation time tested on a variety of tissues and cell lines

| Fixatives | Fixation time | |
|---|---|---|
| | Cells | Tissue |
| 1. Ethanol/acetic acid (3:1 v/v) + ethanol | 15 min R.T.[a] 5 min R.T. | 30 min R.T. 10 min R.T. |
| 2. Ethanol/acetic acid (95%:5% v/v) | 15 min R.T. | 30 min R.T. |
| 3. Methanol/acetone (1:2 v/v) | 4 min −20° C | 20 min −20° C |
| 4. Methanol/acetone (1:1 v/v) | 4 min −20° C | 20 min −20° C |
| 5. Bouin's solution | 30 min R.T. | 1 h R.T. |
| 6. 4% Paraformaldehyde (PF) | 30 min R.T. | 1 h R.T. |
| 7. 2% Glutaraldehyde | 30 min R.T. | 1 h R.T. |
| 8. Paraformaldehyde–lysine–periodate (PLP) | 15 min R.T. | 1 h R.T. |
| 9. 4% Paraformaldehyde/methanol (1:9) | 30 min R.T. | 1 h R.T. |
| 10. 4% Glutaraldehyde in 20% ethylene glycol | 5 min 4° C | 30 min 4° C |

[a] R.T., room temperature

**Table 2.** Each tissue and cell line was fixed with the full panel of reagents of Table 1, and hybridised with a suitable probe. In each system the best hybridisation results were obtained following fixation with a specific reagent, as listed below

| Tissue/cells | Probe | Best fixative |
|---|---|---|
| Placenta | Human placenta lactogen (HPL) | 2% Glutaraldehyde |
| Pituitary | Pro-opiomelanocortin (POMC) | 4% Paraformaldehyde |
| Brain | Neuropeptide Y (NPY) | 4% Paraformaldehyde |
| A 431 | Actin | Bouin's |
| T 15 | N-ras | 4% Paraformaldehyde |
| ZNR | Actin | Ethanol/acetic ac. (95:5) |
| Ψ2PD | Actin | Ethanol/acetic ac. (95:5) |
| Chicken embryo fibroblasts | Actin | Bouin's |

animal tissues, and on cultured cells using a variety of probes. The hybridisation results were evaluated in terms of both morphology preservation and hybridisation efficiency, this being assessed as signal-to-noise ratio (Table 2). Although all fixatives present good morphological details of the tissue before hybridisation, it is apparent that following the rigorous proce-

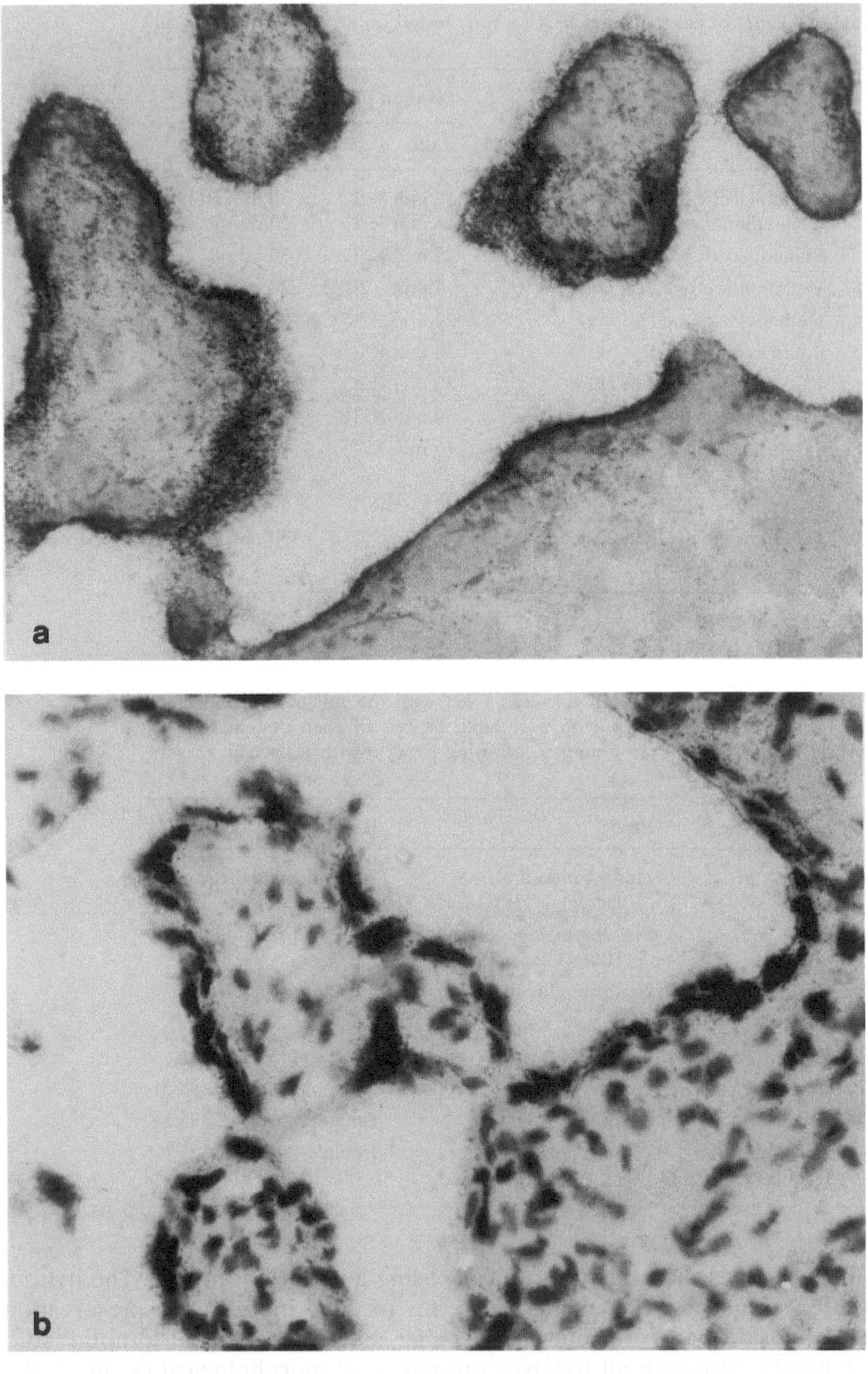
a
b

dure required for ISH the results are discordant for different tissues, as shown for placenta in Fig. 1. When using methanol/acetone fixatives, the tissue morphology is unrecognisable and the hybridisation signal is practically absent, most probably due to loss by diffusion of target mRNA. In contrast, tissue fixed in 2% glutaraldehyde shows a very strong hybridisation signal and good morphological detail. Between these two extremes there is a range of results with different fixatives, but generally the best results are seen with cross-linking fixatives, consistent with the results seen on pituitary and brain. Similar differences were also seen in cultured cells.

Some general indications can be gained from the results of this study. The use of cross-linking fixatives appears to give better final results on tissue blocks, where the complexity and heterogeneity of the cell population might not be sufficiently preserved with precipitating reagents. However, in monolayers of cultured cells precipitating fixatives might be equally suitable, which offers the possibility of avoiding laborious permeabilisation steps before ISH (see Sect. 4.1).

At present it appears that there is not a unique fixation procedure which could be recommended for ISH, and the final choice of fixative depends on the tissue investigated, the rate of target degradation and possibly the type of probe and its length. Hence the choice of fixative for a given tissue should be carried out experimentally by screening using a panel of different fixatives. When this is not possible (e.g. in retrospective clinical studies on formalin-fixed material) the hybridisation protocol should be optimised in order to obtain the maximum hybridisation efficiency possible. It is also notable that prolonged exposure to fixative might reduce the hybridisation signal (Brigati et al. 1983; Wilcox et al. 1986), and proteolytic treatment of the section might be necessary, particularly if strong cross-linking fixatives and paraffin embedding have been used or if deproteination of viral capsid is necessary (Gowans et al. 1981; Brigati et al. 1983; Larsson et al. 1988). Also, for a rapid and uniform fixation of samples such as brain where the fixative penetration is poor, it is advisable, when possible, to fix the tissue by perfusion.

◀ **Fig. 1 a, b.** Cryostat sections of human placenta fixed in 2% glutaraldehyde (**a**) (× 210) and paraformaldehyde-lysine-periodate (PLP) (**b**) (× 210). The sections have been hybridised with human placenta RNA probes labelled with $^{35}S$ and counterstained with haematoxylin and eosin. The glutaraldehyde-fixed placenta shows strong hybridisation signal in the syncytial trophoblasts surrounding the chorionic villi, while in PLP-fixed tissue the cells show a weaker signal intensity

## 2.3 Tissue Processing

The fixed material can be stored safely in washing buffer (e.g. PBS) at 4° C for 1–2 months before sectioning. However, it is suggested that cryostat blocks should be prepared from the fixed tissue as soon as possible, to minimise the risk of bacterial contamination and target degradation. The frozen blocks can be stored for much longer in liquid nitrogen or at −70° C. Cryostat sections are collected onto RNase-free clean glass slides, coated with poly-L-lysine (PLL) (HUANG et al. 1983), air dried and can be stored dry at −70° C for up to 1 year without appreciable loss of nucleic acid target.

An alternative method of tissue processing consists in freezing fresh tissue, the fixation being carried out on glass-mounted frozen sections (HAASE et al. 1984; COGHLAN et al. 1985; GOWANS et al. 1989). Although this method offers the advantage that frozen tissue can be used also for RNA extraction, the morphology of the tissue after ISH is not satisfactory. Furthermore, the handling of unfixed material is more laborious, and a sudden rise of temperature is more likely to compromise the nucleic acid stability.

In situ hybridisation has been carried out successfully on wax-embedded tissue (BRIGATI et al. 1983; PRZEPIORKA and MYERSON 1986; MORLEY and HODES 1988). It has been claimed that this type of tissue processing does not alter significantly the sensitivity of the ISH procedure (HAASE et al. 1984; COGHLAN et al. 1985; TOURNIER et al. 1987; GOWANS et al. 1989; LARSSON et al. 1988), but a comprehensive study comparing hybridisation on wax and frozen sections has not yet been carried out.

Cultured cells can be grown directly onto glass slides or coverslips placed at the bottom of tissue culture petri dishes. The cells are rinsed in isotonic buffer to remove serum and then fixed by placing the slides directly into the fixative. Following fixation the preparations are rinsed in buffer, and dried at 37° C for at least 4 h before ISH (HAMID et al. 1987; LAWRENCE and SINGER 1986). With this method there is no time delay before fixation and virtually no risk of nucleic acid degradation. Also the cell monolayer makes identification of the spatial relationship between cells and the distribution of hybridised target easier. Alternatively, cells which grow in suspension or a trypsinised cell suspension can be cytospun onto PLL-coated slides (approx. $2 \times 10^4$ cell/slide) before fixation (see CUMMING and FALLON 1989 for detailed method).

# 3 Probes

There are many options and considerations when choosing a probe and probe label for ISH experiments. Decisions are dependent on experimental requirements for sensitivity, speed, quantitation and resolution. Workers who are unfamiliar with molecular biology should be aware of the types of probe available, the options with respect to label and the appropriate labelling procedures since there is no universal system appropriate for all applications. Pre-labelled probes are commercially available, but these may limit the range of planned experiments. Some workers therefore choose to synthesise and/or clone their own probes whilst others are able to obtain their probes through academic collaborations.

There are particular requirements for probes used in ISH experiments. Since the signal generated does not give any idea of the size of the target nucleic acid, the specificity of probes must be thoroughly checked prior to use (see Sect. 6.2). Probes should also fall within a certain size range to facilitate entry into whole cells or tissue sections (BRAHIC and HAASE 1978; ANGERER and ANGERER 1981; LAWRENCE and SINGER 1985). For maximum sensitivity, the highest specific activity radioactive probes are currently required. In addition, quantitation of signal may be desired (see Sect. 5.1).

General guidelines for choosing the probe sequence and checking the specificity are given whilst detailed methodology for synthesis, cloning and sequencing are beyond the scope of this chapter. In addition, advantages and disadvantages of a range of probe types and probe labels are also discussed.

### 3.1 Choice of Probe Sequence

Some general guidelines are outlined below when selecting a suitable probe sequence from published data for synthesis and cloning (mistakes in published sequences are a real hazard!).

A large fraction of the DNA of most eukaryotes consists of repeated nucleotide sequences (LEWIN 1980). In selecting a specific probe, repetitive sequence should be avoided since the probe may also bind to multiple irrelevant genomic sequences.

Gene sequence homologies exist between species (CLEVELAND et al. 1980; ISENBERG 1979), some oncogenes (RALSTON and BISHOP 1984), certain virus types (GISSMAN and SCHWARZ 1985) and gene families (LEVINE et al. 1988) for example. The greatest variation between related gene sequences usually lies within the 5′ or 3′ untranslated regions, which are less conserved than those encoding protein. Probe sequences selected from these regions can therefore specify a particular species-specific probe for example. Where continuous regions of diverse sequence are too short, high-stringency conditions and/or short oligonucleotide probes may be enough to establish probe specificity.

The melting temperature (Tm) of hybrids is altered according to probe length and the specific sequence. GC base pairs are stabilised by three hydrogen bonds whereas AT base pairs contain two. For example a probe rich in AT residues has a decreased Tm compared with a probe having an equal number of GC residues. Tm is also reduced as a function of probe length whereby the shorter the probe, the lower the Tm, although this is not a linear relationship (BRITTEN and DAVIDSON 1985).

When choosing a DNA template sequence for RNA probes, inverted repeats or transcriptional stop sequences should be avoided since premature termination of transcription will occur during probe synthesis.

A range of probe sizes from 100 to 400 bases are, in our experience, more suitable for ISH using freshly fixed cultured cells (see Sect. 4.2.1 for further discussion). When using RNA probes, it is convenient to select a template sequence within this range of sizes, thus avoiding the need to hydrolyse larger probes such that they fall within this size range (COX et al. 1984). Clearly longer probes hydrolysed to the correct size offer more target coverage and therefore more sensitive detection.

When choosing an effective probe for clinical diagnostic purposes, additional considerations are necessary. In particular, care must be taken to ensure that the probe does not cross-react with non-pathogenic species of the same genus. Ideally, a sequence specifying virulence is desirable (PALMER and FALKOW 1985).

## 3.2 Probe Types

There are three main types of probe currently in use for ISH: DNA probes (double or single stranded), RNA probes and oligonucleotide probes. RNA and DNA probes are produced from a cloned sequence whilst oligonucleotide probes are chemically synthesised using automated DNA synthesisers.

### *3.2.1 DNA Probes*

Double-stranded complementary DNA probes (cDNA) are commonly used for ISH (LAWRENCE and SINGER 1985). Generally, these sequences specify gene coding regions (exons) or untranslated regions although, recently, intervening or intron sequences have been used specifically to detect unprocessed heterologous nuclear RNA within the nucleus (FREMEAU et al. 1986). Probe sequences are cloned into bacterial plasmid or bacteriophage vectors, ensuring a constant supply of probe. Once cloned, these probes are simple to prepare and label using defined protocols or commercially available kits.

DNA probes are generally more robust than RNA probes since nucleases specific for DNA (DNases) are more easily eliminated from preparations than RNases. There is some evidence to show that heterologous double-stranded DNA probes of around 1000–1500 bases are able to form networks in cells resulting in signal amplification (LAWRENCE and SINGER 1985). Networking can occur when a hybridising probe strand has a free end capable of base pairing to a complementary region of another probe molecule such that the amount of probe and therefore signal at that site is increased.

Despite these advantages, double-stranded DNA probes have certain drawbacks which effectively decrease signal-to-noise ratios. Use of larger probes (1000–1500 base pairs) impinges on the efficiency of penetration into whole cells or tissue sections such that there might be a trade-off between efficient probe entry and signal amplification through networking (BRAHIC and HAASE 1978; ANGERER and ANGERER 1981).

Double-stranded probes must be denatured by boiling so that complementary single strands can hybridise to the single-stranded target molecules. Therefore there is competition between the hybridisation reaction with immobilised target nucleic acid and the reannealing of probe DNA strands in solution. There is evidence suggesting that the latter reaction is favoured, thus decreasing the pool of probe available for hybridisation (COX et al. 1984). It is possible to circumvent this problem whilst retaining some advantages of DNA probes by using single-stranded cDNA probes from M13 although this route is much more labour intensive in terms of probe labelling and purification (GOEDERT 1986).

cDNA probes which include labelled vector sequences can increase the background signal on cell preparations either by vector sequences hybridising to parts of the target tissue genome, or through entrapment of extra labelled nucleic acid within the cell matrix. This problem can be eliminated by purifying the cDNA insert away from the vector sequences using gel electrophoresis.

### *3.2.2 RNA Probes*

Single-stranded RNA probes are increasingly favoured particularly for high sensitivity ISH since they have a number of advantages over double-stranded cDNA probes. The use of RNA probes for ISH was pioneered by COX et al. (1984).

RNA probes are asymmetric, single-stranded probes which therefore do not require denaturation prior to use and so reannealing in solution does not occur, ensuring that all of the probe is available for hybridisation with target. A complementary probe sequence or antisense strand will hybridise with cellular mRNA. A labelled sense strand can be used as a good negative control probe since this, by definition, comprises the same sequence as the mRNA of interest. Strand-specific probes have been useful in identifying viral nucleic acid replicative intermediates (GOWANS et al. 1983). An asymmetric RNA probe will also hybridise to denatured DNA.

The melting temperature of hybrids containing RNA is higher than for DNA : DNA hybrids (RNA : RNA > RNA:DNA > DNA : DNA) (CASEY and DAVIDSON 1977; WETMUR et al. 1981; COX et al. 1984). This means that higher hybridisation and wash temperatures can be used, which still allow efficient specific base pairing between homologous sequences but more effectively prevent or remove weak or non-specific probe binding.

RNA probes are transcirbed in vitro from pieces of DNA inserted adjacent to a promoter or RNA polymerase binding sequence within a bacterial plasmid. They are virtually free of contaminating vector sequences because the plasmid is made linear using a restriction endonuclease just downstream from the inserted sequence (Fig. 7). Saturation of target sequences occurs at lower effective probe concentration when the probe does not contain vector sequences, which implies a higher signal-to-noise ratio since background signal is partly dependent on the concentration of probe applied to the section.

Transcripts generated from the plasmid template are of a defined size. This is an advantage for ISH since probe size is an important factor when considering probe entry into intact cells for example; in this case, optimal probe sizes fall within the 100–400 base range. If the insert size is greater than this, transcripts are easily and reproducibly hydrolysed to the correct size (COX et al. 1984).

Limiting nucleotides in the labelling reaction or transcriptional stop sequences within the probe template can result in prematurely terminated transcripts which may impinge on ultimate sensitivity of the ISH. The integrity of transcripts is easily checked using formaldehyde gel electrophoresis (MANIATIS et al. 1982).

RNA transcripts are readily synthesised using RNA polymerase (MELTON et al. 1984; ANGERER et al. 1985), and kits are commercially available. High specific activity probes can be generated ($\sim 10^9$ dpm/μg) and 70%–80% of radioactively labelled nucleotides incorporated. Incorporation of biotin-labelled nucleotides is less efficient (LUEHRSEN and BAUM 1987), but alternative procedures have been proposed (see Sect. 3.2.2).

The ubiquity of ribonucleases means that special precautions should be taken when handling RNA probes (see Sect. 2.1) (CUMMING and FALLON 1989).

These precautions are readily introduced as routine procedure and are necessary anyway when the target nucleic acid is RNA since non-specific digestion or nicking of target sequences could result in changed stringency requirements.

Rates of DNA : RNA hybridisation are slower than those of DNA : RNA hybridisation (CASEY and DAVIDSON 1977; WETMUR et al. 1981). This is not a particular disadvantage since RNA : RNA hybridisations can be largely complete within 1 h (unpublished data), using high probe concentrations.

### *3.2.3 Oligonucleotide Probes*

Reliable automated DNA synthesisers are available based on solid phase synthesis using deoxynucleoside phosphoramidites and insoluble silica supports (CARUTHERS et al. 1982; PENSCHOW et al. 1986). This has increased the range of researchers using ISH since probes can be selected from published gene sequence data, independently of restriction sites, and synthesised quickly and in large amounts as required (LEWIS et al. 1985; COGHLAN et al. 1985). In addition, short oligonucleotide probes are ideal when distinguishing transcripts from gene families, for example where sequence homologies are particularly high (SCOTT YOUNG III et al. 1986).

Oligonucleotide probes consist of single-stranded DNA, which means that competition between reannealing and hybridisation reactions does not occur. Furthermore, these single-stranded probes are not susceptible to ubiquitous RNases.

Probes can be labelled to relatively high specific activities dependent on the procedure. The size range (~ 17–150 bases) facilitates probe entry without the need for permeabilisation of cells. However, the overall sensitivity of these probes is compromised as target coverage decreases and labelling is not uniform, i.e. a single end label or a labelled short "tail" of nucleotides.

### 3.3 Probe Labelling Procedures

Probe labelling procedures are diverse; some general principles are outlined in this section and appropriate protocols referenced. Most labelling reagents are commercially available in convenient kit formats with adequate literature.

#### *3.3.1 Nick Translation of DNA Probes*

The nick translation reaction (RIGBY et al. 1977) involves the simultaneous action of two enzymes. Pancreatic deoxyribonuclease 1 (DNase 1) introduces nicks randomly in each strand along the length of a DNA molecule, resulting in both free 3′ hydroxyl and free 5′ phosphate groups. *E. coli* DNA polymerase 1 (DNA pol 1) has two enzyme activities which act at the nicked sites. A 5′–3′ exonuclease activity progressively removes nucleotides from the 5′ ends whilst a 5′–3′ polymerase activity successively adds nucleotides to the 3′ ends using the complementary DNA strand as a template. The initial nick is therefore translated along the molecule in a 5′–3′ direction (Fig. 2). Since the DNase 1 introduces nicks randomly, the net result is a uniformly labelled probe molecule. One or more labelled nucleotide triphosphate can be included in the reaction, resulting in radioactive probes of high specific activity, e.g. $> 6.6 \times 10^8 - 10^9$ dpm/μg. Most standard protocols result in an average single-strand probe length of 400–500 bases, although the probe length can be varied by altering the concentration of DNase 1 in the reaction (MOENCH et al. 1985).

Non-radioactively labelled nucleotides such as biotin dUTP are readily incorporated in this way (SINGER and WARD 1982). Microgramme quan-

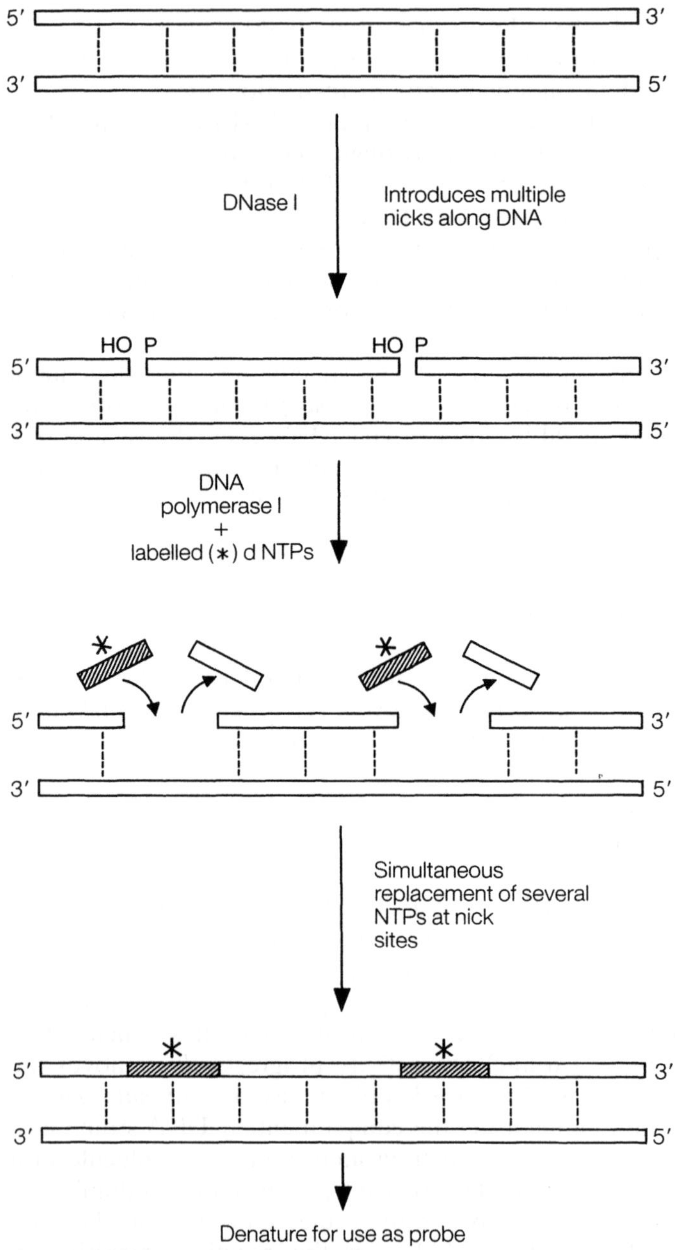


**Fig. 2.** The nick translation reaction

tities of biotin-labelled probe are produced and the probes are stable for many months.

Subcloning is not requiered, and any plasmid containing a sequence of interest can be nick translated. In addition there is no need to linearise the template. Clearly the labelled probe must be denaturted immediately prior to use.

### *3.3.2 Primer Extension of DNA Probes*

There are several approaches to primer extension, utilising different enzymes in combination with unique or random primers. For the purposes of ISH, random priming is increasing in popularity over nick translation largely because it is feasible to incorporate higher specific activity radioactive nucleotides and also because of the ability to label electrophoretically pure probe fragments which may be contaminated with agarose (FEINBERG and VOGELSTEIN 1983, 1984).

This procedure involves the annealing of random hexonucleotides to linear single-stranded DNA template and addition of labelled and unlabelled nucleotides to the 3′ ends of the primers using the Klenow fragment of *E. coli* DNA polymerase I (see Fig. 3).

A single-stranded template is required; therefore double-stranded DNA must be heat denatured prior to annealing the primers. Linear DNA is preferred since complementary circular DNA reanneals more rapidly.

The random primers are generated by DNase 1 digestion of calf thymus or oligonucleotide synthesis. Reagent kits containing the primers, enzyme, reaction buffer etc. are commercially available.

Due to strand displacement, a net synthesis of probe occurs using this technique. Even so, yields of labelled probe are lower than for nick translation; however, since the level of template DNA in the reaction is low, it is feasible to use radioactively labelled nucleotides of higher specific activity, resulting in high specific activity probes, e.g. $5 \times 10^9$ dpm/μg. Biotin-labelled nucleotides can also be incorporated in this way.

The size range of probes generated is generally wider than for a nick-translated probe. Probe size is determined largely by the concentration of labelled nucleotide in the reaction such that the lower the concentration, the shorter the probe fragments. Probe size can be reduced further by sonication. This may be necessary for tritiated probes where the total amount of input label ranges between 20 and 200 pmol with one to four labelled nucleotides and the average size is 400–500 base pairs.

### *3.3.3 Single-Stranded Uniformly Labelled DNA Probes*

Single-stranded uniformly labelled DNA probes avoid the self-reannealing problems discussed previously and are not open to attack by ubiquitous ribonucleases. Technically they are cumbersome to prepare and involve

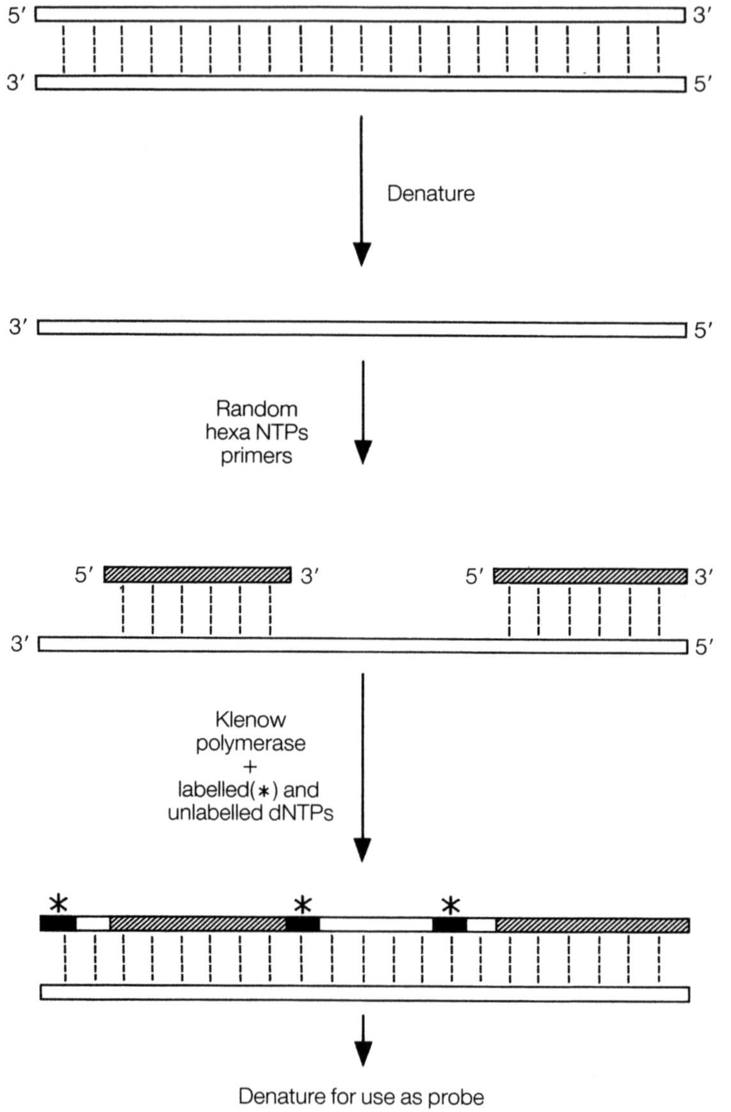


**Fig. 3.** The random primer reaction

subcloning into appropriate vectors. The probes are produced from a single-stranded bacteriophage vector called M 13 which forms a double-stranded replicative intermediate when it infects a bacterial host cell. Both the replicative form and the single-stranded form can be prepared in large quantities (Hu and Messing 1982; Goedert 1986).

The probe sequence of interest is first subcloned into an M 13 replicative form in both orientations, ensuring production of single-stranded phage containing the probe DNA in sense and antisense orientations.

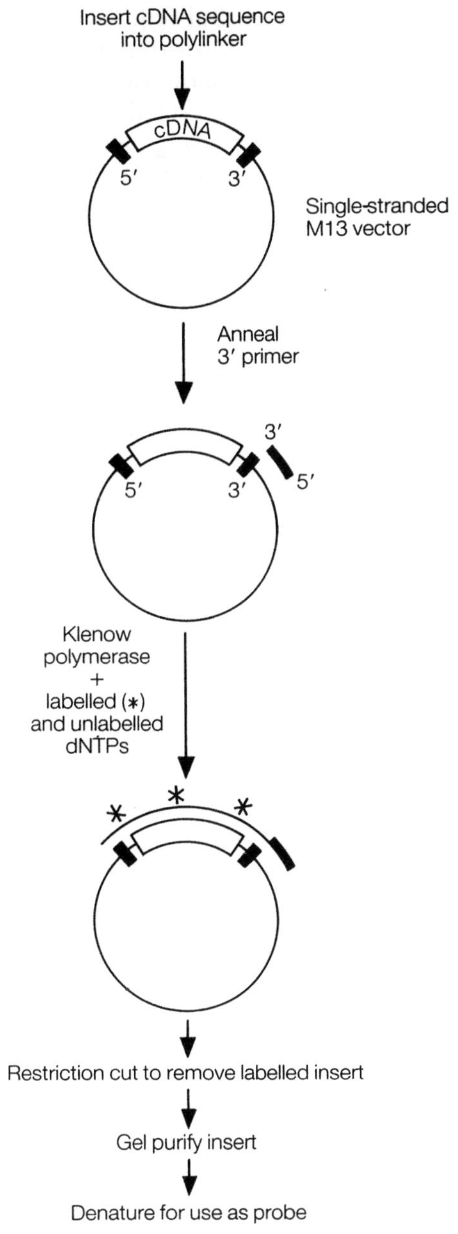


**Fig. 4.** Preparation of strand-specific M 13 probes using the hybridisation primer annealed 3′ to the multiple cloning site

Labelled probes are prepared by annealing an oligonucleotide sequence primer to the single-stranded M 13 5′ to the probe sequence and a mixture of labelled and unlabelled nucleotides are incorporated beginning at the 3′ end of the primer by Klenow polymerase. The amounts of nucleotide supplied to the M 13 template are limited such that only the probe sequence plus a small amount of vector is synthesised. Subsequent digestion with a restriction endonuclease which cuts just downstream of the probe template isolates the probe from vector. Heat denaturation will release the probe from the template. The labelled probe is effectively separated from the probe template and vector sequences using denaturing agarose gel electrophoresis (Fig. 4). The labelled probe fragment can then be purified from the gel and used as a probe.

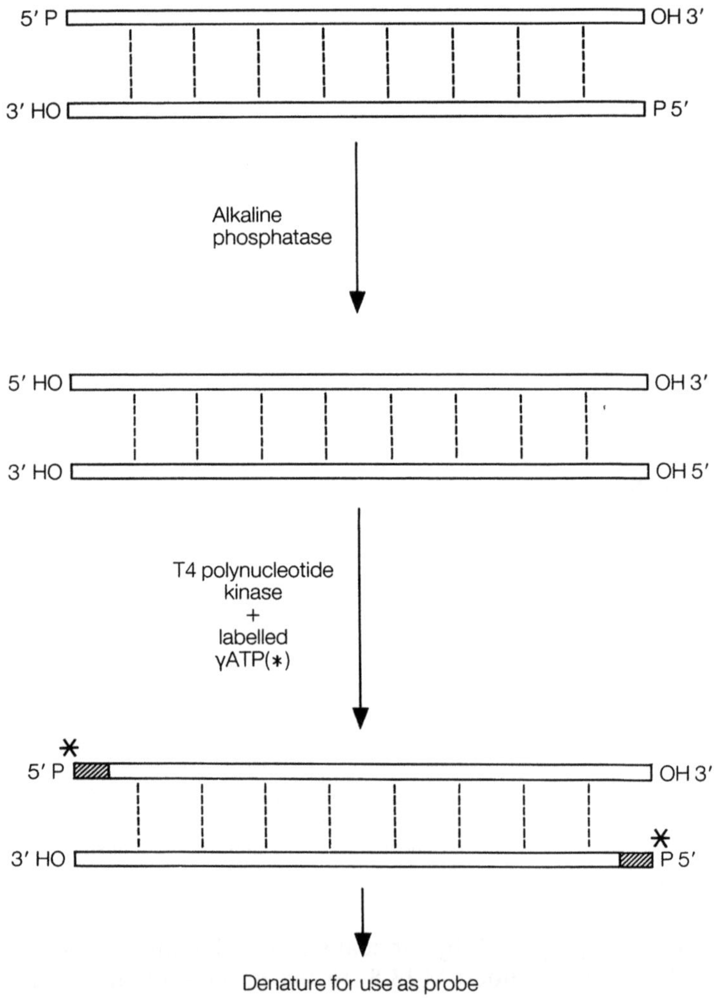


**Fig. 5.** The 5′ end labelling reaction using T4 polynucleotide kinase. NB: Synthetic oligonucleotides do not require treatment with alkaline phosphatase

### 3.3.4 End Labelling of Oligonucleotide Probes

A wide variety of techniques are available for introducing labelled nucleotides to the 5′ or 3′ ends of oligonucleotide probes and reagent kits are commercially available.

T4 polynucleotide kinase catalyses the transfer of a Y phosphate of a ribonucleoside 5′ triphosphate donor to the 5′ hydroxyl group of a synthetic oligonucleotide (MANIATIS et al. 1982) (Fig. 5). γ-ATP with the appropriate label is most frequently used as a donor. This labelling procedure is particularly convenient for synthetic oligonucleotides since they are synthesised with a 5′ hydroxyl group, thus obviating the need to remove a 5′ unlabelled phosphate group.

Alternatively, terminal deoxynucleotidyl transferase (BOLLUM 1974) can be used to add a series of deoxyribonucleotides to the 3′ end of the oligonucleotide (LEWIS et al. 1985) (Fig. 6). This approach results in probes of higher specific activity than 5′ end labelling, which can be an advantage for raising overall sensitivity of ISH reactions towards the level of uniformly labelled probes. Studies have shown that "tailing" specific oligonucleotides in this way does not interfere significantly with stringency requirements (LEWIS et al. 1985).

Modified nucleotides can be introduced during synthesis of an oligonucleotide using commercially available phosphoramidites containing an aminoalkylphosphoramide linker arm. These require subsequent modification to include the desired reporter group such as biotin.

Recently, protocols have become available for producing a specific oligonucleotide which includes a promoter sequence for RNA polymerase (BRYSCH et al. 1988). This approach combines some of the advantages of

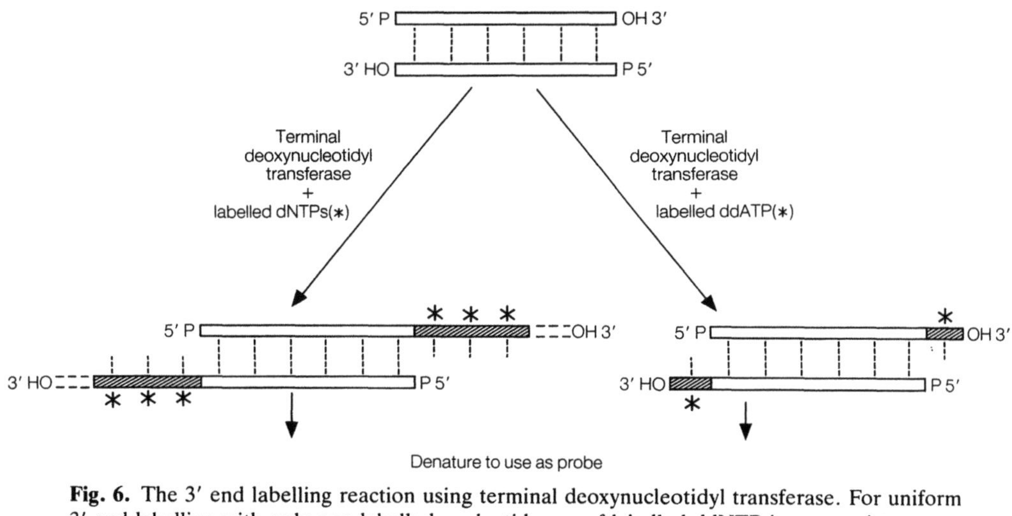


**Fig. 6.** The 3′ end labelling reaction using terminal deoxynucleotidyl transferase. For uniform 3′ end labelling with only one labelled nucleotide use of labelled ddNTP is appropriate

oligonucleotide probes with those of RNA probes in that it is then possible to synthesise uniformly labelled RNA probes from an oligonucleotide template.

### 3.3.5 *In Vitro Transcription of RNA Probes*

Various reagent kits and protocols for labelling RNA probes are commercially available. Kits are also available facilitating the insertion of a sequence of interest into a suitable plasmid vector which contains a promoter sequence or binding site for RNA polymerase.

RNA polymerases catalyse the synthesis of RNA from ribonucleotide triphosphates using a DNA template. In particular, RNA polymerases from bacteriophages are currently in use because of the high specificity for their own promoter sequence in vitro (e.g. salmonella phage SP6 polymerase, coliphage T3 and T7 polymerase (BUTLER and CHAMBERLAIN 1982; DAVANLOO et al. 1984).

Promoter sequences are "built in" to certain plasmids and the probe sequence inserted immediately downstream of the promoter. By digesting the plasmid with a restriction endonuclease which cuts immediately after the sequence of interest, transcription of probe without vector sequences is largely ensured. The basic reaction is illustrated in Fig. 7.

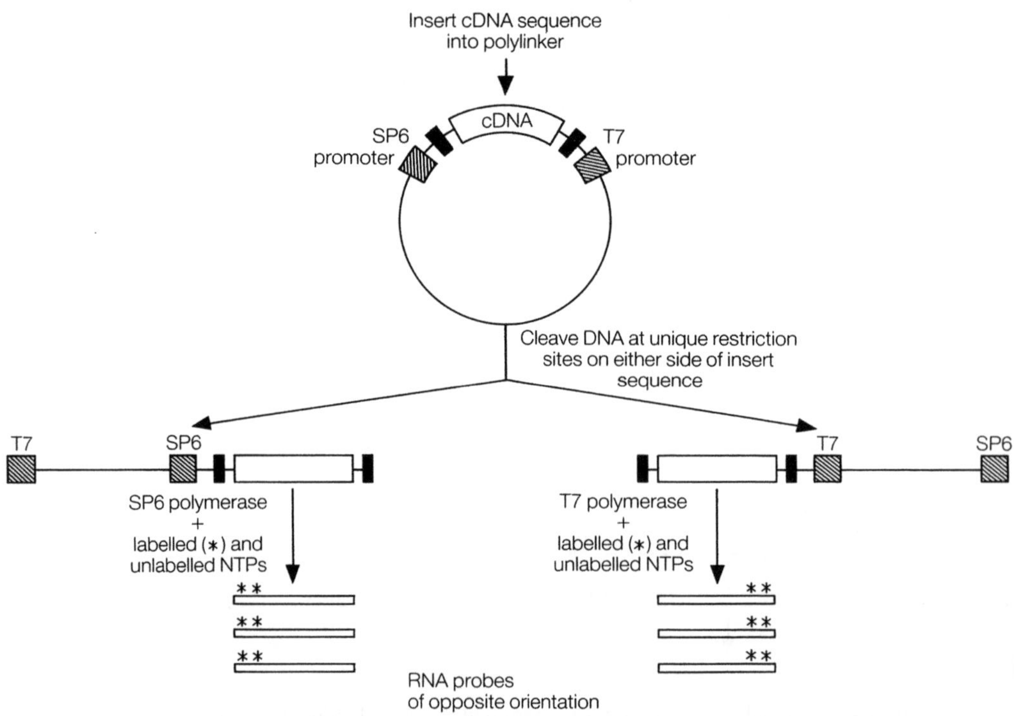


**Fig. 7.** Preparation of RNA probes using a dual promoter vector

The more frequently used vectors contain two phage promoters inserted in opposite orientations and separated by multiple unique restriction sites for cloning. The use of two promoters allows transcription from either strand of template DNA so that antisense or sense RNA may be produced. Therefore these dual promoter vectors are useful since an adequate negative control probe can be synthesised from the same template, thus ensuring that both probes can be subjected to the same stringency conditions since they have identical GC content.

Probe yields are of the order of 120 ng from 1 µg of template using limiting concentrations of radioactively labelled nucleotides. Specific activities of radioactively labelled probes are of the order of $1.3 \times 10^9$ dpm/µg and full length probes of up to 2.5 Kb can be synthesized.

Incorporation of biotin-UTP is in our hands less efficient than incorporation of unlabelled or radioactively labelled nucleotides. However, sufficient biotinylated probe can be generated by using 1 mM biotin-labelled UTP in the final reaction. In this case 4%–30% incorporation can be expected. Allyl UTP is incorporated more efficiently into the RNA probe using RNA polymerase as detailed. The allyl UTP is then modified post-transcriptionally with a biotin reporter molecule (LUEHRSEN and BAUM 1987). In our hands, these probes result in high backgrounds when used for ISH (unpublished results).

### *3.3.6 Direct Labelling*

There are several methods which do not utilise enzymes to incorporate label. Both photoactivatable and chemically reactive groups (VAN DER PLOEG et al. 1986) have been covalently linked to nucleic acids. A photoactivatable analogue of biotin has been synthesised and used to label large quantities of DNA and RNA (FORSTER et al. 1985). This analogue is now commercially available and probes have been used successfully for ISH (CHILDS et al. 1987; BRESSER and EVINGER-HODGES 1987).

## 3.4 Choice of Probe Label

Commonly used labels for nucleic acids can be divided into two types, radioactive and non-radioactive (Table 3). In contrast to other techniques for detecting nucleic acids where macroscopic resolution is sufficient, ISH requires resolution at the microscopic level; hence, the resolution afforded by a probe label is an important consideration. Clearly maximum resolution is not always necessary for a particular experiment, for example it may be sufficient to localise a target sequence merely to a group of cells. However, when requirements relate to where on particular chromosome, in which cell, or the distribution of a target sequence within the cell, then the choice of probe label is influenced by those offering the best resolution.

**Table 3.** Labels for nucleic acids

| | |
|---|---|
| *Radioactive labels* | |
| | Phosphorus 32 ($^{32}P$) |
| | Sulphur 35 ($^{35}S$) |
| | Tritium ($^{3}H$) |
| | Iodine 125 ($^{125}I$) |
| *Non-radioactive labels* | |
| | Biotin |
| | Directly linked enzymes |
| | Sulphonated cytidine |
| | Mercury |
| | Acetylaminofluorene (AAF) |
| | Colloidal gold |

**Table 4.** Characteristics of radionuclides used in nucleic acid labelling

| Radionuclide | Half-life | Type of emission | Maximum energy of emission (MeV) | Specific activity range of nucleotides (Ci/mmol) |
|---|---|---|---|---|
| $^{32}P$ | 14.3 days | β | 1.71 | 400–6000 |
| $^{35}S$ | 87.4 days | β | 1.67 | 400–1500 |
| $^{125}I$ | 60 days | β, γ | 0.035/0.035 | 1000–2000 |
| $^{3}H$ | 12.43 years | β | 0.018 | 25–100 |

Many experiments require high levels of sensitivity, e.g. less than 100 copies of mRNA per cell or detection of single copy genes on metaphase chromosomes. For these experiments the sensitivity of the detection system used for particular probe labels is an important consideration. Some researchers wish to quantitate signal following an ISH experiment (see Sect. 5.1).

In summary, there is a trade-off between resolution and sensitivity in combination with a range of other factors, including speed of development, quantitation, probe stability, safety, ease of use and cost.

### 3.4.1 Radioactive Labels

The ideal radioactive probe label for ISH would be of high specific activity whilst the emitted energy would be low. Detection of such a label would be sensitive and rapid with high resolution. No available radionuclide fully meets these exacting criteria. Selection of a radioactive probe label is based therefore on a balance between sensitivity, speed and resolution. The characteristics of radionuclides available for nucleic acid probe labelling are given in Table 4, and Fig. 8 shows the difference in resolution obtained with various radiolabels.

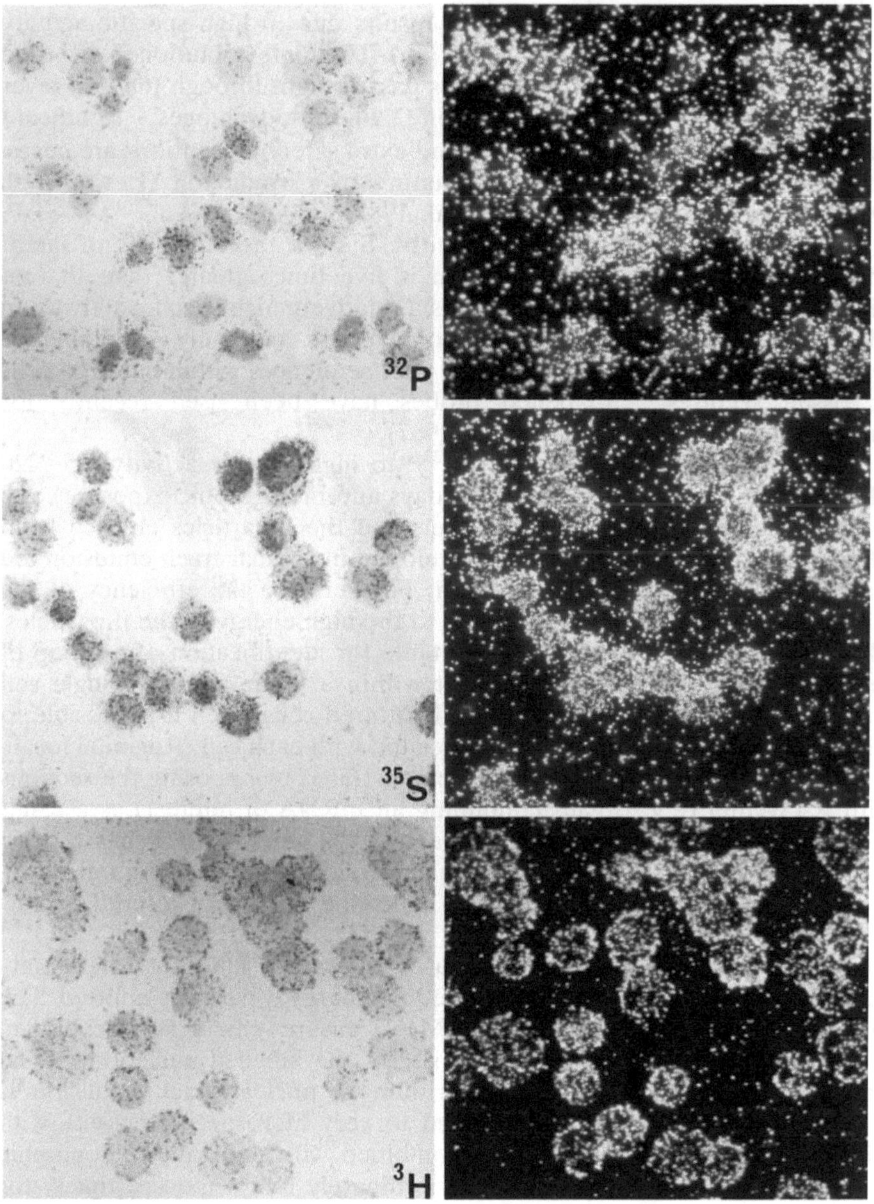


**Fig. 8.** Autoradiography of cultured chicken embryo cells which have been cytospun onto glass slides and hybridised with chicken β-actin RNA probes. The autoradiographs demonstrate the degree of resolution produced by different radioactive labels. The cells are shown in bright field (*left*) and dark field (*right*). Cells hybridised with a $^{32}P$-labelled RNA probe (specific activity $1.3 \times 10^9$ dpm/μg) exposed for 4 days; cells hybridised with a $^{35}S$-labelled RNA probe (specific activity $6.6 \times 10^8$ dpm/μg) exposed for 6 days; cells hybridised with a $^3H$-labelled RNA probe (specific activity $8.6 \times 10^7$ dpm/μg) exposed for 14 days

Use of $^{125}I$ as a label gives rapid results due to high specific activity with relatively good resolution (1–10 μm). The high resolution is afforded by the Auger electrons since the γ-rays pass straight through the thin layer of nuclear track emulsion. Despite these advantages, significant backgrounds are generally observed and extra safety precautions are necessary due to the highly penetrating nature of γ-irradiation (LEWIS et al. 1986; ALLEN et al. 1987; JANSSEN et al. 1987).

Rapid results are also achieved with $^{35}S$ since the efficiency of latent image formation within the emulsion is five times greater than that of tritium, for example, and the label has a relatively high specific activity. In addition, $^{35}S$ offers 15–20 μm resolution which is commonly acceptable for certain applications, e.g. detection of viral sequences within cultured cells or tissue sections. $^{35}S$-labelled probes have been used recently for chromosomal analysis (SOREQ et al. 1987).

Nucleotides can be labelled with $^{32}P$ to high specific activity (15–220 TBq/mmol). This permits rapid (1–7 days autoradiographic exposure) and sensitive detection. However, the energy of the β-particles emitted from $^{32}P$ is high. Therefore silver grains developed in nuclear track emulsion are distant from the site of hybridisation. Furthermore the efficiency of the nuclear emulsion is compromised due to the high energy of the β-particles. $^{32}P$-labelled probes are really only suitable for identification of a group of cells containing the target of interest within a tissue section. Single cell resolution is possible with well-spaced cultured cells when it is possible to distinguish a higher grain density associated with each cell. Regional localisation within tissue sections can be demonstrated by opposing the sections with X-ray film (e.g. organs containing an mRNA of interest) as a result of the high-energy β-particles emitted by $^{32}P$ (Sect. 5.1). $^{32}P$ has a short half-life (~ 2 weeks) which can be inconvenient when carrying out multiple ISH experiments and ultimately limits the sensitivity offered by this probe label.

In contrast, tritium is commonly used as a probe label for ISH largely due to high resolution afforded by the low energy β-particles emitted. The low energy of the β-particles (0.018 MeV) ensures that grains developed in the emulsion relate closely to the site of the labelled nucleotide. (The track length of a β-particle from tritium in nuclear track emulsion is ~ 1 μm.) Nucleotides are not labelled to very high specific activity with tritium (0.9–3.7 TBq/mmol); this, combined with poor autoradiographic efficiency (ROGERS 1979), leads to inordinately long exposure times, for example 1–3 months for unique sequences. Autoradiographic exposure times can be shortened by incorporating more than one tritium-labelled nucleotide into a probe. Tritium has a long half-life (12.43 years) and labelled probes are relatively stable over a period of months.

### 3.4.2 *Non-radioactive Labels*

There is now a wide choice of methods for labelling non-radioactively nucleic acid probes (Table 3) (MITCHEL et al. 1986; VAN DER PLOEG 1986; NIEDOBITEK et al. 1988).

Biotin-labelled probes were first introduced by LANGER et al. (1981), and were detected by using antibodies specific for biotin and labelled secondary antibodies. This method was later modified (HUTCHISON et al. 1982; LEARY et al. 1983) and is still the basis for the methods currently in use. In most of the published studies, biotin-labelled DNA is the probe of choice possibly because of the difficulties encountered initially in biotin labelling RNA probes. In our experience, RNA probes can be successfully labelled with biotin-UTP, although some modifications of the protocol are necessary (see Sect. 3.3.5). Recent reports have shown the successful application of biotinylated RNA (FORSTER et al. 1985; ZABEL and SCHAFER 1988) and oligonucleotide probes (GUITTENY et al. 1988; LARSSON et al. 1988) for ISH on a variety of tissues. This flexibility of use has probably contributed to the widespread use of biotin compared to other non-radioactive labelling methods.

The detection systems for biotinylated probes are based on immunocytochemical methods using streptavidin or antibodies to biotin, which can give rapid detection in a matter of hours. The resolution afforded by these probes is much higher than that obtained with radioactive ISH, and the results are comparable in definition to those of immunocytochemistry (Figs. 9, 10). However, a drawback is the presence of endogenous biotin in most tissues, which might cause background staining and difficulty in evaluating the results (Sect. 4.1.2).

Biotin-UTP is most commonly used for non-radioactive probe labelling. Alternative approaches include direct labelling with photobiotin (FORSTER et al. 1985) and incorporation of allyl-UTP, which is then available to react with biotin (LUEHRSEN and BAUM 1987). In our experience, when probes labelled with photobiotin or derivated allyl-UTP are used for ISH they give a higher background staining than probes labelled with biotin-UTP.

Because of the stability of the probe, the rapidity of detection and the high resolution, non-isotopic probes are generally preferred to radioactive ones, particularly in clinical and diagnostic situations (MCKEATING et al. 1985; UNGER et al. 1986; LEWIS et al. 1987). In addition, the use of non-radioactive probe labels for ISH has produced data on chromosome organisation in interphase nuclei which are not available from radioactive techniques (MANUELIDIS 1985; TRASK et al. 1988). Up to now their widespread application has been curbed by low sensitivity of target detection (SINGER et al. 1986; GILLAM 1987; HÖFLER 1987), although more recent publications have shown the detection of single copy genes and a sensitivity comparable to that of radioactive probes (BHATT et al. 1988; LAWRENCE et al. 1988).

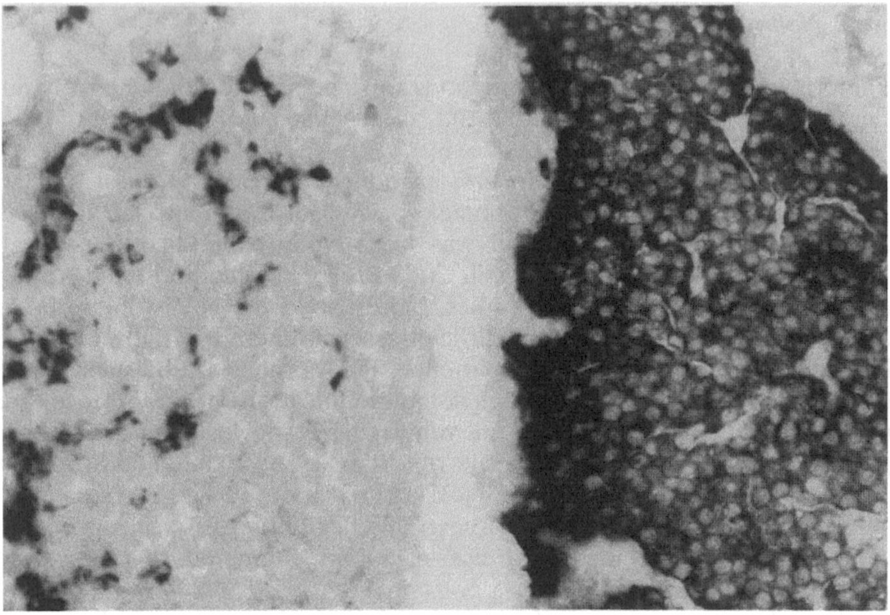

**Fig. 9.** ISH of pro-opiomelanocortin mRNA in cells of intermediate and anterior portions of rat pituitary, using biotinylated antisense RNA probes. The hybridised probes were detected using a streptavidin-biotin-alkaline phosphatase system. Pyronin counterstain, × 170

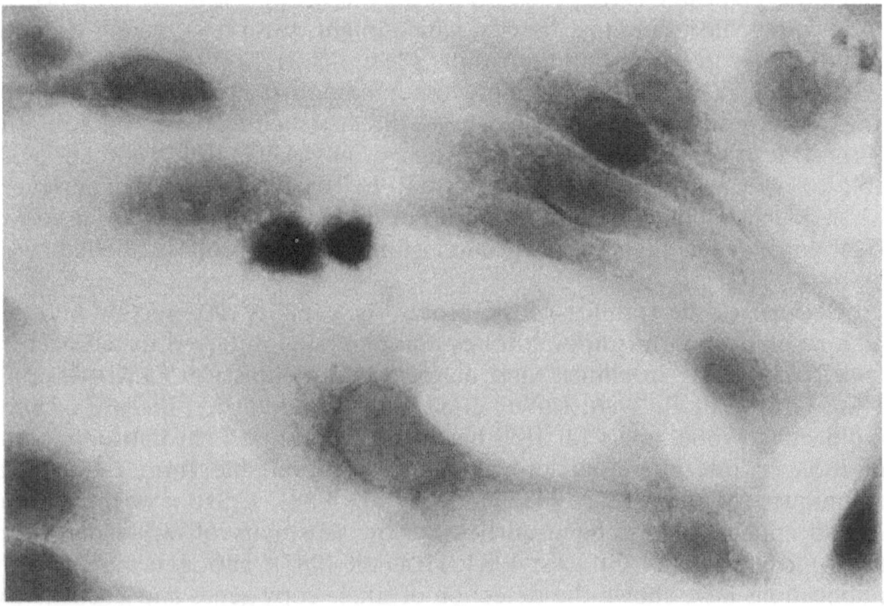

**Fig. 10.** ISH of N-ras mRNA in T15 cell line following dexamethasone stimulation. The antisense RNA probes were labelled with biotin and detected as explained in Fig. 7. × 720

# 4 Hybridisation

## 4.1 Pre-hybridisation Treatments

### *4.1.1 Tissue Permeabilisation*

As already discussed in Sect. 2, tissues or cultured cells should be processed and fixed in order to preserve morphology and prevent loss of target nucleic acids. An additional requirement is that target nucleic acid be rendered accessible to labelled probe. These criteria conflict, since treatments which permeabilise the tissue can also cause loss of morphology and leaching of target nucleic acid. It is therefore necessary to determine empirically the optimal conditions for permeabilisation with respect to tissue type, probe size in respect of tissue fixation and processing. Indeed, the treatment of tissues after fixation represents a major source of variation in the plethora of published protocols.

Limited enzyme digestion with proteases or treatment of tissues with dilute acid to remove basic proteins has been shown to enhance the ISH signal both in our experience and in that of others (GEE and ROBERTS 1983; BRAHIC and HAASE 1978; HÖFLER et al. 1986a; ANGERER and ANGERER 1981) whilst some data show loss of target RNA (SHIVERS et al. 1986b; LAWRENCE and SINGER 1985) and deterioration of cell morphology with excessive deproteination.

The extent of deproteination required depends on a range of interdependent factors including the extent of fixation, the tissue type, probe size, batch of enzyme and whether other permeabilising steps have been utilised.

A range of proteases have been used with or without dilute acid treatment and workers express a preference in the context of the particular system under study, for example, pepsin. Hydrochloric acid has been found to be more effective by some (BURNS et al. 1988) whilst pronase or proteinase K has been most effective for others (SINGER and WARD 1982; HAFEN et al. 1983; COX et al. 1984; ANGERER et al. 1985).

It is important to keep the treatments prior to hybridisation free from nucleases which could introduce nicks or partially degrade target nucleic acid particularly if mild fixation procedures have been used. Random nicking or degradation of target nucleic acids can result in decreased signal due to altered requirements for hybridisation stringency and/or loss of target. To this end, pronase can be rendered nuclease free by predigestion (HOGAN et al. 1986) and proteinase K can be obtained in a nuclease-free state.

The action of proteases can be stopped by incubating the tissue with glycine and/or a post-fixation step (HÖFLER et al. 1986a). The post-fixation step following permeabilisation is additionally advocated by some workers who have demonstrated increased ISH signal (SINGER and WARD 1982; HAFEN et al. 1983; HÖFLER et al. 1986a).

Detergents such as triton X-100 have also been included in tissue treatments prior to hybridisation as an alternative or additional means of permeabilising cells (HÖFLER et al. 1986a). Dehydration or deparaffinisation in solvents such as ethanol or xylene will also permeabilise cells to some extent, as will repeated freezing and thawing of tissues.

### *4.1.2 Probe Non-specific Binding*

The sensitivity of any assay depends on the minimisation of background signal to achieve the best signal-to-noise ratio; this is additionally advantageous following radioactive detection of ISH since a shorter exposure time is required to obtain a cleaner signal. Sources of background signal encountered with ISH arise from entrapment of labelled probe within the cells, electrostatic interaction of the probe with cellular components such as proteins, the formation of imperfect duplexes between certain regions of the probe and irrelevant cellular sequences together with background arising from the detection system employed. The hybridisation itself and post-hybridisation events are also important in achieving the best signal-to-noise ratio and will be discussed subsequently (Sect. 4.2, 4.3).

Electrostatic interactions between the probe and basic proteins within cells has been decreased by pre-hybridisation treatment with 0.25% acetic anhydride, which is thought to acetylate cellular proteins (HAYASHI et al. 1978). Our own data and those of LAWRENCE and SINGER (1985) indicate that acetylation does not decrease backgrounds significantly when utilising short (~ 200 base) probes. However, LAWRENCE and SINGER (1985) do show a significant decrease in background signal when utilising longer probes and with biotin-labelled probes. There are also data to suggest that in addition to preventing electrostatic interactions, the acetic anhydride may serve to block potential binding of probe to polylysine used to coat slides (HAASE et al. 1984).

The majority of published ISH protocols include a pre-hybridisation step designed to decrease the binding of labelled probe by saturating sites within the section or cells and on the glass slide. Pre-hybridisation solutions are analogous to those used in filter hybridisation and are not really tailored to the support or matrix used for ISH. The pre-hybridisation step is commonly carried out at the hybridisation temperature for a period of up to 2 h. Slides and cells can be treated with detergents, Ficoll, polyvinylpyrollidone, heparin and bovine serum albumin to decrease non-specific binding to proteins, polysaccharides and nucleic acids. In addition, sodium pyrophosphate is used to block sites which could interact with the phosphate backbone of the probe. Freshly denatured, sonicated, unlabelled genomic DNA and/or t-RNA are commonly used in prehybridisation solutions as a means of blocking sticky sites within the cells or on the glass slide. Aurintricarboxylic acid (ATA), a potent inhibitor of protein-nucleic acid interactions, has been shown to be effective for ISH (GONZALES et al. 1979).

Biotin-labelled probes are frequently used for ISH. Biotin is a vitamin and a covalent cofactor for several cellular carboxylases. The problem of background signal due the detection of endogenous biotin within cells has been documented both for immunocytochemistry and for ISH (POLAK and VAN NOORDEN 1986; NAOUMOV et al. 1988). Particular tissues such as liver and kidney have been highlighted for their high levels of detectable endogenous biotin compared with other tissues. Background staining caused by biotin is a problem when probing tissues with particularly high levels of endogenous biotin or when probing for a small number of target molecules, for example < 100 copies/cell. Some blocking procedures for endogenous biotin have been described which involve pre-incubation of the tissue or cells with unlabelled detection system and, in the case of avidin or streptavidin, further blocking by a second incubation with free biotin. Although these procedures are generally useful for immunocytochemistry (COGGI et al. 1986), they seem to be less efficient following ISH. There are a number of possible reasons for this; for example, the biotin block may be subsequently removed by the stringent conditions used for ISH, viz. high temperatures in the presence of formamide and detergents, or the stringent conditions expose more endogenous biotin or biotin-like molecules which bind the detection system.

Where alkaline phosphatase or horseradish peroxidase is used in the detection system, endogenous cellular phosphatases or peroxidases can be effectively eliminated from problem tissues using acetic acid or periodic acid and hydrogen peroxide respectively prior to hybridisation (COGGI et al. 1986; ANDREW and JASANI 1987; CHIN-YANG et al. 1987).

## 4.2 Considerations for the Hybridisation Step

Hybridisation is a term which describes the formation of hydrogen bonds between nucleotides resulting in the association of single-stranded nucleic acid molecules by complementary base pairing. Hybrids can be formed between two single-stranded DNA molecules, between single-stranded DNA and RNA or between two single-stranded RNA molecules. Factors affecting the rate of formation of hybrids and their subsequent stability have been elucidated for reactions in solution and similar principles govern ISH. ISH is somewhat different in that the target nucleic acid is immobilised within fixed cells whilst the labelled probe is in a solution phase. In this respect it is broadly analogous to filter hybridisation in respect of the conditions required.

Hybridisation of nucleic acid probes essentially to a solid support (reviewed by MEINKOTH and WAHL 1984) depends on diffusion of the probe to target sequences, the rate of hybrid formation and the stability of resulting duplexes.

### 4.2.1 Probe Diffusion

As examined above, the extent of tissue fixation sets a balance between the preservation of morphology, retention of target sequences, hybridisation efficiency and accessibility of the probe to target by diffusion. There are additional factors governing the rate of diffusion of probe to target when considering ISH. Since the target is held within the cell, probe size also affects the rate of diffusion to target. There have been several studies using different systems to determine the optimal probe size (BRAHIC and HAASE 1978; ANGERER and ANGERER 1981; GEE and ROBERTS 1983; MOENCH et al. 1985; LAWRENCE and SINGER 1985). A consensus probe size range includes 200–500 bases; however, LAWRENCE et al. (1985) present data showing that probes ⩾ 1000 bases and containing vector sequences can enhance signal by a factor of tenfold on average. Under the conditions used in our laboratory for freshly fixed, cultured cells, a probe size of between 200 and 500 bases is optimal. Shorter probes and probes exceeding 500 bases result in decreased signal intensity. Oligonucleotides provide maximal tissue penetration due to their small size but may produce weaker signals than uniformly labelled longer probes.

The probe concentration for ISH reactions is in vast excess over target concentration, for example, ~ 0.5 ng/µl for radioactive probes, ~ 2 ng/µl for non-radioactive probes. The optimal concentration is set to provide the best signal-to-noise ratio since the level of background may be linearly related to probe concentration (COX et al. 1984). Although most of the probe does not take part in the hybridisation reaction, it has been suggested that probe is removed by non-specific interaction with the tissue during hybridisation. Clearly, the hybridisation temperature and extent of cell permeability will also affect the diffusion of probe.

### 4.2.2 Hybridisation Kinetics and Optimal Temperature

The probe concentration is in vast excess for ISH reactions to facilitate diffusion of the probe to target. Under these conditions, the rate of hybrid formation follows first-order kinetics. The time required for half the probe to hybridise ($t_{1/2}$) is related to the probe complexity (N), i.e. the number of base pairs in a non-repeating sequence, probe length (L) and probe concentration (C) in mol nucleotides/litre.

$$t_{1/2} = \frac{N \ln 2}{3.5 \times 10^5} \times L^{0.5} \times C$$

Rates of hybridisation are approximately tenfold lower for ISH than predicted by this equation, which is based on solution hybridisation (BRAHIC and HAASE 1981; HAASE et al. 1984; COX et al. 1984; LAWRENCE and SINGER 1985). Knowledge of the time course is important when maximising the sensitivity of the ISH reaction. Most of the ISH reaction is complete within a few hours (Cox et al. 1984; LAWRENCE and SINGER 1985); however, there is a much slower second phase of hybridisation (ANGERER et al. 1987). The hybridisation buffer is a good solvent for proteins, and with prolonged hybridisation times, target RNA can be released from the tissue. Although very long hybridisation times have been published, such as 1–4 days (EDWARDS and WOOD 1983; HAASE et al. 1982; HAYASHI et al. 1978; SINGER and WARD 1982), shorter hybridisation times of a few hours to overnight are recommended (LAWRENCE and SINGER 1985; BRESSER and EVINGER-HODGES 1987). In our hands, hybridisation for 1 h is sufficient to detect less than a 100 copies of mRNA per cell.

The equation detailed above implies that long probes of low complexity hybridise faster. However, longer probes may result in lower rates of hybrid formation in situ since they are likely to take longer to diffuse into a tissue section. In addition, when double-stranded probes are used, there is a competition between the hybridisation reaction and the reannealing of complementary probe strands in solution. It has been suggested that the latter is the more favourable reaction and that removal of probe from the hybridisation buffer through reannealing lowers the effective probe concentration, thus slowing hybrid formation some eightfold (Cox et al. 1984).

Addition of polymers to the hybridisation solution such as dextran sulphate or polyethylene glycol can enhance the rate of hybrid formation, particularly for double-stranded probes (WAHL et al. 1979; RENZ and KURZ 1984). Network formation is promoted between overlapping sequences and probe concentration is increased by volume exclusion of probe within the hybridisation buffer. The inclusion of a polymer in the hybridisation buffer can therefore increase both the rate of hybrid formation and the amount of labelled at a target site. We have also found inclusion of a polymer necessary for short RNA : RNA hybridisations.

The maximum hybridisation rate occurs at about 25° C below the temperature at which half the hybrids dissociate or melt (Tm). The maximum rate in situ occurs at approximately 30° C below Tm due to the large number of interdependent variables involved, including the possibility that only partial duplexes can form with the probe due cross-linking of target molecules (ANGERER et al. 1987). Since there is a broad optimum range, the hybridisation rate is not affected greatly within 5° C of the Tm (WETMUR and DAVIDSON 1968; BRAHIC and HAASE 1978; Cox et al. 1984). The Tm can be calculated but there are different equations for DNA : DNA duplexes as opposed to RNA : RNA duplexes, as detailed below (THOMAS and DANCIS 1973; BODKIN and KNUDSON 1985; ANGERER et al. 1987).

Probe length contributes to hybrid stability. Long complementary probes form a greater number of hydrogen bonds with the target and are therefore more stable, while in this respect short oligonucleotide probes can be less stable. In addition, the effect of a mismatched base on Tm is more pronounced for short oligonucleotides than for longer probes. Despite these contra-indications, short oligonucleotide probes are increasingly used for ISH since they are readily available and by virtue of their size are more easily accessible to target nucleic acid within tissue samples.

Formamide is used to lower the temperature required for hybridisation reactions (KOURILSKY et al. 1971). For ISH in particular, the lower hybridisation temperature afforded by addition of formamide preserves tissue morphology and minimises tissue loss from the glass slide. Each percentage of formamide included in the hybridisation buffer decreases the effective hybridisation temperature required by 0.35° C for RNA : RNA duplexes (COX et al. 1984) and by 0.65° C for DNA : DNA duplexes (MCCONOUGHY et al. 1969).

RNA : RNA hybrids are about 10°–15° C more stable than DNA : DNA hybrids of the same base composition (WETMUR et al. 1981; COX et al. 1984) whereas DNA : RNA hybrids are intermediate in Tm (CASEY and DAVIDSON 1977).

In summary, hybridisation reactions are generally carried out under conditions of low stringency which favour the formation of nucleic acid duplexes. The temperature is generally set at 25°–30° C below the Tm, the formamide concentration at 50% and the salt concentration around 0.75 M.

### 4.3 Post-hybridisation Treatments

The aim of the post-hybridisation steps is to remove non-specifically bound probe from the tissue or cells, thus enhancing the signal-to-noise ratio as much as possible. Washes are designed to be more stringent than hybridisation conditions with a view to dissociating weakly complementary hybrids and further reducing non-specific association of the probe with cell components. Stringency conditions can be altered as described in Sect. 4.2 bearing in mind the fact that hybrid stability decreases progressively from RNA : RNA to RNA : DNA and DNA : DNA duplexes.

In addition to stringency washes, limited digestion with nucleases can be used to decrease background binding of labelled probe. The often high background binding of RNA probes can be decreased using post-hybridisation treatment of the preparation with RNase, which degrades single-stranded RNA and leaves hybrid molecules intact (JOHN et al. 1969; LYNN et al. 1983; COX et al. 1984). Most published protocols detailing the use of cDNA probes do not utilise post-hybridisation enzyme treatments although GODARD (1983) demonstrate the effectiveness of limited S 1 nuclease treatment. High S 1 nuclease concentration, however, can lead to artefactual relocalisation of labelled hybrids in situ.

Of particular importance when using $^{35}S$-labelled probes is the inclusion of 10–100 mM dithiothreitol (DTT) in both the hybridisation buffer and the stringency washes to prevent background binding of the sulphur to the cells. Use of unlabelled thio-UTP has also been shown to reduce background binding of $^{35}S$-labelled probes in this way (BANTTLOW et al. 1987), although this approach was not particularly effective in our hands.

# 5 Detection

## 5.1 Radioactive Detection

Following post-hybridisation treatments, tissue preparations or cells are dehydrated in graded ethanols, air dried and subjected to autoradiography. The importance of this step in the ISH procedure is sometimes overlooked, but an erroneous methodology at this stage could lead to false negative results. A good understanding of autoradiography in theory and in practice is vital to be able to distinguish artefacts or rectify a poor result (ROGERS 1979, for a comprehensive guide to autoradiography).

Autoradiography is essentially the recording of radioactive emission within a photographic film. Micro-autoradiography is most commonly used for ISH whereby a thin layer of nuclear emulsion is applied to the tissue, cells or chromosome preparations using either a liquid emulsion or stripping film product. The resultant silver grains in the emulsion layer are examined for spatial information and/or quantitative information (ROGERS et al. 1987) in respect of the cells beneath.

It is occasionally useful to use a higher energy radioisotope, such as $^{32}P$ or $^{35}S$ in conjunction with X-ray film, because of the rapidity offered by this method in checking the results of hybridisation. This approach can be particularly useful in rapidly assessing the parameters of ISH in cultured cells by relating the grain density to cell number, or when carrying out ISH on sections of whole brain or whole embryos, since the size of the preparations might preclude an even coating with liquid emulsion.

The nuclear emulsions are suspensions of silver bromide crystals in gelatin. The interaction of radioactive emission or light with the silver bromide causes the formation of sensitivity specks of silver (the latent image) which can be visualised by treatment with a chemical developing agent. The track length of β-particles in nuclear emulsion from $^{32}P$, $^{35}S$ and $^{3}H$ is dependent on their respective energies (see Table 4). As described in Sect. 3.4.1, the low energy of β-particles from $^{3}H$ ensures that grains are formed closest to the site of hybridisation compared with the β-particles from $^{35}S$ and $^{32}P$ for example. The thickness of the emulsion layer is important for microautoradiography. Generally, the thinner the emulsion layer, the better the resolution. Clearly, increasing the emulsion thickness above the maximum range of β-particle path length in this medium has no effect on resolution. Both liquid nuclear emulsion and stripping film are commercially available. One of the advantages of using stripping film over liquid nuclear emulsion is that the emulsion layer is of constant thickness over the whole preparation. The majority of workers choose to use liquid emulsion as opposed to stripping film for ISH probably due to ease of handling. In addition, the liquid emulsion forms a more intimate contact with the cells or tissue sample presented on the slide. The thickness of liquid emulsion coating a slide can be regulated up to a point by dilution. There is, however, an inherent variation in the thickness of emulsion over the whole slide, and dipping procedures require practice in order to achieve some degree of reproducibility (CUMMING and FALLON 1989; ROGERS 1979).

Liquid emulsions are available in a form which will produce a variety of grain sizes offering different resolution. The differences between commercially available products are often not appreciated (ROGERS 1979). For example, the thickness of coating over a dipped slide may be completely different for two given emulsions. In addition, the sensitivity to latent image fading, safe-light conditions and background grain formation as a result of drying vary dramatically. There is some evidence that dilution of emulsions with 0.6 M ammonium acetate as opposed to water maintains the integrity of hybrids at this stage (BRAHIC and HAASE 1978). This might be an important consideration when working at high sensitivity. Some commercially available liquid emulsions are not compatible with ammonium acetate although this does not fully preclude their use for ISH. (NB: see manufacturer's literature for precise details.)

The efficiency of nuclear emulsions is also dependent on a whole variety of factors which generally conflict with those required for high resolution (ROGERS 1979).

The conditions of drying and exposure of liquid emulsion are critical. The emulsion should be allowed to gel quickly at low temperature followed by slow and gentle drying in a humid atmosphere. For exposure, slides should be packed in a light tight box in the presence of dessicant and stored at 4° C.

The developer, stop fixer and counterstaining solutions should be at the same temperature to avoid loss or cracking of emulsion. In addition the choice of staining procedure is important. The use of prolonged acid destaining can bleach grains (ROGERS 1979).

Clearly a range of controls should be adopted for each experiment to check for artefacts of autoradiography; these are discussed in Sect. 6.4.1.

Quantitative studies based on ISH have been carried out (ROGERS et al. 1987). Grain counting is a convenient way of quantitating the extent of hybridisation in a given preparation and sophisticated image analysis has been used in comparative studies. Absolute quantitation of target within cells or direct comparison of quantities between cells is difficult due to variation introduced at each stage of the ISH procedure. For example, there might be some lack of uniformity in probe diffusion, the efficiency of fixation, target loss, the efficiency of hybridisation and the thickness of emulsion. Although the combination of these factors precludes absolute quantitation through grain counting, statistical analysis of grains is effective when assigning a gene to "particular regions" on a chromosome and in broad-based comparative physiological studies (SCOTT YOUNG III et al. 1986).

Colour micro-autoradiography and double labelling by ISH has been described (HAASE et al. 1985). However, this technique has not been widely adopted to date.

## 5.2 Non-radioactive Detection

Currently the methods for non-radioactive detection are based on those used in immunocytochemistry. Generally these include the streptavidin-biotin method (COGGI et al. 1986), the indirect antibody detection method using a variety of reporter labels (POLAK and VAN NOORDEN 1986), or a combination of the two in order to achieve greater sensitivity. The majority of these methods are based on the use of biotinylated probes, although other ways to label the probes non-radioactively have been proposed (MITCHELL et al. 1986; VAN DER PLOEG et al. 1986; NIEDOBITEK et al. 1988).

To detect the biotinylated probes, streptavidin is most commonly used either as a reporter labelled molecule or as a link for biotin molecules bound to a detectable marker. The avidin-biotin method is based on the high affinity between avidin and biotin molecules, the possibility of coupling biotin to a variety of reporter molecules through a simple chemical reaction, and the ability of avidin to act as a bridge between two different biotinylated molecules such as probes and alkaline phosphatase.

Because of its small size, biotin does not modify the chemical or physical properties of the molecules to which it is bound. Also multiple biotin molecules can be bound to an enzyme, each biotin being able to bind specifically to avidin. The preferred labels are usually fluorochrome (i.e. fluorescein) (LAWRENCE et al. 1988) or an enzyme (i.e. peroxidase or phosphatase) HÖFLER 1987; LEWIS et al. 1987; BHATT et al. 1988; ZABEL and SCHAFER 1988; LARSSON et al. 1988).

The immunocytochemical methods using avidin-biotin have similar requirements to the ones based on antibodies. The biotinylated probes can be visualised with avidin acting as a bridge with biotinylated enzymes, or with an avidin-biotin-enzyme/fluorochrome complex (COGGI et al. 1986). The reaction between avidin and biotin is very rapid and extremely stable, but one must be aware of likely non-specific binding of streptavidin to endogenous biotin, and of the possible presence of endogenous alkaline phosphatase. The latter can be blocked by using acetic or periodic acid (COGGI et al. 1986). Binding of streptavidin to endogenous biotin, or other tissue components such as glycoproteins which can be unmasked by the hybridisation procedure, is more difficult to control and can cause serious problems of background noise (see Sect. 4.1.2). Various blocking procedure have been recommended to decrease this type of non-specific binding (COGGI et al. 1986; SINGER et al. 1986; GILLAM 1987), but a satisfactory answer to the problem has not yet been found.

Biotinylated probes can be detected also by using antibodies to biotin. These can then be visualised with labelled secondary antibodies (indirect method), or with a bridging antibody and an enzyme-anti-enzyme complex (unlabelled antibody-enzyme method). Both polyclonal and monoclonal antibodies have been used, in combination with a variety of labelled molecules, using essentially traditional immunocytochemistry methods

(CUELLO 1983; POLAK and VAN NOORDEN 1986). However, these seem to have lower sensitivity of detection than streptavidin-biotin methods, and their application is more limited (unpublished results).

## 6 Specificity

In line with immunocytochemical methods, suitable controls are necessary during every ISH experiment to assess the reproducibility, specificity and sensitivity of the reagents and procedures. Also a correct set of controls will confirm that the tissue presents the target of interest, which is identified by a suitable probe and hence the hybrid identified at the microscope is the correct one. In some cases it may be difficult to decide whether the signal is specific, and it is only by applying a full set of positive and negative controls that non-specific signal can be eliminated. It is convenient to examine the different controls separately, as several approaches have been developed.

### 6.1 Tissue Controls

It is most important that the morphology of the tissue is retained throughout the ISH procedure so that the hybridisation signal is distributed to recognisable cell subpopulations, which can be identified in relation to other anatomical structure. Staining of sections before and after the in situ procedure with a routine haematoxylin/eosin method can provide an adequate control. Indeed, the tissue morphology can appear intact at the beginning of the hybridisation procedure, but it may be grossly altered at the end of it if the fixation is inadequate.

Hybridisation can occur when there are sufficient, but unexpected, homologies between the probe and the numerous sequences present in any cell. A useful negative control is a tissue known not to contain the target nucleic acid investigated. Hybridisation on this tissue should be negative, a positive result being indicative of a non-specific signal.

The reproducibility of the technique can be checked by carrying out the hybridisation in parallel on positive cell/tissue models with a well-characterised probe. Alternatively, a probe recognising ubiquitous sequences (e.g. actin mRNA) can be applied to the tissue under investigation as positive control. In this way it is possible to check at the same time the condition of the tissue and the methodology.

When antibodies to the gene product are available, a comparison between ISH and immunocytochemistry could further confirm the correct distribution of the signal. The two methods can be carried out on adjacent (HÖFLER et al. 1986a; TERENGHI et al. 1987; STEEL et al. 1988) or on the same sections (BRAHIC et al. 1984; SHIVERS et al. 1986a). It is noteworthy that although there might be a consistent distribution, there could be a numerical discrepancy between cells identified by the two methods. Indeed, gene transcription might not be a continuous event for all cells, whereas different amounts of transcribed and translated products are shown by the two techniques.

### 6.2 Probe Controls

An appropriate set of controls will confirm whether the probe is hybridising to the intended sequence. A Northern blot analysis of total RNA using the specific probe is necessary to establish specificity for a given target (MANIATIS et al. 1982).

Target RNA examined by Northern blot analysis is extracted from a large and, in the case of tissues, heterogeneous population of cells. If the number of cells containing the target sequences is a small percentage of the totality, the level of target RNA may be at variance with the current limits of sensitivity for Northern blotting because of the dilution effect. This may be the case when only a few cells express the message of interest, which could still be picked up by ISH.

When using RNA probes, transcripts with the same orientation as the target sequence, or sense probes, can effectively be used as negative control. Similarly, a non-homologous probe hybridised to the test tissue should give a negative result. When ISH is carried out with probes which contain the vector sequence, hybridisation with labelled sequence vector should also be included, since homology to the target sequence may produce false positive results.

Useful information can be gained by carrying out competition controls, where a mixture of labelled and unlabelled probes are hybridised to the section (JILBERT et al. 1986). Since there is only a determinate amount of specific target compared to potential non-specific hybridisation sites, the intensity of the hybridisation signal, or the decrease of it, should be in direct relation to the ratio of the two probes in the mix. An unchanged signal intensity would then indicate a non-specific hybrid.

### 6.3 Hybridisation Controls

These controls verify more broadly the specificity of the hybridisation reaction. Negatively charged probes may "stick" non-specifically to positively charged tissue proteins. Also probes might bind weakly to non-related target sequences if there are regions with a high proportion of C-G bases (JONES and HYMAN 1983; PATIENT 1984). Non-specific "sticky" regions may also be uncovered in the tissue by the hybridisation procedure itself. Acetylation of the tissue or increased stringency of hybridisation and wash conditions (see Sect. 4) can prevent these non-specific reactions. A control preparation processed routinely, but incubated in the absence of probe, should yield a negative result.

The stability of the signal with washes of increasing temperature is indicative of specificity, as mismatched probe binding is less stable than that of true hybrid because of a lower melting temperature (Tm); hence a non-specific hybrid is removed more readily with increasing temperature (SZABO et al. 1977).

Nuclease digestion of the tissue prior to hybridisation can also be included as a control to confirm that the probe is hybridising to nuclei acid sequences. DNase or RNase digestion can be carried out on a separate preparation, which is then hybridised with the chosen probe. The purpose of this control is to obtain a decrease of signal intensity, rather than a complete deletion, which could be achieved by careful titration of nuclease concentration and digestion time. It is worth noting that this control will only confirm that the probe is hybridising to digestable nucleic acid; it will not confirm the specificity of the hybrid. Care must also be taken to eliminate all traces of nuclease after the reaction, as these could compromise the integrity of the probe and lead to mistaken conclusions. For example, RNA probes might be degraded by traces of RNase not blocked following nuclease digestion (GOWANS et al. 1989).

## 6.4 Detection Controls

Detection controls ensure that the detection system is working correctly. The types of control are different according to whether radioactively or non-radioactively labelled probes have been employed.

### *6.4.1 Autoradiography Detection*

The main problem can be a high background deposit of silver grains, which may be caused by mechanical stress during emulsion dipping and/or drying, exposure to light or prolonged exposure. Furthermore, deposition of silver grain can be caused by chemography, a chemical reaction between tissue components and radiographic emulsion (ROGERS 1979). These eventualities can be easily checked by processing a tissue preparation for autoradiography which either underwent hybridisation in the absence of probe or was not processed for ISH. Also a blank slide should be dipped in emulsion and exposed alongside the other preparations to check the condition of the emulsion. If this control shows unacceptable amount of silver grain deposits, the emulsion should be discarded and substituted with a new one. A further emulsion-coated blank slide should be exposed briefly to light prior to development. This will ensure that possible negative ISH results are not due to failure of latent image formation.

### 6.4.2 Non-radioactive Detection

At present this type of detection is mainly based on immunocytochemical staining methods (see 6.2), and it involves similar type of controls. If in separate tissue preparations one or more steps of the detection reaction are omitted, a negative result should be obtained. In addition, the full detection system, or part of it, is applied to a preparation which has gone through hybridisation in absence of probe. The combination of controls should pinpoint the cause of false positivity. For a full review on the subject the reader should refer to more comprehensive immunocytochemical publications (POLAK and VAN NOORDEN 1986; CUELLO 1983).

## 7 Some Applications

One of the major advantages of ISH is the ability to examine morphologically the functional aspect of cells and tissues. From the early stages ISH has been applied to embryology and other areas of cell differentiation, as the localisation of gene expression could be detectable before any functional development (HAFEN et al. 1983; MCALLISTER et al. 1983; COX et al. 1984; BLOOM et al. 1988).

Peptides and proteins have been widely studied with immunocytochemical techniques (POLAK and VAN NOORDEN 1986). However, the demonstration of an intracellular peptide or protein antigen does not supply any information on intracellular function. For example, atrial natriuretic peptide (ANP) has been localised immunocytochemically in both atria and ventricles of mammalian heart. While the presence of ANP in atrial cells was well known to be the result of intracellular synthesis, it was difficult to exclude a possible uptake of circulating peptide in ventricular cells. Using RNA probes complementary to ANP mRNA it has been possible to confirm ANP gene expression in both atrial and ventricular cells, in either tissue sections or cultured cells (HAMID et al. 1987) (Fig. 11).

Using probes to different peptides it has been possible to compare their transcription sites with the known distribution of the mature peptide. Neuropeptide Y (NPY) gene transcripts have been shown in neuronal cells of the human cerebral cortex (Fig. 12) (TERENGHI et al. 1987). Although the distribution of neurons positive for ISH was consistent with that of NPY-immunoreactive cells, there is a small numerical difference between cells identified with the two methods. This might be the reflection of different metabolic states of the cells, as gene transcription might not be a continuous event for all the cells, which show a different profile of mature peptide. In different situations cells that show a strongly positive hybridisation signal may appear negative by immunocytochemistry, due to a rapid peptide degradation or secretion (HÖFLER et al. 1986b, 1987).

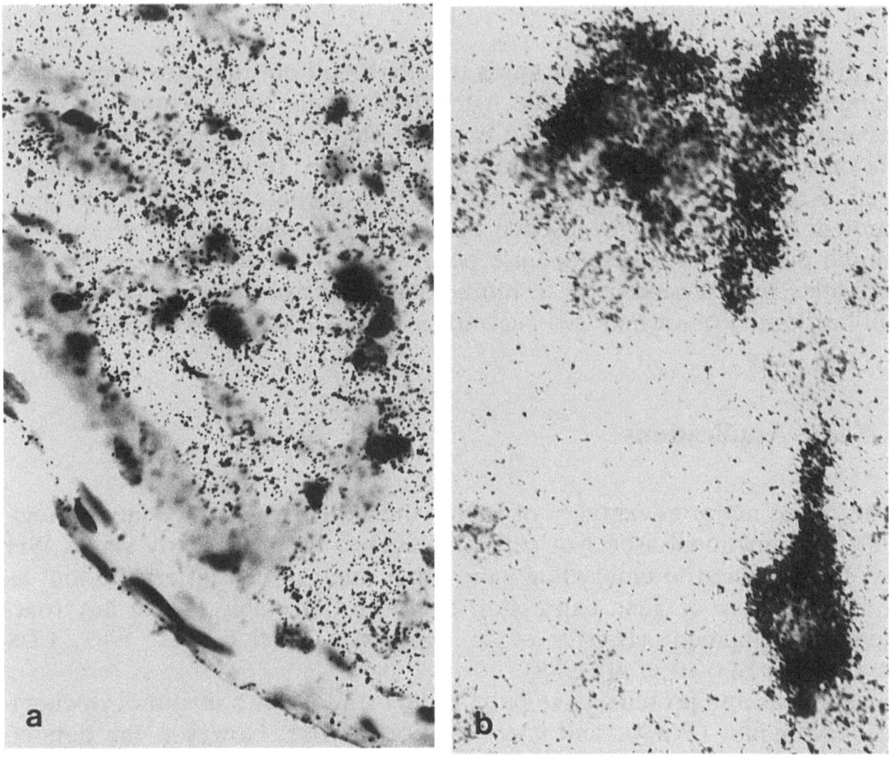


**Fig. 11 a, b.** ISH of ANP mRNA in sections of rat atrium (**a**) (× 250) and culture myocytes of rat atrium (**b**) (× 490), using a $^{32}$P-labelled antisense RNA probe. Haematoxylin counterstain

Hybridisation of peptide mRNAs has also been applied to the study of physiological and endocrine functions such as pregnancy and lactation (STEEL et al. 1988), and of pathological conditions (GOEDERT et al. 1986; ZAJAC et al. 1986; HÖFLER et al. 1987; GUITTENY et al. 1988; RUDA et al. 1988).

In situ hybridisation has been widely used to investigate gene expression (see COGHLAN et al. 1985 for review). A field of particular interest is the detection of oncogene expression, as oncogenes have become increasingly important in the study of tumour development and prognosis. Although oncoproteins can be localised morphologically by immunocytochemistry (GASTL et al. 1986), the availability of specific antibodies is still limited. ISH has allowed a wider investigation in this field and now there are several reports linking the expression of oncogenes in relation to tumour progression (SLAMON et al. 1984; CHAN and MCGEE 1987), in particular in neuroblastoma (SCHWAB et al. 1984; GRADY-LEOPARDI et al. 1986), lung small cell carcinoma (NAN et al. 1986) and medullary thyroid carcinoma (HÖFLER et al. 1986c). These are just few examples in a continuously growing list, which indicates the importance of

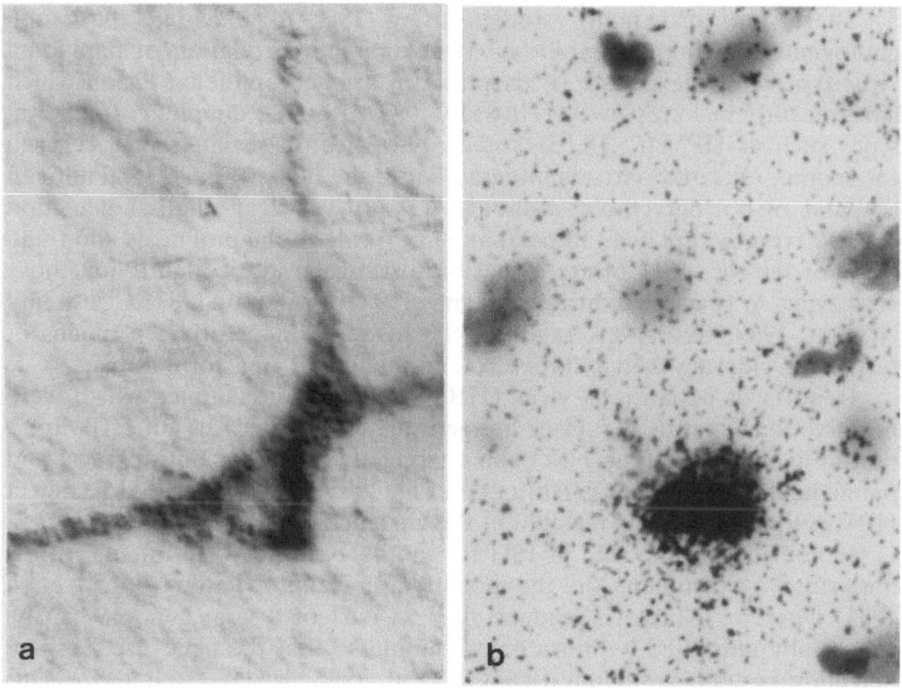


**Fig. 12 a, b.** Neurons in layer VI of human temporal cortex from a post-mortem case identified by immunocytochemistry using antibodies for NPY (**a**) (× 580), and by ISH using $^{32}P$-labelled antisense RNA probes for NPY (**b**) (× 580). Immunocytochemistry was carried out according to the peroxidase-antiperoxidase method. The autoradiograph is counterstained with cresyl fast violet

ISH in the investigation of these components and its potential application to clinical diagnosis.

The detection of intracellular viral genomes using ISH has been of interest for some time in the study of viral pathogenesis (HAASE et al. 1977; STOWRING et al. 1985; FORHANI et al. 1985). An advantage of viral detection by ISH is that it allows identification of one or a few positive cells in the midst of a large number of negative cells. For this reason, the detection and typing of viruses by ISH has potential for clinical applications (FLEMING 1987) and has recently been reviewed (HAASE 1986; MCDOUGALL et al. 1986; MAITLAND et al. 1986).

Immunohistochemistry is not always useful in detecting latent viral infections where there is restricted expression of viral genes in small numbers of cells and a block on synthesis of viral antigens during this stage of the disease process. Indeed, this realisation has led to an appreciation of the longevity of the virus-host relationship, for example the detection of herpes simplex viruses in brain (FORHANI et al. 1985). More recently, latent HIV infection in less than 0.01% of peripheral blood mononuclear cells has been detected using ISH (RANKI et al. 1987).

There are also many studies using ISH to elucidate both the pathogenesis of human papilloma virus and the correlation of papilloma virus type with the degree of dysplasia in a given epithelial lesion which have recently been reviewed (HOWLEY 1987). In this example, certain viral types such as HPV 16, 18, 31 and 35 have been associated with cervical carcinoma. Once the virus is integrated into the host genome, viral antigen is sparse or not detectable. Although it is not yet clear whether detection of HPV types will play an important clinical role in the prognosis and treatment of lesions, ISH offers a feasible practical way of identifying those viral types in cervical biopsies and smears (GUPTA et al. 1987; CUBIE and NORVAL 1988; SYRJÄNEN et al. 1988). In addition, it offers a means of further elucidating their putative role in malignant transformation.

Reports of hepatitis B virus (HBV) detection in samples which were negative for hepatitis B surface antigen suggest a wider role for HBV infection in liver disease than had been envisaged, which has both clinical and public health implications (BRECHOT et al. 1985). The use of ISH to detect HBV in liver biopsies is feasible in this instance.

In considering the role of ISH for the detection of viral infection, it is important to comment on the polymerase chain reaction (SACKI et al. 1985) whereby a target sequence such as part of a viral genome can be specifically amplified within a given sample so that the need for sensitive detection systems is obviated (SACKI et al. 1985; KWOK et al. 1987). This is a cornerstone technique for molecular biology and has important implifications for clinical diagnosis, including the detection of viral genomes. The next few years should see the emergence of some important data in respect of the relationship between viral load, the immune system and progression of infection. ISH will probably continue to have a role in staging viral disease with respect to cell type and the number of cells infected in a given population.

*Acknowledgement.* The authors would like to thank Christiane Whitehead for excellent typing assistance, and all their colleagues for helpful suggestion and discussion.

## References

Allen JM, Sasek CA, Martin JB, Heinrich G (1987) Use of complementary $^{125}$I-labelled RNA for single cell resolution by in situ hybridisation. Biotechniques 5:774–777

Andrew SM, Jasani B (1987) An improved method for the inhibition of endogenous peroxidase nondeleterious to lymphocyte surface markers. Application to immunoperoxidase studies on eosinophil rich tissue perparations. Histochem J 19:426–430

Angerer LM, Angerer RC (1981) Detection of poly A$^+$ RNA in sea urchin eggs and embryos by quantitative in situ hybridisation. Nucleic Acids Res 9:2819–2840

Angerer RC, Cox KH, Angerer LM (1985) In situ hybridisation to cellular RNAs. Genet Eng 7:43–65

Angerer LM, Stoler MH, Angerer RC (1987) In situ hybridisation with RNA probes: an annotated recipe. In: Valentino K, Erberwine JH, Barchas JD (eds) In situ hybridisation. Applications to neurobiology. Oxford University Press, Oxford pp 42–47

Banttlow CE, Heumann R, Schwab ME, Thoenen H (1987) Cellular localisation of nerve growth factor synthesis by in situ hybridisation. EMBO J 6:891–899

Belasco JG, Nilsson G, von Gabain A, Cohen SN (1986) The stability of *E. coli* gene transcripts is dependent on determinants localised to specific mRNA segments. Cell 46:245–251

Bhatt B, Burns J, Flamery D, McGee J (1988) Direct visualisation of single copy genes on banded metaphase chromosome by non-isotopic in situ hybridisation. Nucleic Acids Res 16:3951–3961

Bloom FE, Naus CCG, Miller FD, Morrison JH (1988) Immunocytochemical and in situ hybridisation analysis of the development of the rat somatostatin-containing neocortical neuronal system. J Comp Neurol 269:448–464

Bodkin DK, Knudson DL (1985) Assessment of sequence relatedness of double stranded RNA genes by RNA-RNA blot hybridisation. J Virol Methods 10:45–52

Bollum FJ (1974) Terminal deoxynucleotidyl transferase. In: Boyer PD (ed) The enzymes. Academic Press, New York

Brahic M, Haase AT (1978) Detection of viral sequences of low reiteration frequency by in situ hybridisation. Proc Natl Acad Sci USA 75: 6125–6129

Brahic M, Haase AT, Cash E (1984) Simultaneous in situ detection of viral RNA and antigens. Proc Natl Acad Sci USA 81:5445–5448

Brawerman G (1987) Determinant of messenger RNA stability. Cell 48:5–6

Brechot C, Degas F, Lugassy C (1985) Hepatitis B virus DNA in patients with chronic liver disease negative tests for hepatitis B surface antigen. N Engl J Med 312:270–276

Bresser J, Evinger-Hodges MJ (1987) Comparison and optimisation of in situ hybridisation procedures yielding rapid, sensitive mRNA detections. Gene Anal Techn 4:89–104

Brigati DJ, Myerson D, Leary JJ et al. Detection of viral genomes in cultured cells and paraffin-embedded tissue sections using biotin-labelled hybridisation probes. Virology 126:32–50

Brahic M, Haase AT (1981) Lentivirinae: Maedi/uisna virus group infections. Comparative aspects and diagnosis. In: Kurstak E, Kurstak C (eds) Comparative diagnosis of viral diseases, vol IV, part B. Academic Press, New York, pp 619–643

Britten RJ, Davidson EH (1985) Hybridisation strategy. In: Hames BD, Higgings SJ (eds) Nucleic acid hybridisation – a practical approach, IRL Press, pp 3–15

Brysch W, Hagendorff G, Schlingensiepen (1988) RNA probes transcribed from synthetic DNA for in situ hybridisation. Nucleic Acids Res 16:2333

Burns J, Graham AK, McGee J O'D (1988) Non isotopic detection of in situ nucleic acid in cervix: an updated protocol. J Clin Pathol 41:897–899

Butler ET, Chamberlain MJ (1982) Bacteriophage SP6-specific RNA polymerase. J Biol Chem 257:5772–5778

Caruthers MH, Beaucage SL, Efcavitch JW et al. (1982) Chemical Synthesis and biological studies on mutated gene control regions. Cold Spring Harbor Symp Quant Biol 47:411–418

Casey J, Davidson N (1977) Rates of formation and thermal stabilities of RNA : RNA and DNA : DNA duplexes at high concentrations of formamide. Nucleic Acids Res 4:1539–1552

Chan VTW, McGee J (1987) Cellular oncogenes in neoplasia. J Clin Pathol 40:1055–1063

Childs GV, Lloyd JM, Unabia G, Ghanb SD, Wierman ME, Chin WW (1987) Detection of luteinising hormone β messenger ribonucleic acid (RNA) in individual gonadotropes after castration: use of a new in situ hybridisation method with a photobiotinylated complementary RNA probe. Molec Endocrinol 1:926–932

Chin-Yang L, Zeismer SC, Lazcano-Villareal O (1987) Use of azide and hydrogen peroxide as an inhibitor for endogenous peroxidase in immunoperoxidase method. J Histochem Cytochem 35:1457–1460

Cleveland DW, Lopata MA, MacDonald RJ, Cowan NJ, Rutter WJ, Kirschner MW (1980) Number and evolutionary conservation of α- and β-tubulin and cytoplasmic β and γ-actin genes using specific cloned cDNA probes. Cell 20:95–105

Coggi G, Dell'Orto P, Viale G (1986) Avidin-biotin methods. In: Polak JM, van Noorden S (eds) Immunocytochemistry – modern methods and applications. Wright, Bristol, pp 54–70

Coghlan JP, Aldred P, Haralambidis J, Niall HD, Penschow JD, Tregear GW (1985) Hybridisation histochemistry. Anal Biochem 149:1–28

Cox KH, De Leon DV, Angerer LM, Angerer RC (1984) Detection of mRNAs in sea urchin embryos by in situ hybridisation using asymmetric RNA probes. Dev Biol 101:485–502

Cubie HA, Norval M (1988) Synthetic oligonucleotide probes for the detection of human papilloma viruses by in situ hybridisation. J Virol Methods 20:239–249

Cuello AC (1983) Immunocytochemistry. J. Wiley, Chichester

Cumming RDF, Fallon RA (1989) Subcellular localisation of biological molecules. In: Slater R (ed) Radioisotopes in Biology. IRL Press

Davanloo P, Rosenberg AH, Dunn JJ, Studier FW (1984) Cloning and expression of the gene for bacteriophage 77 RNA polymerase. Proc Natl Acad Sci USA 81:2035–2039

Edwards MK, Wood WB (1983) Location of specific messenger RNAs in *Caenorhabditis elegans* by cytological hybridisation. Dev Biol 97:375–390

Feinberg AP, Vogelstein B (1983) A technique for radiolabelling DNA restriction endonuclease fragments to high specific activity. Anal Biochem 132:6–13

Feinberg AP, Vogelstein B (1984) A technique for radiolabelling DNA restriction endonuclease fragments to high specific acitivity. Addendum, Anal Biochem 137:266–267

Fleming K (1987) In situ hybridisation – a role in clinical pathology. J Pathol 153:201–202

Forhani B, Dupuis KW, Schmidt NJ (1985) Rapid detection of herpes simplex virus DNA in human brain tissue by in situ hybridisation. J Clin Microbiol 22:656–658

Forster AC, McInnes JL, Skingle DC, Symous RH (1985) Non-radioactive hybridisation probes prepared by the chemical labelling of DNA and RNA with a novel reagent, photobiotin. Nucleic Acids Res 13:745–761

Fremeau RT, Lundblad JR, Pritchett DB, Wilcox JN, Roberts JL (1986) Regulation of proopiomelanocortin gene transcription in individual cell nuclei. Science 234:1265–1269

Gall G, Pardue HL (1969) Formation and detection of RNA-DNA hybrid molecules in cytological preparations. Proc Natl Acad Sci USA 63:378–381

Gastl G, Ward JH, Rapp UR (1986) Immunocytochemistry of oncogenes. In: Polak JM, Van Noorden S (eds) Immunocytochemistry – modern methods and applications. Wright, Bristol, pp 273–283

Gee CE, Roberts JL (1983) A technique for the study of gene expression in single cells. DNA 2:157–163

Gillam IC (1987) Non-radioactive probes for specific DNA sequences. Tib Tech 5:332–334

Gissman L, Schwarz E (1985) Cloning of papillomavirus DNA In: Becker Y (ed) Recombinant DNA research and viruses, Martinus Nijhoff, Boston, p 173

Godard CM (1983) Improved method for detection of cellular transcripts by in situ hybridisation: detection of virus specific RNA in Rous sarcoma virus-infected cells by in situ hybridisation to cDNA. Histochemistry 77:123–131

Goedert M (1986) Single-stranded DNA probes using an M 13 template In: Uhl G (ed) In situ hybridisation in brain. Plenum, New York, pp 236–237

Goedert M, Fine A, Hunt SP, Ullrich A (1986) NGF-mRNA in peripheral and central rat tissue and in the human central nervous system: lesions effects in the rat brain and levels in Alzheimer's disease. Mol Brain Res 1:85–92

Gonzalez RG, Blackburn BJ, Schleich T (1979) Fractionation and structural elucidation of the active components of aurintricarboxylic acid, a potent inhibitor of protein nucleic acid interactions. Biochim Biophys Acta 562:534–545

Gowans EJ, Burrell CJ, Jilbert AR, Marmion BP (1981) Detection of hepatitis B virus sequencing in infected hepatocytes by in situ cytohybridisation. J Med Virol 8:67–78

Gowans EJ, Burrell CJ, Jilbert AR, Marmion BP (1983) Patterns of single- and double-stranded hepatitis B virus DNA and viral antigen accumulation in infected liver cells. J Gen Virol 64:1229–1239

Gowans EJ, Jilbert AR, Burrell CJ (1989) Detection of specific DNA and RNA sequences in tissues and cells by in situ hybridisation. CRC series, in press
Grady-Leopardi EF, Schwab M, Ablin AR, Rosenau W (1986) Detection of N-*myc* oncogene expression in human neuroblastoma by in situ hybridisation and blot analysis relationship to clinical outcome. Cancer Res 46:3196–3199
Guitteny AF, Fouque B, Mongin C, Teoule R, Boch B (1988) Histological detection of mRNAs with biotinylated synthetic oligonucleotide probes. J Histochem Cytochem 36:563–571
Gupta JW, Gupta PK, Rosenshein N, Shah K (1987) Detection of human papillomavirus in cervical smears. A comparison of in situ hybridisation, immunocytochemistry and cytopathology. Acta Cytol 31:387–396
Haase AT (1986) Analysis of viral infections by in situ hybridisation. J Histochem Cytochem 34:27–32
Haase AT, Stowring LS, Narayan O, Griffin D, Price D (1977) Slow persistent infection caused by visna virus: role of host restriction. Science 195:175–177
Haase AT, Stowring L, Harris JD, Traynor B, Ventura P, Peluso R, Brahic M (1982) Visna DNA synthesis and the tempo of infection in vitro. Virology 119:399–410
Haase AT, Brahic M, Stowring L (1984) Detection of viral nucleic acids by in situ hybridisation. In: Maramorosch K, Koprowski H (eds) Methods in virology, VII. Academic Press, New York, pp 189–226
Haase AT, Walker D, Stowring L et al. (1985) Detection by hybridisation of viral infection of the human central nervous system. Ann NY Acad Sci 436:103–108
Hafen E, Levine M, Garber RL, Gehring WJ (1983) An improved in situ hybridisation method for the detection of cellular RNAs in *Drosophila* tissue sections and its application for localising transcripts of the homoeotic antennapedia gene complex. EMBO J 2:617–623
Hamid Q, Wharton J, Terenghi G et al. (1987) Localisation of atrial natriuretic peptide mRNA and immunoreactivity in the rat heart and human atrial appendage. Proc Natl Acad Sci USA 84:6760–6764
Hayashi S, Gillam IC, Delaney AB, Tener GM (1978) Acetylation of chromosome squashes of *Drosophila melanogaster* decreases the background in autoradiographs from hybridisation with [$^{125}$I]-labelled RNA. J Histochem Cytochem 26:677–679
Höfler H (1987) What's new in "in situ hybridisation". Pathol Res Pract 182:421–430
Höfler H, Childers H, Montminy MR, Lechan RM, Goodman RH, Wolfe HJ (1986a) In situ hybridisation methods for the detection of somatostatin mRNA in tissue sections using antisense RNA probes. Histochem J 18:597–604
Höfler H, Childers H, Dayal Y et al. (1986b) Detection of neuroendocrine gene expression of tumour cells by combined in situ hybridisation and immunocytochemistry. Verh Dtsch Ges Pathol 70:211–216
Höfler H, Childers H, Mobtaker H, Tischer AS, De Lellis RA, Wolfe HJ (1986c) Fos oncogene expression in a mouse medullary thyroid carcinoma cell line demonstrated by blot and in situ hybridisation. Lab Invest 54:15A
Höfler H, Putz B, Ruhri C, Wirnsberger G, Klimpfinger M, Smolle J (1987) Simultaneous localisation of calcitonin mRNA and peptide in a medullary thyroid carcinoma. Virchows Arch [Cell Pathol]54:144–151
Hogan B, Costantini F, Lacy E (1986) Manipulating the mouse embryo: a laboratory manual. Cold Spring Harbor Laboratory
Howley PM (1987) The role of papillomaviruses in human cancer 4. In: Devita VT Jr, Hellman S, Rosenberg SA (eds) Important advances in oncology. Lippincott, Philadelphia
Hu N, Messing J (1982) The making of strand specific M13 probes. Gene 17:271–277
Huang WM, Gibson SJ, Facer P, Gu J, Polak JM (1983) Improved section adhesion for immunocytochemistry using high molecular weight polymers of L-lysine as a slide coating. Histochemistry 77:275–279
Hutchison NJ, Langer-Safer PR, Ward DC, Manikalo BA (1982) In situ hybridisation at the electron microscopic level: hybrid detection by autoradiography and colloidal gold. J Cell Biol 95:609–618
Isenberg I (1979) Histones. Ann Rev Biochem 48:159–191

Janssen HP, van Loon AM, Meddens MJ, Herbrink P, Linderman J, Quint WGV (1987) Comparison of in situ DNA hybridisation and immunological staining with conventional virus isolation for the detection of human cytomegolavirus infection in cell cultures. J Virol Methods 17:311–318

Jilbert AR, Burrel CJ, Gowans EJ, Rowland R (1986) Histological aspects of in situ hybridisation. Histochemistry 85:505–514

John HA, Burnstiel ML, Jones KW (1969) RNA:DNA hybrids at the cytological level. Nature 223:582–587

Johnson SA, Morgan DG, Finch CE (1986) Extensive postmortem stability of RNA from rat and human brain. J Neurosci Res 16:267–280

Jones TR, Hyman RW (1983) Specious hybridisation between herpes simplex virus DNA and human cellular DNA. Virology 131:555–560

Kourilsky P, Leidner J, Tremblay GY (1971) DNA : DNA hybridisation, on filters at low temperature in the presence of formamide or urea. Biochimie 53:1111–1114

Kwok S, Mack DH, Mullis KB et al. (1987) Identification of human immunodeficiency virus sequences by using in vitro enzymatic amplification and oligomer cleavage detection. J Virol 61:1690–1694

Langer PR, Waldrop A, Ward D (1981) Enzymatic synthesis of biotin labelled polynucleotides: novel nucleic acid affinity probes. Proc Natl Acad Sci USA 78:6633–6637

Larsson L-I, Christensen T, Dalboge H (1988) Detection of POMC mRNA by in situ hybridisation using a biotinylated oligodeoxynucleotide probe and avidin-alkaline phosphatase histochemistry. Histochemistry 89:109–116

Lawrence JB, Singer RH (1985) Quantitative analysis of in situ hybridisation methods for the detection of actin gene expression. Nucleic Acids Res 13:1777–1799

Lawrence JB, Singer RH (1986) Intracellular localisation of mRNA for cytoskeletal proteins. Cell 45:407–415

Lawrence JB, Villnave CA, Singer RH (1988) Sensitive, high-resolution chromatin and chromosome mapping in situ: presence and orientation of two closely integrated copies of EBV in a lymphoma line. Cell 52:51–61

Leary JJ, Brigati DJ, Ward DC (1983) Rapid and sensitive colorimetric method for visualizing biotin-labelled DNA probes hybridized to DNA or RNA immobilized on nitrocellulose. Proc Natl Acad Sci USA 80:4045–4049

Levine BJ, Liv T, Marzluff WF, Skoultchi AI (1988) Differential expression of individual members of the histone multigene family due to sequences in the 5' and 3' regions of the genes. Mol Cell Biol 8:1887–1895

Lewin BM (1980) Eukaryotic genomes. In: Gene expression 2:503–569, 861–930, Wiley, New York

Lewis ME, Sherman TG, Watson SJ (1985) In situ hybridisation histochemistry with synthetic oligonucleotides; strategies and methods Peptides 6 [Suppl 2]: 75–87

Lewis ME, Arentzen R, Baldino F (1986) Rapid, high-resolution in situ hybridisation. Histochemistry with radioiodinated synthetic oligonucleotides. J Neurosci Res 16:117–124

Lewis FA, Griffith S, Dunncliff R, Wells M, Dudding N, Bird CC (1987) Sensitive in situ hybridisation technique using biotin-streptavidin-polyalkaline phosphatase complex. J Clin Pathol 40:163–166

Luehrsen KR, Baum MP (1987) In vitro synthesis of biotinylated RNA probes from A-T rich templates: problems and solutions. Biotechniques 5:660–670

Lynn DA, Angerer LM, Bruskin AM, Kleen WH, Angerer RC (1983): Localisation of a family of mRNAs in a single cell type and its precursors in sea urchin embryos. Proc Natl Acad Sci USA 80:2656–2660

Maitland NJ, Cox MF, Lynas C, Prime S, Crane I, Scully C (1986) Nucleic acid probes in the study of latent viral disease. J Oral Pathol 16:199–211

Maniatis T, Fritsch EF, Sambrook J (1982) Molecular cloning. Cold Spring Harbor Laboratory

Manuelidis L (1985) Indications of centromere movement during interphase and differentiation. Ann NY Acad Sci 450:205–221

McAllister LB, Scheller RH, Kaudel ER, Axel R (1983) In situ hybridisation to study the origin and fate of identified neurones. Science 222:800–808

McAllister HA, Rock DL (1985) Comparative usefulness of tissue fixatives for in situ viral nucleic acid hybridisation. J Histochem Cytochem 33:1026–1032
McConaughy BL, Laird CD, McCarthy BJ (1969) Nucleic acid reassociation in formamide. Biochemistry 8:3289–3295
McDougall JK, Myerson D, Beckman AM (1986) Detection of viral DNA and RNA by in situ hybridisation. J Histochem Cytochem 34:33–38
McKeating JA, Al-Nakib W, Greenaway PJ, Griffiths PD (1985) Detection of cytomegalovirus by DNA-DNA hybridisation employing probes labelled with $^{32}P$ or biotin. J Virol Methods 11:207–216
Meinkoth J, Wahl G (1984) Hybridisation of nucleic acids immobilized on solid supports. Anal Biochem 138:267–284
Melton D, Kneg P, Rebagliati M, Maniatis T, Zinn K, Green MR (1984) Efficient in vitro synthesis of biologically active RNA and RNA hybridisation probes from plasmids containing a bacteriophage SP6 promoter. Nucleic Acids Res 12:7035–7056
Mitchell AR, Ambros P, Gosden JR, Morten JEN, Porteous DJ (1986) Gene mapping and physical arrangements of human chromatin in transformed hybrid cells: fluorescent and autoradiographic in situ hybridisation compared. Somatic Cell Mol Genet 12:313–324
Moench TR, Gendelman HE, Clements JE, Narayan O, Griffin DE (1985) Efficiency of in situ hybridisation as a function of probe size and fixation technique. J Virol Methods 11:119–130
Morley DJ, Hodes ME (1988) Amylase expression in human parotid neoplasms: evidence by in situ hybridisation for lack of transcription of the amylase gene. J Histochem Cytochem 36:487–492
Nan MM,Brooks BJ, Carney DN, Gazdar AF (1986) Human small cell lung cancers shows amplification and expression of the N-*myc* gene. Proc Natl Acad Sci USA 83:1092–1096
Naoumov NV, Alexander GJM, Eddleston ALWF, Williams R (1988) In situ hybridisation in formalin fixed, paraffin wax embedded liver specimens: method for detecting human and viral DNA using biotinylated probes. J Clin Pathol 41:793–798
Niedobitek G, Finn HH, Bornhoft G, Gerdes J, Stein H (1988) Detection of viral DNA by in situ hybridisation using bromodeoxyuridine labelled DNA probes. Am J Pathol 1313:1–4
Palmer L, Falkow S (1985) Selection of DNA probes for use in the diagnosis of infectious disease. In: Kingsbury DT, Falkow S (eds) Rapid detection and identification of infectious diseases. Harcourt Brace Jovanovich, New York, pp 211–233
Patient R (1984) DNA hybridisation – beware. Nature 308: 15–16
Penschow JD, Haralambidis J, Aldred P, Tregear GW, Coghlan JP (1986) Location of gene expression in CNS using hybridisation histochemistry. Methods Enzymol 124:534–548
Polak JM, van Noorden SV (1986) Immunocytochemistry – modern methods and applications. Wright, Bristol
Pontecorvi A, Tata JR, Phyllaier M, Robbins J (1988) Selective degradation of mRNA: the role of short-lived proteins in differential destabilization of insulin-induced creative phosphokinase and myosin heavy chain mRNAs during rat skeletal muscles L6 cell differentiation. EMBO J 7:1489–1495
Przepiorka D, Myerson D (1986) A single step silver enhancement method permitting rapid diagnosis of cytomegalovirus infection in formalin-fixed paraffin embedded tissue sections by in situ hybridisation and immunoperoxidase detection. J Histochem Cytochem 34:1731–1734
Ralston R, Bishop JM (1984) Evolutionary relationships among oncogenes of DNA and RNA tumour viruses: myc, myb and adenovirus EIA In: Vande Woude GF, Levine AJ, Topp WC, Watson JD (eds) Cancer and cells 2: oncogenes and viral genes. Cold Spring Harbor Laboratory, pp 165–172
Ranki A, Krohn M, Allain J et al. (1987) Long latency precedes overt seroconversion in sexually transmitted human-immunodeficiency-virus infection. Lancet II: 589–593
Renz M, Kurz C (1984) A colorimetric method for DNA hybridisation. Nucleic Acids Res 12:3435–3444
Rigby PWJ, Dieckmann M, Rhodes C, Berg P (1977) Labelling deoxyribonuclei acid to high

specific activity in vitro by nick translation with DNA polymerase 1. J Mol Biol 113:237–251

Rogers AW (1979) Techniques of autoradiography. Elsevier, Amsterdam

Rogers WT, Schwaber JS, Lewis ME (1987) Quantitation of cellular resolution in situ hybridisation histochemistry in brain by image analysis. Neurosci Lett 82:315–320

Ruda MA, Iadarola MJ, Cohen LV, Young III WS (1988) In situ hybridisation histochemistry and immunocytochemistry reveal an increase in spinal dynorphyn biosynthesis in a rat model of peripheral inflammation and hyperalgesia. Proc Natl Acad Sci USA 85:622–626

Sacki RK, Scharf S, Faloona F, Mullis KB, Horn GT, Erlich HA, Arnheim N (1985) Enzymatic amplification of β-globin genomic sequences and restriction site analysis for diagnosis of sickle cell anaemia. Science 230:1350–1354

Schwab M, Ellison J, Busch M, Rosenau W, Varmus HE, Bishop JM (1984) Enhanced expression of the human gene N-*myc* consequent to amplification of DNA may contribute to malignant progression of neuroblastoma. Proc Natl Acad Sci USA 81:4940–4944

Scott Young III W, Mezey E, Siegel RE (1986) Vasopressin and oxytocin mRNAs in adrenalectomized and Brattleboro rats: analysis by quantitative in situ hybridisation histochemistry. Mol Brain Res 1:231–241

Shaw G, Kamen R (1986) A conserved AU sequence from the 3' untranslated region of GM-CSF mRNA mediates selective mRNA degradation. Cell 46:659–667

Shivers BD, Harlan RE, Pfaff DW, Schachler BS (1986 a) Combination of immunocytochemistry and in situ hybridisation in the same tissue section of rat pituitary. J Histochem Cytochem 34:39–43

Shivers BD, Schachter BS, Pfaff DW (1986 b) In situ hybridisation for the study of gene expression in the brain. Methods Enzymol 124:497–510

Singer RH, Ward DC (1982) Actin gene expression visualized in chicken muscle culture by using in situ hybridisation with a biotinated nucleotide analog. Proc Natl Acid Sci 79:7331–7335

Singer RH, Lawrence JB, Villnave C (1986) Optimization of in situ hybridisation using isotopic and non-isotopic detection methods. Biotechniques 4:230–250

Slamon DC, de Kernion JB, Verma IH, Cline H (1984) Expression of cellular oncogenes in human malignancies. Science 224:256–262

Soreq H, Zamur R, Zevin-Sonkin D, Zakut H (1987) Human cholinesterase genes localised by hybridisation to chromosomes 3 and 16. Hum Genes 738:1–4

Stanssens P, Remaut E, Fiers W (1986) Inefficient translation initiation causes premature transcription termination in the Iac Z gene. Cell 44:711–718

Steel JH, Hamid Q, Van Noorden S et al. (1988) Combined use of in situ hybridisation and immunocytochemistry for the investigation of prolactin gene expression in immature, pubertal, pregnant, lactating and ovariectomised rats. Histochemistry 89:75–80

Stopa EG, Uhl GR, Mobtaker H et al. (1989) Somatostatin gene expression in human brain: in situ hybridisation studies in postmortem tissue. Neuroscience, in press

Stowring L, Haase AT, Petursson G et al. (1985) Detection of visna virus antigens and RNA in glial cells in foci in demyelination. Virology 141:311–318

Syrjänen S, Partanen P, Mäntyjärvi R, Syrjänen K (1988) Sensitivity of in situ hybridisation techniques using biotin and $^{35}$S labelled human papilloma virus (HPV) DNA probes. Virol Methods 19:225–238

Szabo P, Elder R, Steffensen DM, Uhlenbeck OC (1977) Quantitative ISH of ribosomal RNA species to polytene chromosomes of *Drosophila melanogaster.* J Mol Biol 115:539–563

Taylor GR, Carter GI, Crow TJ, Johnson JA, Fairbairn AF, Perry EK, Perry RH (1986) Recovery and measurement of specific RNA species from postmortem brain tissue: a general reduction in Alzheimer's disease detected by molecular hybridisation. Exp Mol Pathol 44:111–116

Terenghi G, Polak JM, Hamid Q et al. (1987) Localisation of neuropeptide Y mRNA in neurons of human cerebral cortex by means of in situ hybridisation with a complementary RNA probe. Proc Natl Acad Sci USA 84:7315–7318

Thomas CA, Dancis BM (1973) Ring stability. J Mol Biol 77:44–55

Tournier I, Bernau D, Poliard A, Schoevaret D, Feldman G (1987) Detection of albumin mRNAs in rat liver by in situ hybridisation: usefulness of paraffin embedding, and comparison of various fixation procedures. J Histochem Cytochem 35:453–459

Trask B, van den Engh G, Pinkel D, Mullikin J, Waldmann F, van Dekken H, Gray J (1988) Fluorescence in situ hybridisation to interphase cell nuclei in suspension allows flow cytometric analysis of chromosome content and microscopic analysis of nuclear organization. Hum Genet 78:251–259

Unger ER, Budgeon LR, Myerson D, Brigati DJ (1986) Viral diagnosis by in situ hybridisation. Description of a rapid simplified colorimetric method. Am J Surg Pathol 10:1–8

van der Ploeg, Landegent JE, Hopman HHN, Raap AK (1986) Non-autoradiographic hybridocytochemistry. J Histochem Cytochem 34:126–133

Wahl GM, Stern M, Stark GR (1979) Efficient transfer of large DNA fragments from agarose gels to diazobenzyloxymethyl-paper and rapid hybridisation by using dextran sulphate. Proc Natl Acad Sci USA 76:3683–3687

Wetmur JG, Davidson N (1968) Kinetics of renaturation of DNA. J Mol Biol 31:349–370

Wetmur JG, Ruyechan WT, Douthart RJ (1981) Denaturation and renaturation of *Penicillinin chrysogenum* mycophage double-stranded ribonucleic acid in tetraalkylammonium salt solutions. Biochemistry 20:2999–3002

Wilcox JN, Gee CE, Roberts JL (1986) In situ cDNA-mRNA hybridisation: development of a technique to measure mRNA levels in individual cells. Methods Enzymol 124:510–533

Zabel M, Schafer H (1988) Localisation of calcitonin and calcitonin gene-related peptide mRNAs in rat parafollicular cells by hybridocytochemistry. J Histochem Cytochem 36:543–546

Zajac JD, Penschow J, Mason T, Tregear G, Coghlan J, Martin TJ (1986) Identification of calcitonin and calcitonin gene-related peptide messenger ribonucleic acid in medullary thyroid carcinomas by hybridisation histochemistry. J Clin Endocrinol Metab 62:1037–1043

# Subject Index

## Index of Volumes 78–81 Current Topics in Pathology

ɴᴛ, H. J. HÖHLING, A. ROESSNER, E. GRUNDMANN, ·ormation in Different Osteosarcomas: Topological, ıl, and Histochemical Analyses

ʀ, A. ROESSNER, P. WUISMAN, A. HÄRLE, E. GRUNDMANN, . roliferation Behavior of Bone Tumors Investigated with the ɔnoclonal Antibody Ki-67

.MELLIN, W. DIERSCHAUER, W. HIDDEMANN, A. ROESSNER, G. EDEL, P. WUISMAN, A. HÄRLE, E. GRUNDMANN, Flow Cytometric DNA Analysis of Bone Tumors

L. E. WOLD, T. SPELSBERG, N. JIANG, F. SIM, Steroid Receptors and Giant Cell Tumor of Bone

J. J. BROOKS, J. Q. TROJANOWSKI, V. A. LIVOLSI, Chondroid Chordoma: A Low-Grade Chondrosarcoma and Its Differential Diagnosis

V. BOUROPOULOU, A. BOSSE, A. ROESSNER, E. VOLLMER, G. EDEL, P. WUISMAN, A. HÄRLE, Immunohistochemical Investigation of Chordomas: Histogenetic and Differential Diagnostic Aspects

A. ROESSNER, E. VOLLMER G. ZWADLO, C. SORG, M. KOLVE, D. B. v. BASSEWITZ, P. WUISMAN, A. HÄRLE, E. GRUNDMANN, The Cytogenesis of Macrophages and Osteoclast-like Giant Cells in Bone Tumors with Special Emphasis on the So-Called Fibrohistocytic Tumors

**Volume 81: Gastrointestinal Pathology.** Edited by G. T. WILLIAMS

M. F. DIXON, Progress in the Pathology of Gastritis and Duodenitis

F. POTET, V. DUCHATELLE, Barrett's Oesophagus

P. SIPPONEN, Gastric Dysplasia

A. H. QIZILBASH, Duodenal and Peri-ampullary Adenomas

C. FENGER, Intra-epithelial Neoplasia in the Anal Canal and Peri-anal Area

M. M. MATHAN, V. I. MATHAN, Gastrointestinal Biopsy Diagnosis in the Tropics

D. A. LEVISON, P. A. HALL, A. J. BLACKSHAW, The Gut-Associated Lymphoid Tissue and Its Tumours

J. M. SLOAN, D. C. ALLEN, P. W. HAMILTON, P. C. H. WATT, The Place of Quantitation in Diagnostic Gastrointestinal Pathology

W. V. BOGOMOLETZ, Collagenous Colitis

A. B. PRICE, Ischaemic Colitis

P. S. TEGLBJÆRG, Intestinal Spirochaetosis

PH. U. HEITZ, P. KOMMINOTH, Biopsy Diagnosis of Hirschsprung's Disease and Related Disorders

P. HERMANEK, Malignant Polyps – Pathological Factors Governing Clinical Management

J. R. JASS, Prognostic Factors in Colorectal Cancer

N. A. SHEPHERD, H. J. R. BUSSEY, Polyposis Syndromes – An Update

MIX
Papier aus verantwortungsvollen Quellen
Paper from responsible sources
FSC® C105338

If you have any concerns about our products,
you can contact us on
**ProductSafety@springernature.com**

In case Publisher is established outside the EU,
the EU authorized representative is:
**Springer Nature Customer Service Center GmbH**
**Europaplatz 3, 69115 Heidelberg, Germany**

Printed by Libri Plureos GmbH
in Hamburg, Germany